Satoshi Koizumi, Christoph Nebel, and Milos Nesladek

Physics and Applications of CVD Diamond

Related Titles

Berakdar, J.

I. Concepts of Highly Excited Electronic Systems / II. Electronic Correlation Mapping from Finite to Extended Systems

approx. 565 pages in 2 volumes
2006
Hardcover
ISBN: 978-3-527-40412-4

Riedel, R. (ed.)

Handbook of Ceramic Hard Materials

1091 pages in 2 volumes with 622 figures and 45 tables
2000
Hardcover
ISBN: 978-3-527-29972-0

Satoshi Koizumi, Christoph Nebel, and
Milos Nesladek

Physics and Applications of CVD Diamond

WILEY-VCH Verlag GmbH & Co. KGaA

The Editors

Dr. Satoshi Koizumi
NIMS
Japan

Dr. Christoph Nebel
Diamond Research Center
Japan

Dr. Milos Nesladek
CEA-LIST
Frankreich

Library of Congress Card No.:
applied for

British Library Cataloguing-in-Publication Data
A catalogue record for this book is available from the British Library.

Bibliographic information published by the Deutsche Nationalbibliothek
Die Deutsche Nationalbibliothek lists this publication in the Deutsche Nationalbibliografie; detailed bibliographic data are available in the Internet at <http://dnb.d-nb.de>.

Typesetting SNP Best-set Typesetter Ltd., Hong Kong
Printing Strauss GmbH, Mörlenbach
Binding Litges & Dopf Buchbinderei GmbH, Heppenheim

Printed in the Federal Republic of Germany
Printed on acid-free paper

ISBN: 978-3-527-40801-6

Contents

Physics and Applications of CVD Diamond. Satoshi Koizumi, Christoph Nebel, and Milos Nesladek

ISBN: 978-3-527-40801-6

Preface

Diamond related research is emerging again, in laboratories and universities all over the world, after nearly 15 years of slumber, a period of time that has been dominated by other wide band gap semiconductors, most prominently GaN and related alloys. These optoelectronic semiconductors diverted a significant amount of research activity away from diamond. Nonetheless, it was still possible to achieve significant progress during this period. Now, more and more researchers recognize the unique properties of diamond, which promise new device applications in the twenty-first century. These applications are in the fields of room temperature quantum computing, biosensing, biointerfaces, microelectromechanical systems (MEMS), color centers and particle detectors, to name a few. For more than 20 years, diamond has been known to be a perfect material for a variety of applications, as it shows the highest electron and hole mobilities, a high electric field breakdown strength, and a low dielectric constant. In combination with its unmatched thermal conductivity and hardness, many applications have been proposed. However, the breakthrough never occurred, as diamond was not available as semiconductor grade material in areas required for technological processing.

In 1996, it his been shown that diamond can be n-type doped by phosphorus. In combination with the already established p-type doping with boron, diamond became one of the wide band gap semiconductors which are now candidates for deep UV light emitting devices (LED), bipolar diodes, and transistors for high field switching applications. In addition to achievements with respect to doping, plasma enhanced chemical vapor deposition of diamond has been optimized, during recent years, to a level which allows the growing of crystals with ultra low defect densities, with controlled isotope mixtures, or isotopically enriched, and with atomically smooth surfaces.

This book is dedicated to presenting reviews of progress in diamond research and technology. After some critical opening statements by Marshall Stoneham, three chapters about the growth of diamond will demonstrate that diamond can be grown with high perfection. The authors of these chapters span the field, from nanocrystalline diamond films, to homoepitaxial growth of single crystalline diamond, including discussion of single crystalline growth by heteroepitaxial methods.

Physics and Applications of CVD Diamond. Satoshi Koizumi, Christoph Nebel, and Milos Nesladek

ISBN: 978-3-527-40801-6

Bioelectronics and MEMS applications are discussed by three contributors. They show that diamond is truly the best working electrode in electrochemical applications, and therefore, a very promising material for biosensor devices. The surface conductivity of undoped diamond is discussed based on the "transfer doping model". During the last five years, the diamond surface has been utilized as an ideal platform for chemical modifications with different organic linker molecular layers. Two types are introduced in detail, namely amine- and phenyl-linker molecules, including a discussion of growth and layer formation. The realization of DNA sensors, one of several applications of diamond in biotechnology, is also described. This shows that diamond is available as a promising device for biochemical sensing, with unmatched chemical stability.

While diamond doping is long established, still new effects arise from the incorporation of extrinsic atoms into the diamond lattice. We present two summaries. One examines theoretical views of defects and dopants in diamond, and the other is concerned with the latest experimental achievements, demonstrating the high quality of n-type doping by phosphorus. Controlled defect formation adds two new fields of application to diamond. One is related to quantum computing at room temperature using the nitrogen/carbon–vacancy complex (NV-center). The other is the field of color centers where the nickel/nitrogen complex has been discovered as a new light emitting center for single photon emission applications in quantum cryptography.

In addition, undoped diamond shows very promising excitonic properties. In combination with its high thermal conductivity, this may result in excitonic Bose–Einstein condensates. As the binding energy of excitons is about 80 meV, experiments are currently performed at room temperature to investigate this phenomenon in even more detail. Due to these properties, and due to its radiation hardness, undoped diamond has also been very successfully introduced as a sensor for high energy particles.

Although Boron doping of diamond is long established, recently, it has attracted significant attention as, in case of metallic doping with densities above $10^{21}\,cm^{-3}$, superconducting properties have been revealed in the temperature regime below 8–10 K. This is discussed in a chapter summarizing related discoveries and introducing experiments that have been applied in this context.

From our point of view, this book presents hot topics taking into account recently achieved progress in the field of diamond research and development. We hope that it is stimulating for readers, and that it will contribute to the progress of diamond, which is still a "promising wide band gap semiconductor" which needs to mature into a technologically relevant material of the future.

S. Koizumi, Christoph E. Nebel, Milos Nesladek

1
Future Perspectives for Diamond

A. Marshall Stoneham
University College London, Department of Physics and Astronomy, London Centre for Nanotechnology, London WC1E 6BT, UK

1.1
The Status Diamond and the Working Diamond

Diamond has inspired both legends and industries. Its special properties have also spawned what can legitimately be called a wealth of niche applications. But should we expect more, new *major* applications, ones that inspire new legends or transform an industry?

Today, there are just two ways in which diamond is both unique and dominant: the status diamond, and the working diamond. Status diamonds are symbols of power. Their beauty, rarity and cost limits ownership. When kings, queens, emperors and presidents are crowned or invested, it is diamond, more than rich gold and rubies, that hints most effectively at power and continuity of tradition. Even a model who wears a special stone for a few minutes feels privileged by the transient sense of ownership. The status diamond will be expensively set and worn by the richest people. Working diamonds are entirely different. The stones and their mountings are chosen for economy and effectiveness. Even before the industrial revolution, Diderot's 1751 *Encyclopaedia* showed a tradition for diamonds as tools, with many diverse applications for these working diamonds. Even small and ugly diamonds have value. Their mechanical properties dominate within significant niche applications such as thermal sinks.

The *major* applications for diamond thus exploit only a fraction of diamond's special properties: visual for status diamonds, and mechanical for working diamonds. Could there be other major applications, perhaps exploiting different diamond properties? The materials developments of the last 50 years include silicon becoming the semiconductor of choice, many new and better developed polymers, and the transformation of communications by silica-based optical fibers. Diamond has been synthesized. Could there be new, major, radical opportunities for it?

Physics and Applications of CVD Diamond. Satoshi Koizumi, Christoph Nebel, and Milos Nesladek

ISBN: 978-3-527-40801-6

1.2
On Diamond's Future

Diamond, as both a source of power and an object of beauty, is already a remarkably successful material. Its mechanical properties still win through because of diamond's impressive performance and diamond's track record as being tougher than new materials, despite the claims of "superhard" alternatives. Diamond's wealth of materials properties, better than those of any common material, have led to many niche applications, though other materials remain dominant. Could diamond's special virtues yield major new opportunities? Its optical properties are exceptional, usually in desirable ways (although high refractive indices can create indirect problems). The mechanical properties of diamond are truly exceptional, again usually in desirable ways (although adhesion can be challenging). Its thermal properties are similarly exceptional, with a thermal conductivity that stands comparison with copper. Diamond withstands aggressive environments, including extremes of pH. Its electron mobility can be phenomenal, and electron emission can be excellent. Moreover, diamond can be compatible with silicon electronics, even if the involvement of a second material is inconvenient. But here the problems start. Control of diamond's properties is difficult. For instance, it is difficult and barely practical to create n-type diamond. Electrical contacts can be tricky to fabricate. The perception of diamond as unacceptably expensive is a further issue. Yet, thin film diamond is not expensive: it is the gemstone associations that prove a barrier to some major applications.

1.3
The Electron in Carbon Country

For almost all measures of performance, there is some carbon-based material that performs better than silicon. Yet, just as for diamond, it has proved tough to exploit these carbons in electronics, apart from niche applications. Might hybrid carbon-based materials be more successful? Should we think less about "diamond" and more about the *integration* of diamond as one component of carbon electronics [1]? Silicon only became the semiconductor of choice in the 1950s, when the winning feature was an oxide that passivates the surface, acts as an impressive dielectric, and enables high-resolution lithography. But even silicon does not stand alone as a material. Device fabrication needs lithography optics and resists, and processing at the anticipated smaller scales may well exploit new electronic excitation methods. Alternative dielectrics and interconnect materials introduce new compatibility issues, and there are further varied constraints from displays, spintronic components, electron emitters or transparent conductors. Could the many carbon-based materials with interesting functional properties lead to a new class of alternative systems?

Carbon-based electronic materials are strikingly diverse. They include diamond, graphites, buckey structures, amorphous carbons, and nanodiamond. Add hydrogen and one has a range of diamond-like carbons and the wealth of organics. Such carbon-based materials include small molecules and polymers: impressive insulators, semiconducting and conducting polymers, switchable forms, superconducting and magnetic forms, and intercalates with better electrical conductivity than copper (Cu) for a given weight. Their impressive properties may be electronic or photonic, mechanical or thermal. Whilst diamond is superhard, diamond-like carbons can have controllable mechanical properties from the viscoelastic to the highly rigid. Photochemistry brings opportunities for novel processing methods. Even water-based processing may be possible sometimes (alas, not for diamond), and additional tools like self-organization of organic molecules on surfaces have been demonstrated.

Current semiconductor technology achieves its goals by controlling bandgaps and mobilites through alloying and stress. For carbon-based systems, these simple band pictures may mislead, though there is scope for control by molecular design and choices of sidegroups or terminations. Blending can create systems that have coexisting n-type (electron transporting) and p-type (hole transporting) behavior [2]. Texture can be controlled to optimize a mesostructure [3]. The best carbons display impressive, sometimes superior, performances, including the mobility and optical properties of diamond, spin-conserving transport in buckeytubes, and electron emission. Despite these virtues, carbon-based electronics is not in the same league as established silicon technology. But there is no reason why carbons should compete with silicon in those areas where silicon is supreme. Diamond is not just another semiconductor with different standard properties. An alternative view is that carbon electronics should judiciously combine carbons and organics to take the very best properties, avoiding weaker features. At the least, this would provide a basis for niche applications, such as combinations of good mechanical and optical behavior with resistance to corrosion or radiation, or ingenious micromachines.

1.4 Social Contexts: Twenty-First Century Needs

Any major new applications will have to match demands from social pressures. There are three clear technology drivers for the twenty-first century. One comes from the biomedical technologies, where there is wide expectation that novel molecular scale science will transform the quality of many individual lives. A second stems from the demand for energy, especially for the developing world, where one must not ignore global constraints or industrial impacts on the biosphere. Thirdly, there are the information technologies, which are the major enabling technologies, and for which new ways to do things will be needed.

1.5
The Biomedical and Life Sciences Context

Diamond's good *electrochemical* properties should give it a niche. Good as they are, diamond's *functional* properties are not unique. So, whilst nanodiamond might be used in nanobiomedicine, biomolecules may help diamond more than vice versa. Nature is already adept at self-organization on surfaces, photochemical manipulation and other means for materials modification. Nature is even rather good at managing efficient nanoscale energy transport along protein α-helices. So, even if there are plenty of niches, it is hard to see a really major opportunity exploiting diamond's functional properties.

What about human scale biomedical science or technology? The more obvious opportunities, like replacing hip joints, seem niche, even though diamond and diamond-like carbons have excellent tribological properties, and even though carbon fiber composites can match bone elastic constants and anisotropy to reduce bone wear. Surgeons all seem to have their favorite choices of implant, and major benefits from materials improvements may need to wait for a new generation of surgeons.

1.6
Fusion: Opportunity and Challenge

It is an understatement to say that the fusion reactor first wall presents materials challenges. Dramatic pictures of plasma disruption events show large local energy releases (Figure 1.1). Could one improve on the graphites used in the UK based

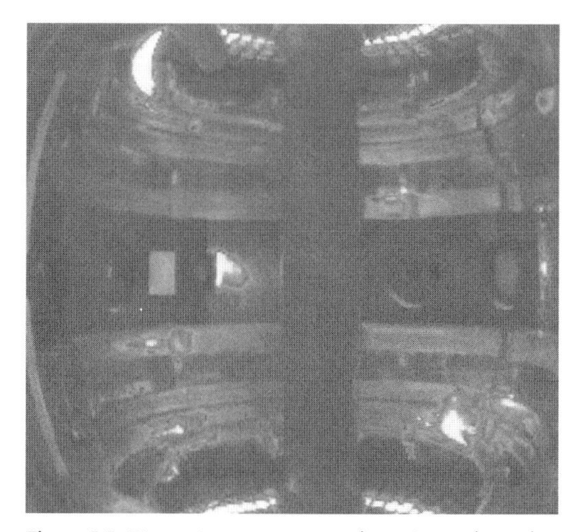

Figure 1.1 Disruption occurring in the MAST spherical tokamak, showing dramatic large local energy releases. Figure courtesy Euratom–UKAEA Fusion Association.

Joint European Torus (JET), the world's leading fusion experimental facility? Ignoring routine engineering problems, there are four substantial challenges for that critical first inch [4].

First, the wall must *survive*, even under disruptions. This means that a low ablation rate is desirable (the erosion problem). For erosion resistance, a high atomic binding energy is usually beneficial. There should be a low fragmentation and spalling rate (the dust problem). This requires resistance to thermal shock. Diamond's high thermal conductivity and low thermal expansion are good, but its brittleness is unhelpful. However, diamond-like carbons can be viscoelastic, with their degree of viscosity and rigidity controlled partly by their hydrogen concentration and partly by the proportions of sp^2 and sp^3 bonded carbons. Current demands for a conceptual power plant would have a surface energy flux at the divertor of 50 MW/cm^2; that elsewhere at the first wall is less, around 10 MW/cm^2. The power density at the first wall is about 55 MW/cm^2.

The second major issue is *radiation damage* from 14 MeV neutrons, α and β radiation, photons, and bombardment from hydrogenic species H, D, T in the plasma. The total neutron flux anticipated is 2.10^{11} neutrons per m^{2}s. Also expected are 2300 atomic ppm H from (n, p) reactions per calendar year and 500 atomic ppm He from (n, α) reactions per calendar year. There will be of order 50 displacements per atom (depending on the material chosen) per full power year. Amazingly, many radiation related properties of diamond are not known. Some of these complexities are discussed in [6], with interesting suggestions that radiation might stabilize diamond [7]. The views of the "informed" field are extraordinarily diverse. One common concern is chemical erosion [8].

Thirdly, there are strong legal limits concerning the tritium inventory, and these constrain tritium retention and release. The tritium should be recoverable for reuse, if possible. Fourthly, material from the wall will enter the plasma, but such material should not harm it. In practice, atoms with a high atomic number are bad news, since they can cool (take energy from) the plasma locally. Carbon has $Z = 6$, which is one reason why graphite and carbon composite systems are used today.

So what might be done? One suggestion [4] is to create a "designer carbon", constructed from various carbon forms, and with a chosen microstructure. Thus, the plasma facing surface might be diamond, to exploit its superb thermal and mechanical properties; there might be a subsurface layer of diamond-like carbon, whose viscoelastic properties might balance the brittleness of diamond and inhibit dust formation and spalling. If tritium is to find its way back into the plasma, one might need to exploit the hydrogen transport properties of cheap buckeytubes. Various approaches have been outlined [4]. Diamond's cost in a designer carbon is not prohibitive. Whilst materials for the immediate next generation of fusion devices have been settled [8], many of these suggestions are close to 1970s–1980s choices. Such choices were very sensible, but could not recognize the alternatives now available. The power plant of 2050 is still at the concept stage. There is time for a serious look at novel options.

1.7
Extending the Information Technologies

Silicon electronics is firmly established, and developing so rapidly that it is unlikely to be replaced in the foreseeable future. Thinking of diamond as an alternative to silicon electronics is far fetched. But does the diversity of carbons promise new classes of application alongside silicon? Could one link the functionalities of different carbons, so as to use the best properties of each of the carbon forms used?

Diamond plus compatible organic materials could meet many significant challenges, so long as one could exploit the convenience of organics to eliminate diamond's problems. So why are molecules special for electronic applications? Molecules have the right size (1–100 nm) to create functional nanostructures. The interactions between molecules enable a degree of nanoscale self-assembly, albeit with modest accuracy. Molecules can be switched in structure, enabling novel functionalities (cf. retinal); molecules can be tailored to control electron transport and optical properties (e.g. color based on charge transfer). However, most molecules cannot stand high temperatures.

The anticipated challenges to silicon stem from demands that current silicon technology is finding hard to meet. These demands reflect technological needs for greater miniaturization, higher device speed, lower power use, and heat dissipation. Despite developments like porous silicon, silicon-based displays are unlikely to be leaders. Nor is silicon the clear material of choice for new opportunities like electronic books. Silicon is challenged in hostile environments, whether at high temperatures or under electrochemical conditions. Silicon's key advantage is in know-how: there is reliable experience at an industrial level and at the state of the art. Present day carbon-based systems often use electrons in mundane ways. Devices based on small organic molecules and on semiconducting and conducting polymers can perform strongly, though often with some ingredient for commercial success still missing. Yet, there is a wealth of proposals [9]. Polymeric electrical insulators have long been standard in everyday life, with scope for science-based improvements. Designer organics with controlled electronic and ionic processes are the basis of battery and allied systems, and growing in their roles, but more substantial ideas are needed for carbon-based electronics to go beyond niches.

Do carbon systems offer more sophisticated options? Earlier, in ref [1], I noted five possible ideas. First, interface engineering promises effective control of electron transfer between different (perhaps carbon-based) media. Diamond shows good cold cathode behavior, and certainly clever things are possible with carbon-based nanotubes, including effective electron emission. A big problem is buckeytube reproducibility: if you find the ideal buckeytube, it may take a very long time to produce another like it. Secondly, there is plenty of scope to construct mesostructures [3, 10] to form photonic structures, whether by molecular design [11], texturing, blending, self-organization, or some combination of crosslinking or scission [3]. In understanding such mesostructures, average properties may be

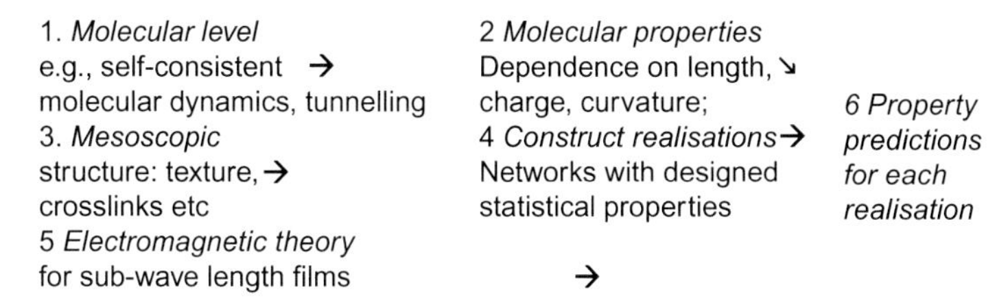

Figure 1.2 The linking of energy scales in mesoscopic modeling of carbon electronics (after ref [10]).

misleading, and many findings must be examined. It is crucial that features at the atomic, mesoscopic and larger scales are integrated (Figure 1.2).

Thirdly, there could be viable ideas for field-effect devices. One carbon-based material might supply the field, and another supply carriers, an especially powerful approach for highly correlated systems [12]. Fourthly, we think of wires to carry power (like interconnects on a chip) and optical fibers to carry signals, but we may forget how efficiently the α-helix in protein transfers energy, whether by solitons or otherwise. Mimics of such biological performance might offer value in physics-based technologies, for the ways in which carbon is exploited in life processes are remarkably ingenious. Fifthly, carbons can offer superb optical properties (apart from laser action, which still seems elusive), not just for conventional devices. Indeed, several proposed quantum computer gates are based on diamond. In all these options, there will be issues at several scale lengths: most critical action may be at the atomic and molecular scale; there will be macroscale optical wavelengths components or contacts, and the mesoscale is relevant whenever the topology of polymer components or similar are involved.

So it is still an open question as to whether carbon-based materials can find success in areas that silicon technology cannot address. For diamond to enable such developments, the key need is for simple ways to control dopants: diamond is still far less *convenient* than current semiconductor technology. Whilst diamond needs the advantages of the other carbons, it must be said that carbon electronics may not need diamond.

1.8
Can the Quantum be Tamed?

Quantum information processing raises two implicit questions. First, why should we bother with quantum computing when classical silicon-based computing is so powerful? Secondly, what can diamond do that silicon cannot do better? Quantum information processing [13] encompasses secure communication, database searching, computing, and other information technologies. Quantum information processing has emerged partly from push and partly from pull. The push comes from the success of silicon-based microelectronics and consequent technical demands from the drives to speed, miniaturization, low energy use, and heat dissipation.

These demands are reaching interesting limits, for example, the Semiconductor Industry Roadmap predicts the number of electrons needed to switch a transistor will fall to only one electron by 2020, making extrapolation tricky.

The pull to quantum information processing comes from basic questions and opportunities raised by quantum physics. How do you describe the state of a physical system? How does the state change if it is not observed? How do you describe observations and their effects? These questions provide significant reasons for pursuing quantum information processing. How do you describe the state of a physical system? Whereas a classical state is often simply state $|1>$ or state $|0>$, the quantum state is usually given as an amplitude, and can be a superposition of the sort $\alpha|0> + \beta|1>$. The coefficients α, β can embody quantum information. How does the state change if it is not observed? It evolves (deterministically) following the Schrödinger equation, which is a linear equation. Just because the equation is linear, in essence, we have multiple parallelism. Quantum parallelism relies on interference to extract a joint property of all solutions of some computational problem. How do you describe observations and their effects? Ignoring the subtleties, the system will collapse with some probability into a single eigenstate; one can know that a measurement has been made. The system "knows" it has been looked at, and this underlies strategies to frustrate eavesdroppers.

The minimum requirements for a quantum computer demand (i) some well-defined quantum states as qubits; (ii) the ability to initialize, that is, prepare acceptable states within this set; (iii) the means to carry out a desired quantum evolution (run the device), generally using *universal gates* to manipulate single qubits and to control the entanglements of two or more qubits (control their "quantum dance"); (iv) the avoidance of decoherence long enough to compute; (v) a means to read out the results (read process). Challenges concerning the means of avoiding or compensating for errors, and scalability, namely the capability to link enough qubits to do useful calculations, are implicit.

Many systems have been proposed for quantum computing. Yet, it is still not certain that a quantum *computer* is possible in the usual sense. Just as friction defeated nineteenth century mechanical computer makers, so decoherence (loss of quantum entanglement) might defeat large scale quantum computing, and long decoherence times imply long switching times [14], though some cases seem achievable [15].

Could diamond offer a seriously practical route to quantum information processing? Many important ingredients have been demonstrated, often using the NV^- nitrogen vacancy center, sometimes in combination with the substitutional nitrogen center N_s. The earliest experiments date from the 1980s [16]; subsequent important results [17–19] include work at room temperature. Some advantages of diamond are clear. Optical transitions in diamond can be very sharp in energy. If electron or nuclear spins are to be used as qubits, then spin–lattice relaxation could lead to decoherence, but this is slower when the velocity of sound is high and the spin–orbit coupling small, as in diamond. The spin lattice relaxation time at room temperature of N_s is of the order of a millisecond. One should stress that quantum behavior is not restricted to low temperatures. Certainly quantum statistics only differ from classical statistics at low temperatures, since the key factor is $h\omega/k_BT$, but quantum statistics relates primarily to systems at or near equilibrium. In quantum information processing, one exploits quantum dynamics, and aims to keep well away from equilibrium. Quantum dynamics involves h in several ways, and quantum phenomena are not directly limited by temperature.

Rather than examine a number of the many approaches, I shall describe the optically controlled spintronics ideas of Stoneham, Fisher and Greenland [20] (the SFG approach), who introduced three key concepts. The first idea was to use spins controlled by laser pulses to give universal gates. Secondly, entanglement achieved through stronger interactions in the electronic excited state of a "control" dopant, with only weak interactions in the ground state. This second new concept separates the *storage* of quantum information in qubit spin states from the *control* of quantum interactions. The third new concept is the exploitation of the disordered distribution of dopant atoms that is ordinarily found.

Dopants do not have to be placed at precise sites: the disorder is needed for *system* reasons. The original proposal suggested devices based on deep donors in silicon. Use of a random distribution of dopants has advantages because it turns out that the fabrication and operational steps are almost all ones that have been demonstrated by someone, somewhere in the world. This, of course, does not mean creating a working device is trivial. Implementation in silicon seems possible, and would exploit the power of silicon fabrication technology. Diamond could be even better. Indeed, with N_s as qubits, diamond offers a system with potential for room temperature operation that is an acceptably silicon-compatible alternative to devices based directly on silicon is possible.

Is diamond promising in other respects? Are there suitable qubits? The simple N_s donor seems excellent. Can we initialize the qubits? The NV^- center can be initialized, and quantum information exchanged with N_s even at room temperature. Are there suitable excited states for control dopants? The simple P donor may prove effective. Whilst excited state *energies* of many diamond defects and impurities are known very precisely, there is far less information on wave functions; indeed, it is surprising how little is known. Acceptor effective mass theory is recognized as working well, whereas donor effective mass theory is largely untested. There is evidence (albeit limited and indirect) for effective mass states for some deep defects. The GR2-GR8 lines of the neutral vacancy appear to be describable as excitations to effective mass states of an electron bound to the positive vacancy.

Will the entanglement survive enabling useful quantum information processing: can a quantum information processor, even a single gate, be run without being overwhelmed by decoherence? Here we must assess likely decoherence mechanisms. Anything enabling entanglement can cause decoherence (the fluctuation-dissipation theorem implies that faster switching means faster decoherence [14]) so *spontaneous emission* can cause loss of entanglement. Spin–lattice relaxation in the ground state would give decoherence (it would affect quantum information storage in the SFG approach) and here, as already observed, diamond is very good, with very long spin–lattice relaxation times for N_s. Excited state processes, like thermal ionization and two-photon ionization, could be problems, but probably not for diamond, given the energies. Entanglement may also be introduced via the control atoms in the SFG approach, but this can be avoided with the right pulse shapes. Readout can be done in several ways, for example, as suggested in ref [20]. Other approaches using NV^- have been demonstrated [17–19]. Thus one possible quantum information processor might be an optically controlled device with electron spins (say N_s) as qubits, P donors as the control dopants whose optical excitation controls the "quantum dance" (entanglement) of the qubits, and NV^- centers as a means to initialize and perhaps for readout. This approach, one of many solid state proposals, has some special features. Quantum information storage (as N_s spin states) is separated from quantum manipulations (adjustments of entanglement by electronic excitation of controls such as substitutional P). The selection of gates and qubits during processing cannot be done by mere focusing of the light used to excite the controls, which might achieve a resolution of about a micron, whereas the qubits and controls must be separated by about an excited orbital radius, say 10 nm. It is here that the disorder in the system is *desirable*, as indicated in Figure 1.3. Excitation energies for each group of control plus qubit groups will vary from one place to another, simply because dopant spacings and relative orientations vary. So individual gates might be oper-

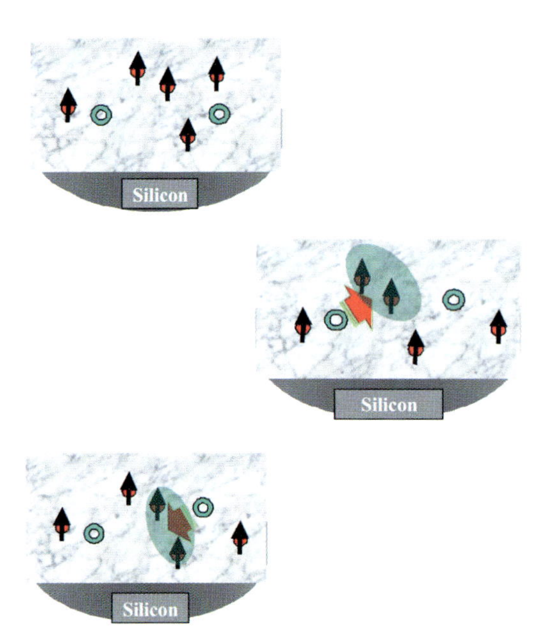

Figure 1.3 The Stoneham–Fisher–Greenland quantum gate. (a) With the control particle in its ground state, the qubit spins are isolated from one another. (b) Laser excitation over a region a few wavelengths across creates a delocalized bound state, causing indirect interaction between the spins. (c) Laser excitation at a different frequency in the same spatial region operates a different gate. Classically, the spins precess; quantally, their entanglement is changed, an example of a universal two-qubit gate.

ated by a combination of spatial (optical focus) and spectral (choice of wavelength) means. More detailed studies suggest that a likely architecture involves “patches” of perhaps 20 gates near the center of optically resolvable regions (perhaps 1–2 μm across), linked to other patches by so-called flying qubits. An approach to flying qubits is currently being developed as a patent application, so details are not available for this paper. What is important is that the device could be scalable, that is, one can recognize ways that quite large numbers of qubits, perhaps several hundreds, might be linked and operated at room temperature.

1.9 Conclusions: Beyond Those Niche Applications

Diamond does not lack applications. However, new, major applications are not easily found. One route may be *mix and match* carbon-based electronics. The flexibility, control, and cheap processing of organics, whether small molecules or polymers, can only be made effective by taking advantage of the truly impressive performance parameters of other carbons. So far, the efforts to bring these significant advantages together have been limited and short-term, and it is rare to see attempts to draw together the full wealth of carbons in a concerted way. There could be an opportunity to remove one of the potential problems for fusion reactors. To succeed, it may be essential to *design* a carbon, taking the best features of different forms in a complementary way to balance resistance to radiation, energy flux, tritium inventory, and other factors effectively. Solid state quantum information processing is one area where diamond seems an undeniable leader, possibly good enough to win through.

Acknowledgments

This work was supported by the Engineering and Physical Sciences Research Council (EPSRC), through its Basic Technologies and Carbon-Based Electronics programs, and also by Culham Laboratory, UK. I am indebted to many of my colleagues for their input. The paper is based on invited talks at the 2006 Institute of Materials, Minerals and Mining (IoMMM) Materials Congress, and at the November 2006 Materials Research Society (MRS) meeting. An earlier version of this paper is given in ref [21].

References

1 Stoneham, A.M. (2004) *Nature Materials*, **3**, 3.
2 Cacialli, F. and Stoneham, A.M. (2002) *Journal of Physics: Condensed Matter*, **14**, V9.
3 Stoneham, A.M. and Harding, J.H. (2003) *Nature Materials*, **2**, 77.
4 Stoneham, A.M., Matthews, J.R. and Ford, I.J. (2004) *Journal of Physics: Condensed Matter*, **16**, S2597.

5 Banhart, F. (1999) *Reports on Progress in Physics*, **62**, 1181.
6 Zaiser, M. and Banhart, F. (1997) *Physical Review Letters*, **79**, 3680.
7 (a) Roth, J. *et al.* (2001) *Nuclear Fusion*, **41**, 1967. (b) Roth, J. *et al.* (1999) *Journal of Nuclear Materials*, **266–269**, 51. (c) Roth, J. *et al.* (2004) *Nuclear Fusion*, **44**, L21. (d) Mech, B.V., Haasz, A.A. and Davis, J.W. (1998) *Journal of Applied Physics*, **84**, 1655.
8 Bolt, H. *et al.* (2002) *Journal of Nuclear Materials*, **307–*11***, 43.
9 Heath, J.R. and Ratner, M.A. (2003) *Physics Today*, **43**.
10 Stoneham, A.M. and Ramos, M.M.D. (2001) *Journal of Physics: Condensed Matter*, **13**, 2411.
11 Cacialli, F. *et al.* (2002) *Nature Materials*, **2**, 160.
12 Ahn, C.H., Triscone, J.-M. and Mannhert, J. (2003) *Nature*, **424**, 1015.
13 Williams, C.P. and Clearwater, S.H. (2000) *Ultimate Zero and One: Computing at the Quantum Frontier*, Copernicus (Springer Verlag), New York.
14 (a) Fisher, A.J. (2003) *Philosophical Transactions of the Royal Society of London*, **A361**, 1441. (b) Fisher, A.J. (2002) Lower Limit on Decoherence Introduced by Entangling Two Spatially-Separated Qubits, http://arXiV.org/quant-ph/0211200 (accessed 26 March 2008).
15 Plenio, M.B. and Knight, P.L. (1997) *Philosophical Transactions of the Royal Society of London*, **453**, 2017.
16 Oort, E., Manson, N.B. and Glasbeek, M. (1988) *Journal of Physics C*, **21**, 4385.
17 (a) Charnock, F.T. and Kennedy, T.A. (2001) *Physical Review B*, **64**, 041201. (b) Kennedy, T.A., Charnock, F.T., Colton, J.S., Butler, J.E., Linares, R.C. and Doering, P.J. (2002) *Physics of the Solid State*, **B233**, 416. (c) Kennedy, T.A., Colton, J.S., Butler, J.E., Linares, R.C. and Doering, P.J. (2003) *Applied Physics Letters*, **83**, 4190.
18 (a) Jelezko, F. and Wrachtrup, J. (2004) *Journal of Physics: Condensed Matter*, **30**, R1089. (b) Jelezko, F., Gaebel, T., Popa, I., Domhan, M. Gruber, A. and Wrachtrup, J. (2004) *Physical Review Letters*, **93**, 130501. (c) Childress, L., Dutt, M.V.G., Taylor, J.M., Zibrov, A.S., Jelezko, F., Wrachtrup, J., Hemmer, P.R. and Ludkin, M.D. (2006) *Science*, **314**, 281. (d) Wrachtrup, J. and Jelezko, F. (2006) *Journal of Physics: Condensed Matter*, **18**, S807.
19 (a) Hanson, R., Gywat, O. and Awschalom, D.D. (2006) *Physical Review B*, **74**, 161203. (b) Hanson, R., Mendoza, F.M., Epstein, R.J. and Awschalom, D.D. (2006) *Physical Review Letters*, *97*, 087601.
20 (a) Stoneham, A.M., Fisher, A.J. and Greenland, P.T. (2003) *Journal of Physics: Condensed Matter*, **15**, L447. (b) Stoneham, A.M. (2005) *Physics of the Solid State*, **C2**, 25.
21 Stoneham, A.M. (2007) MRS Proceedings, Symposium J. vol 956 page 1 Diamond Electronics: Fundamentals to Applications.

2
Growth and Properties of Nanocrystalline Diamond Films

Oliver A. Williams,[1] *Miloš Nesládek,*[2] *Jiří J. Mareš,*[3] *and Pavel Hubík*[3]
[1]Hasselt University, Institute for Materials Research (IMO), Wetenschapspark 1, B-3590 Diepenbeek, Belgium
[2]CEA-LIST (Recherche Technologique), CEA-Saclay, 91191 Gif sur Yvette, France
[3]Institute of Physics ASCR v. v. i., Cukrovarnická 10, 162 53 Prague 6, Czech Republic

2.1
Introduction

Recent progress in nanotechnology has motivated research into small grain sized diamond films known as nanocrystalline (NCD) and ultrananocrystalline diamond (UNCD) [1–6]. These materials have rather unusual properties compared to their single-crystal diamond counterpart and can be deposited from submicron thicknesses on a wide range of substrates including glass, Si and several metals. NCD and UNCD are both intended for applications in emerging nano/microelectromechanical systems (N/MEMS) types of devices, optical coatings, thermal management and finally electrochemistry and bioelectronics [3, 4]. However NCD and UNCD are two entirely different forms of diamond: UNCD is grown from Ar-rich plasma and it is a fine-grained material (5–15 nm) [1], with grain sizes practically independent of the film thickness, due to the fact that UNCD renucleates continuously during film growth. Consequently in excess 10% of the total volume fraction of UNCD is composed of grain boundaries, leading to a greater proportion of nondiamond or disordered carbon in the films. In contrast, NCD films are prepared using more conventional diamond plasma chemistry under atomic hydrogen-rich conditions, resulting in grain sizes which scale with the film thickness due to the columnar type of film growth [5]. The grain sizes of NCD materials are usually quoted as below 100 nm, but grain sizes and film thicknesses from about 30 nm are nowadays possible [6]. Based on the continuous etching of sp^2 carbon by atomic hydrogen during growth, the renucleation is suppressed during NCD growth, thus NCD has far lower proportions of sp^2 carbon than the UNCD films. This makes the NCD advantageous for applications to electrochemistry, where sp^2 carbon often influences the electrode stability.

The electrical conduction mechanism in these two types of film is entirely different. In UNCD, by the addition of nitrogen into the gas phase, it is possible to

Physics and Applications of CVD Diamond. Satoshi Koizumi, Christoph Nebel, and Milos Nesladek

ISBN: 978-3-527-40801-6

generate a high n-type conductivity material[7]. However this type of electrical transport is dominated by grain boundary conduction with a practically 2D type of conduction, which we demonstrate in this paper by magnetotransport measurements at low temperatures, and which leads to weak localization effects [8]. Using an admixture of 10–20% nitrogen (N_2) into the gas phase, high electric conductivity (~$1–10 \times 10^2\,S\,cm^{-1}$) can be achieved [2, 7]. On the other hand, in the NCD films conduction relies on substitutional doping of diamond grains with impurities such as boron [9]. The bulk doping of NCD with boron can yield superconductivity [10] if metallic doping is achieved. The main aim of this review is to compare the structural, optical and electrical properties of NCD and UNCD.

2.2 Growth

Undoped and boron-doped nanocrystalline diamond films can be grown in a variety of conditions, with common features of relatively low pressures and high nucleation densities. To achieve these high nucleation densities, several approaches can be adopted. Nucleation densities up to around $10^{11}\,cm^{-2}$ can be realized using vibration seeding, a process where the silicon wafer is placed on a polishing cloth impregnated with diamond powder on a vibration table. A weight is placed on top of the silicon wafer and the vibration table switched on for around 30 min. This process has the advantage of being very reproducible but has several disadvantages. Firstly, the maximum nucleation density is somewhat less than the state of the art. Secondly, as the process is used repeatedly, build up of damaged silicon/foreign particles from previously seeded substrates can lead to scratching of new wafers. Thirdly, some agglomerates of diamond powder may be left on the surfaces of seeded wafers, resulting in higher surface roughness. An alternative process is ultrasonic seeding in diamond suspensions of alcohol and diamond powders, a technique which is best described as a "black art", relying on many proprietary protocols. In this process diamond powder, usually ultradispersed diamond (UDD) [11] is mixed with an alcohol to form a pseudo-stable suspension. The concentration of this solution is crucial and pretreatments of diamond powders can also influence the stability and effectiveness of the suspension. In particular, bead milling can produce monodisperse colloids that are highly effective in seeding substrates [12]. The wafer is placed in this suspension and in an ultrasonic bath or horn. Figure 2.1 shows an example of the high nucleation densities obtainable by seeding with these colloids.

Ultimately, seeding in solution can yield higher nucleation densities (over $10^{11}\,cm^{-2}$), can treat three dimensional topographies and does less damage to silicon wafers than the vibration table technique. The principal problem with seeding in solution is the aggregation of fine particles into much bigger sizes than the core particles. This can leave hazy or opaque regions on the wafer due to the presence of residual powder. UNCD and NCD can both be grown on a plethora of substrates, Si, SiO_2, SiC, Al, Cu, refractory metals, low alkaline glasses such as

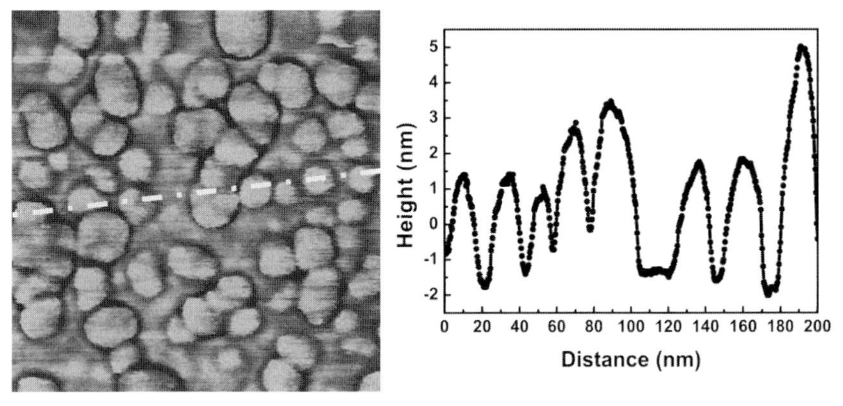

Figure 2.1 High nucleation densities obtained by seeding silicon with a nanodiamond colloid. The area is 200 × 200 nm. The particles appear somewhat bigger than expected due to the convolution effect with the AFM tip. The line scan shows the particles are less than 10 nm [12].

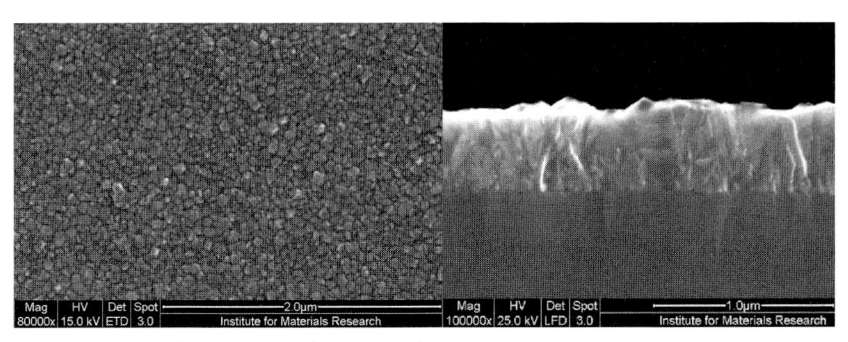

Figure 2.2 Top and cross-sectional scanning electron micrograph views of ~400 nm thin nanocrystalline diamond (NCD) films grown from hydrogen-rich plasma on Si substrate.

Corning Eagle 2000 and so on. Nucleation densities tend to be highest on silicon and refractory metals. Following seeding procedures, wafers were cleaned in a pure alcohol solution and placed inside a microwave plasma chemical vapor deposition reactor. Growth procedures were as follows. For NCD, films were grown in a hydrogen-rich plasma with a small percentage of methane. For doping, trimethylboron was also added to the gas phase. Typical microwave powers were between 1000 and 2500 W depending on substrate size and process pressure. Temperatures between 500 and 800 °C were monitored by dual wavelength pyrometry, the actual value again chosen for the relevant substrate. An example of the morphology of a ~400 nm thick NCD films is shown on the left of Figure 2.2.

It can be seen from this figure that the average grain size is below 100 nm, with the occasional larger grain and there are no macroscopic pin holes visible in the film. Thus the film is totally closed at a thickness of 100 nm: in fact we are able to grow close films as thin as 60 nm. Further optimization of seeding procedures will allow this minimum thickness of coalesced films to be reduced. An example of a cross-section of a much thicker film is shown on the right side of Figure 2.2. Here, the evolution of the microstructure is clearly visible, the grains increasing in size with increasing distance from the silicon substrate. This example was chosen as a particularly poor film grown with high CH_4 admixture. This film is clearly approaching a microcrystalline structure rather than NCD. Effect of the grain enlargement during the growth is a key issue when growing NCD, as it fundamentally represents a form of van der Drift growth, becoming microcrystalline as the bigger grains overtake the smaller ones. Thus the term "survival of the largest", is a good description of the overall process. This means that high nucleation densities are essential to ensure early coalescence of films. The high density of initial grains prevents the formation of large grains which can otherwise dominate the morphology from the beginning. Low methane levels also help ensure that the distribution of grain size remains narrow.

In stark contrast, UNCD films were grown in hydrogen-poor plasmas. Here, low percentages of methane (around 1%) were diluted in Ar. Typical microwave power and pressure levels were 1400 W and 150 torr respectively. The substrate temperature was monitored by a thermocouple attached to the substrate holder and was maintained at 800 °C. The conductivity of UNCD can also be controlled by the addition of nitrogen into the gas phase [7], reducing the Ar rate to maintain a constant total gas flow. Transmission electron microscopy has shown that these films have grain sizes of 3–5 nm with abrupt grain boundaries [13]. The films appear highly reflective but are absorbing in the visible spectrum. SEM images of a typical UNCD film can be seen in Figure 2.3. The left image is the plan view

Figure 2.3 Ttop and cross-sectional scanning electron micrographs of ~1000 nm thin ultrananocrystalline diamond (UNCD) film grown from Ar-rich plasma on Si substrate.

SEM. It can be seen that there is no direct evidence of crystallinity at this length scale as the grains are too small. The cross-section in the right of Figure 2.3 shows that this material shows no columnar structure and is thus fundamentally different to the van der Drift growth mode seen in NCD. This is due to the high renucleation rate of the hydrogen-poor Ar/CH_4 plasma, leading to continuous nucleation of new crystals and thus limiting the maximum grain size. One advantage of this is that the surface roughness is independent of the film thickness, as unlike in NCD, the grains are not expanding and competing. A disadvantage of this is that this material really consists of two phases, the material in the grains and that at the grains boundaries. Of course NCD also contains grain boundary material but the volume fraction is considerably less, in fact minimal relative to the 10% of UNCD carbon at the grain boundaries. These key differences in carbon configuration can be characterized by Raman spectroscopy as shown in Figure 2.4.

2.3 Raman Spectra of NCD and UNCD Films

The Raman spectra in Figure 2.4 show very typical Raman spectral shapes for UNCD and NCD samples. There are major differences in the Raman spectra of UNCD and NCD films. The Raman spectra for various N-doped UNCD films, detailed in Refs. [14, 15], exhibit two broad features at about 1370 cm^{-1} and 1550 cm^{-1} with no clear zone-center peak for diamond at 1332 cm^{-1}. These characteristic spectra can be explained by the fact that UNCD is comprised of nanograins of sp^3 carbon surrounded by grain boundaries of nondiamond carbon, consisting of several bonding configuration (sp, sp^2) and also disordered carbon. The grain boundary carbon occupies around 10% of the total film volume and thus little evidence of diamond is seen in visible Raman due to the enhancement of sp^2 peaks at this wavelength. The broadened zone center Raman line can be detected with UV excitation. The Raman spectra of UNCD films are also practically independent of nitrogen content in the gas phase, suggesting that the non-sp^3 components dominate the Raman spectra. However, the Raman spectra for NCD films which are nominally undoped are entirely different, reflecting the fact that NCD films are similar to columnar microcrystalline diamond, just with reduced grain sizes and thickness. To illustrate this difference Figure 2.4 compares the Raman spectra for undoped UNCD and NCD films with thicknesses of about 500 nm. In NCD films the zone-center phonon line at 1332 cm^{-1} is clearly visible together with a graphitic peak at about 1500 cm^{-1}. It should be noted that the Raman spectral shape also depends on the NCD thickness, that is, for very thin NCD layers in the range of 50 nm thickness, D and G peaks appear under red light excitation, related to disordered carbon present in a larger relative proportion than that for thicker films. For such thin films, peaks at 1120 cm^{-1} and 1450 cm^{-1} associated with transpolyacetylene at the grain boundaries are often detected. An interesting phenomenon occurs upon boron (B) doping of NCD films, detailed in Figure 2.5. For high B-concentrations, in the range around 2×10^{20} cm^{-3} or higher, we can see the so

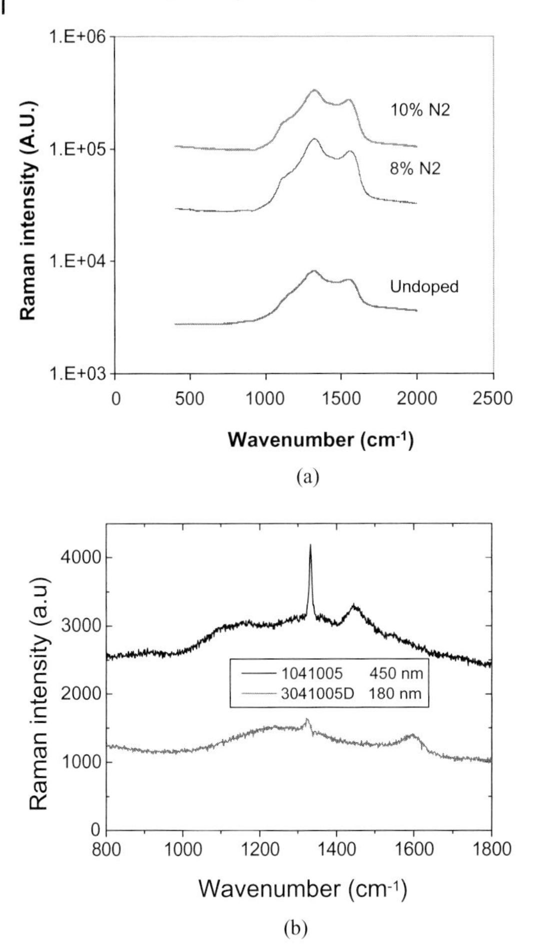

Figure 2.4 Comparison of Raman spectra of undoped and N-doped ultrananocrystalline (UNCD) thin films, prepared from Ar-rich plasma (a) and undoped nanocrystalline diamond (NCD) (b), prepared from the H-rich plasma. Specifically, the UNCD films exhibit broad features centered at 1370 cm^{-1} and 1550 cm^{-1}. For low thickness the NCD films show quite similar features to UNCD. For greater thicknesses a clear zone center diamond peak at 1332 cm^{-1} can be established. The position of the main Raman lines is discussed in the text.

called Fano resonance, which will be discussed below, confirming a substitutional B-incorporation in the diamond lattice. Figure 2.5 shows Raman spectra for two heavily B-doped samples containing 5×10^{19} cm^{-3} and 4×10^{20} cm^{-3} boron, as calibrated by SIMS measurements.

The position of the central zone phonon at 1332 cm^{-1} is marked by a dashed line in Figure 2.5. Upon increasing B-concentration, an asymmetric shoulder at about 1230 cm^{-1} appears with a broad tail towards lower wavenumbers. Similar results have been found for heavily B-doped single-crystal homoepitaxial layers with B-concentrations close to the Metal–Insulator Transition (MIT) transition ($N_B > 2 \times$

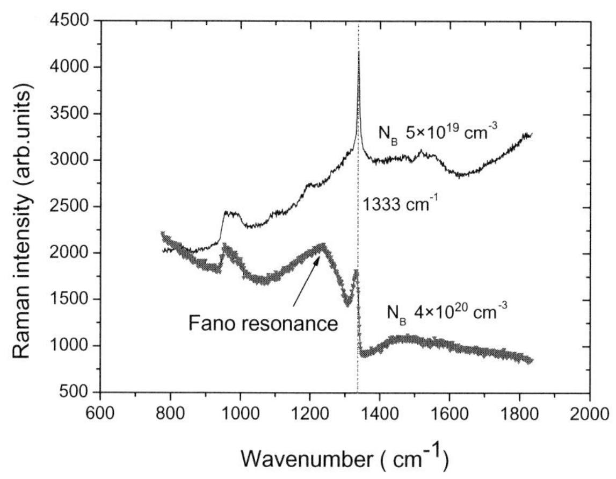

Figure 2.5 Raman spectra of various boron-doped nanocrystalline diamond films with B : C gas doping ratios of 800 and 3000 ppm indicated in figure. The position of the central zone phonon at 1332 cm^{-1} is marked by a dashed line. The Fano resonance is observable as a broad peak at about 1230 cm^{-1} and as a deformation of the zone-center phonon line.

10^{20} cm^{-3}). The peak at 1230 cm^{-1}, typical for heavily B-doped diamond with a B-concentration above MIT was explained by a Fano resonance, for example, quantum interferences between the continuum of electronic excitations and a zone-center phonon [16]. An additional interesting characteristic in Figure 2.5 is that sp^2-related Raman resonances in heavily B-doped films are significantly reduced and for the highest doping levels become almost invisible. This fact suggests that the grain boundaries of heavily B-doped NCD films are partially compensated in such way that sp^2 carbon in the grain boundaries is changed to a Raman inactive mode. This hypothesis needs further TEM study but one possible explanation is that segregation of boron at grain boundaries occurs which leads by subsequent interactions with grain boundary carbon atoms to generation of Raman-inactive B-rich B_xC_{1-x} phases. The segregation of boron onto the grain boundaries can easily be seen based on comparison of SIMS and Hall measurement data. In the metallic regime (there is 1 : 1 ratio between the boron concentration and free carrier concentration above MIT) we can clearly see that the B-concentration obtained from SIMS is a factor of two higher than the carrier concentration obtained from Hall measurements, suggesting the existence of some additional nonactive boron in the B-NCD grain boundaries.

2.4 Optical Properties of UNCD and B-NCD Films

Higher optical transparency of thin NCD films compared to UNCD is an important advantage for several applications requiring a combination of high conduction with full optical transparency. In general, the intrinsic optical absorption is related to the presence of nondiamond carbon such as sp^2 bonded carbon, present mainly at grain boundaries. Figure 2.6 compares the optical absorption of thin UNCD and NCD films, deposited on quartz substrate with the optical absorption of highly transparent CVD microcrystalline diamond [17]. Figure 2.6 demonstrates that the intrinsic absorption in diamond is sensitive to the grain size and thus the concentration of nondiamond carbon located at grain boundaries. However, the character of the broad absorption continuum, which decays towards the IR, is essentially the same. It has been suggested by Zapol *et al.* [18] based on molecular dynamic simulations, that a highly disordered subsystem at grain boundaries leads to an increase of the subgap density of states (DOS), located near the midgap. If nitrogen is introduced into the gas phase, the nitrogen further promotes the formation of midgap π-states. The essence of this picture is that the increased DOS in nanocrystalline diamond relates to sp^2-bonded carbon in grain boundaries as described in our previous work on microcrystalline diamond [19]. In ref. [19] we have used a two Gaussian distribution for π-π^* states to model such broad optical absorption in diamond films. Figure 2.6 thus clearly points towards the fact that in microcrystalline diamond, nanocrystalline diamond and ultrananocrystalline diamond the absorption spectra are related to the same type of nondiamond carbon defects of π-type state origin. However, its concentration increases progressively from microcrystalline diamond towards the UNCD structure. In UNCD films, the absorption coefficient is close to that expected for metallic types of behavior ($10^4\,cm^{-1}$). Compared to UNCD, NCD films have lower overall optical absorption. This is consistent with the columnar NCD growth, which leads to the development of large grains, identical to microcrystalline diamond. Thus the NCD films can be truly considered as an early stage of microcrystalline diamond growth. As an estimation, the optical absorption coefficient of about $10^3\,cm^{-1}$ for a 100 nm thick layer

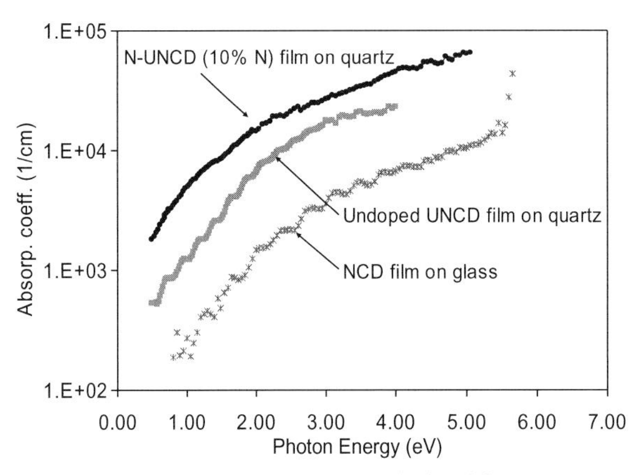

Figure 2.6 Optical absorption spectra calculated from photothermal deflection spectroscopy (PDS) for nanocrystalline (NCD) and ultrananocrystalline diamond (UNCD) films.

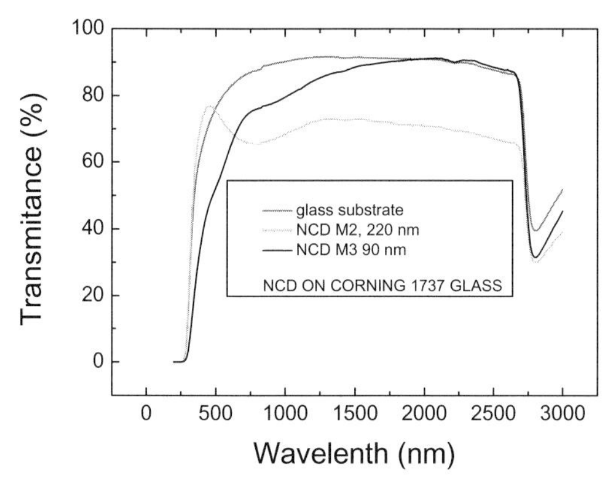

Figure 2.7 Optical transmission data measured in the range from IR to near UV for two undoped nanocrystalline (NCD) films, deposited on Corning #1737 glass substrate. The thickness of the films measured was 80–240 nm. Transmission data of Corning glass substrate is shown for comparison.

leads to the absorbance of only about 1%. This is the main reason why NCD films show very high transparency despite the fact that they have a relatively high absorption coefficient compared to natural or single-crystal CVD diamond. Figure 2.7 shows the optical transmission data for several undoped NCD films of various thicknesses. Figure 2.7 clearly demonstrates high optical transparency of NCD films. The transmission spectra are modulated by an interference pattern, which gives the samples a characteristic color. Typically, the color of a "just coalesced" layer, that is, a layer with grain sizes equal to the film thickness with nucleation density at about $10^{11}\,cm^{-2}$, and about 70 nm thick, is a blue. B-doping has a significant influence on the absorption of NCD films. When boron is incorporated in the diamond lattice it leads to a blue background color, independent of interference effects as discussed above. The influence of boron on the electrical transport properties is discussed in detail in following section.

2.5 Doping and Transport Measurements

One of the aims of this chapter is to compare the transport properties of doped UNCD and NCD films. In summary, in nitrogenated UNCD films, nitrogen does not act as a donor (nitrogen is a deep n-type dopant 1.7 eV from the conduction band minimum) and therefore does not significantly influence the conductivity of the UNCD diamond films. Rather, the high incorporation of nitrogen atoms in UNCD leads to the genesis of a transport path related to a highly conductive subsystem of π-states located in the grain boundaries. A rough estimate for the conductivity of such a subsystem can be made using semi-empirical formula for

Mott's MIT threshold cunductivity [8]: $\sigma_M = (8\pi\kappa\varepsilon_0/\hbar)\varepsilon E_A$, where the relative permittivity $\varepsilon = 5.5$ and the dimensionless constant κ ranges from 0.025 to 0.05 for various possible arrangements of doping centres.)This leads to reasonable limits for σ_M at room temperature of about $\sigma_M \approx 10\,\text{S cm}^{-1}$ with the activation energy of conductivity E_A, directly deduced from the slope of the temperature dependence plotted in Arhenius coordinates. The calculated value is close to typical experimentally measured values of $10–100\,\text{S cm}^{-1}$. Thus we attribute the metallic conduction in nitrogenated UNCD to transport in the grain boundary π-states, inducing a metallic conductivity. The details can be found in ref. [8]. On the other hand, in the case of boron-doped NCD films, boron is incorporated substitutionaly into the diamond lattice. Boron incorporates rather easily due to the fact that the boron atom has a covalent radius which is close to that of carbon, and therefore fits into substitutional sites. An isolated boron atom belongs to the third group of the periodic table and forms in diamond an acceptor state (p-type impurity), having a binding energy of ~0.37 eV. This is obviously a much larger acceptor activation energy than in silicon; however B is the shallowest impurity in diamond. Bohr's radius ($a_B \approx 3.1 \times 10^{-10}\,\text{m}$) for boron is comparable with the lattice constant of diamond ($a \approx 3.56 \times 10^{-10}\,\text{m}$) and boron therefore behaves like a typical shallow acceptor. B-NCD films show the same properties as bulk diamond with boron substitutionaly incorporated into the diamond lattice. There are several substantiations for this statement, starting with the intense blue color of B-NCD films, the observed Fano resonances and the observed superconductivity, similar to microcrystalline or epitaxial B-doped diamond [16]. Figure 2.8 shows for comparison, a set of magnetoresistance (MR) curves measured at magnetic fields perpendicular to the sample and at various temperatures ranging from 1.28 to 4.0 K for both N-UNCD and B-NCD films.

For N-UNCD films (see Figure 2.8a), the resistance increases as the temperature decreases at zero magnetic fields. This transport is atypical for ordered metals and it could be interpreted as a symptom of disordered transport. Further on, for UNCD samples the magnetoresistance data show a reduction of electrical resistivity with applied magnetic field. When looking carefully at the magnetoresistance curves in Figure 2.8a, after passing a certain magnetic field B_k, the MR dependences starts to quickly decrease (negative MR, NMR). The total decrease of resistance due to a magnetic field of 8 T reaches a value of 28% at ~1.28 K. As far as we know, the only effect, which can account for such a huge (giant) NMR could be the effect of weak localization (WL). Recently we have explained the observed NMR curves under assumption that their shape is controlled by a competition of the electrostatic and magnetic Aharonov-Bohm's effects [21, 22]. Our modeling [8] led to the derivation of a formula relating the critical magnetic field B_k to the film conductivity G:

$$\sqrt{B_k} = \gamma(\hbar/e)^{1/2}(G - G_0). \tag{2.1}$$

This formula, derived by Mares *et al.* in ref. [8] relates a critical magnetic field B_k for the transition to delocalized conduction, induced by the magnetic field perturbation. The term $G–G_0$ is usually called the "quantum correction" with G_0 and

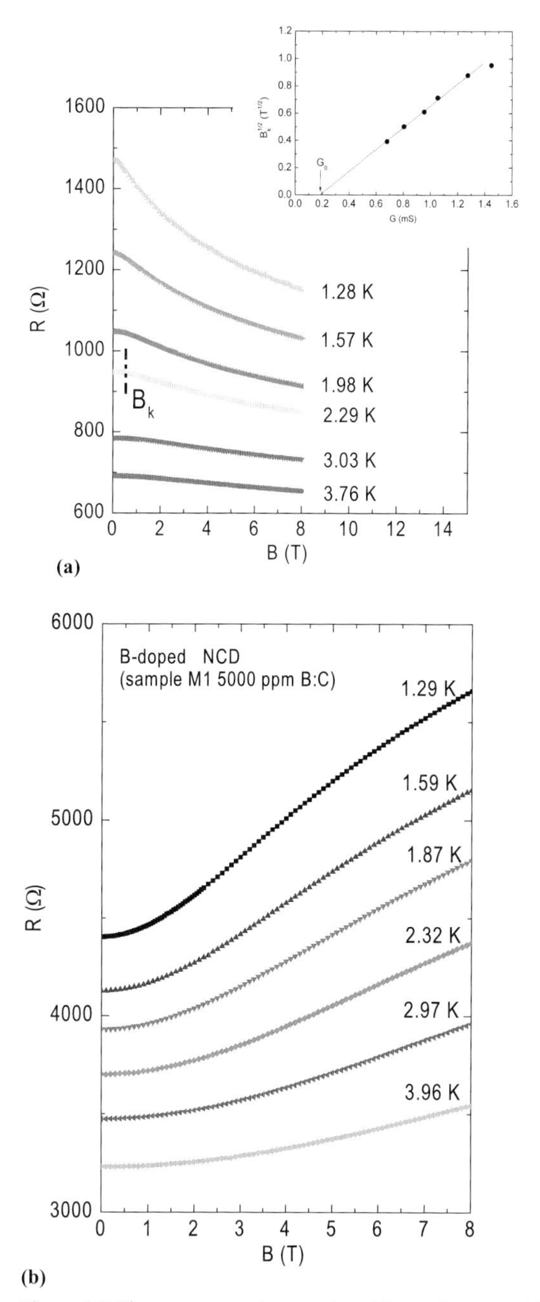

Figure 2.8 The magnetoresistance data (a) measured at low helium temperatures for 20% N-doped UNCD films, prepared at Argonne Laboratories. USA. The inset shows the dependence of conductivity on the critical magnetic field B_k for which the delocalization of the wavefunction occurs according to Equation 2.1, discussed in the text (starting at the B_k critical magnetic field onset); (b) magnetorestistance data for a B-doped NCD sample with B-concentration of $5 \times 10^{19}\,cm^{-3}$ showing a positive magnetoresistance based on the shrinkage of the wave function with applied magnetic field.

γ being empirical constants. When using the experimental B_k values from Figure 2.8a, an excellent agreement between experimental data and formula (1) suggests validity of our assumption of weak localization transport at low temperatures (see the inset oinFigure 2.8a, where the square root dependence of the experimentally determined B_k is plotted as a function of measured conductivities, and fitted using Equation 2.1). It supports the suggestion that the subsystem of grain boundaries has a character comparable to low-dimensional (2D) disordered metals, which is controlled by quantum interference effects of electrons resulting in their weak localization. Estimates for the dimensions of the localization lengths L_φ based on the experimentally determined values of B_k for weakly localized orbits leads to L_φ of about 7 nm. So the extent of phase-coherent wave functions is comparable to the grain sizes, observed by TEM in our UNCD films (10–16 nm). This conduction, similar to that in a disordered metal, explains the negative Hall sign on one side and atypical temperature dependence of conductivity on the other. In contrast, B-NCD films show an entirely different low-temperature transport behavior. To elucidate this transport mechanism Figure 2.8b shows the magnetoresistance at liquid helium temperatures for a B-NCD film where the boron concentration was below the MIT transition (SIMS data show the B-concentration in the range of $5 \times 10^{19}\,cm^{-3}$). In contrast to UNCD, B-NCD films show a positive magnetoresistance (e.g. an increase in resistance with magnetic field). This kind of transport could be explained by localization of the wavefunction on boron atoms [10]. With increasing magnetic field, a shrinkage of the extension of the wave functions leads to a progressive localization, resulting in a decrease of the conductivity with magnetic field, giving rise to a positive magnetoresistance [10].This picture is consistent with the substitutional incorporation of boron into the diamond lattice and reduced effect of conductance via grain boundaries. Recently, superconductivity has been observed for heavily B-doped CVD diamond films with substitutional B-

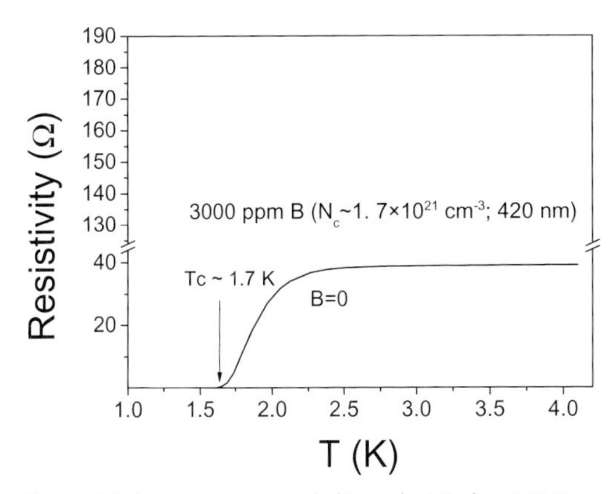

Figure 2.9 Low temperature helium electrical resistivity measurements for boron-doped nanocrystalline diamond (B-NCD) films, with an RT carrier concentration N_B equal to $1.7 \times 10^{21}\,cm^{-3}$. The sample shows a superconductive transition at T_c ~1.7 K.

incorporation [23–25]. Thus for B-NCD we can also expect a superconductive transition at low temperatures. Figure 2.9 depicts low temperature conductivity behavior in a B-doped NCD film of thickness 420 nm, with a room temperature carrier concentration of about $1.6 \times 10^{21}\,cm^{-3}$ and a room temperature mobility of about $1\,cm^2\,V^{-1}\,s^{-1}$. At temperatures of about 1.7 K the resistivity drops below $10^{-4}\,\Omega$, which is the limit of our van der Paw LT Hall setup. For comparison, the epitaxial B-doped layers grown on (100) surfaces in ref. [16] have shown an SC transition at a T_c of 1.4 K for $N = 1.2 \times 10^{21}\,cm^{-3}$: this is close to values found in B-NCD diamond thin films. Thus it is clear that at low temperatures the conduction mechanism in BNCD films is significantly different to UNCD and at the same time almost identical to conduction in heavily substituted B-doped diamond. Finally Figure 2.10 shows the conductivity values for B-NCD layers for three B-concentrations, specifically for 600, 800 and 3000 ppm B : C ratio.

It can be established that heavily boron-doped NCD films indeed show very high conductivity, similar to heavily B-doped epitaxial films, with the MIT transition observed by a strong increase in conductivity at B-concentrations of about $2–3 \times 10^{20}\,cm^{-3}$ [8, 10]. Heavily B-doped diamond single crystals show hopping transport just below the MIT, whilst samples with B-concentration above MIT show semimetal transport behavior [26]. In our case, for concentrations above the MIT transition, the Hall coefficient confirms hole-type conduction with mobilities in the range $0.5–2\,cm^2\,V^{-1}\,s^{-1}$ which are typical for highly disordered Mott's metals. For concentrations below the MIT transition, the conductivity plotted in Arhenius coordinates shows a rather complex behavior. Modeling the transport data by fitting with a variable range hopping formula in 3 dimension ($\sigma \sim T^{(-1/4)}$) or in 2 dimensions ($\sigma \sim T^{(-1/3)}$), or alternatively nearest-neighbor hopping ($\sigma \sim T^{(-1/2)}$) does not give an acceptable fit. Below a level of about $1 \times 10^{19}\,cm^{-3}$ films again tend to show temperature activated behavior and p-type conductivity, however with

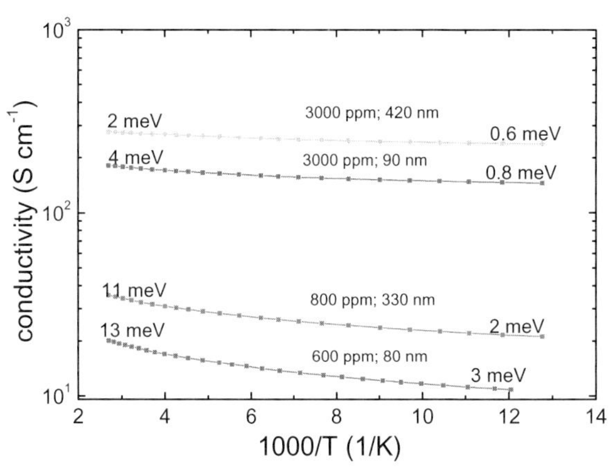

Figure 2.10 Conductivity measurements (a) for a set of boron-doped nanocrystalline (B-NCD) samples for boron concentration N_B of 600, 800 and 3000 ppm B : C ratio for various film thicknesses.

mobilities significantly lower (factor 100–1000) than for B-doped epitaxial diamond [10].

2.6 Conclusions

The aim of this chapter is to summarize the state of the art for the growth and doping of nanocrystalline diamond (NCD) and ultrananocrystalline diamond (UNCD) thin films. Based on experimental data we discuss in detail the NCD and UNCD Raman, optical and electrical transport properties which are important for understanding the nanocrystalline diamond structure and properties related to several applications such as electronics and bioelectronics. NCD films are grown mainly in a H-rich plasma/gas phase and they resemble, because of their columnar growth, microcrystalline diamond with the result that they have similar Raman spectra, exhibiting a 1332 cm^{-1} zone–center phonon Raman line. NCD films can be deposited with thicknesses ranging from about 50 nm to several hundreds of nm on various substrates such as glass, Si and several metals The conduction mechanism in boron-doped NCD layers is of a semiconducting type, which can be converted to a metallic type of conduction when the B-acceptor concentration reaches the Mott metallic conductivity limit of $2 \times 10^{21} cm^{-3}$. At the same time the B-NCD layers are optically transparent. UNCD films, however, have very different structures, due to the growth mechanism in Ar-rich plasmas, leading to continuous diamond renucleation during growth. This leads to a higher proportion of nondiamond carbon in UNCD films compared to NCD films. Consequently, UNCD films have lower optical transparency than NCD films of the same film thickness. Also UNCD films do not show a semiconducting type of behavior; and the UNCD conductivity mechanism is governed by a highly conductive metallic system located at the grain boundaries. However, the advantage of UNCD films is the continuous renucleation during growth which leads to small grain sizes even for thicker UNCD films.

Acknowledgments

The authors would like to thank J. Chevallier and A. Rzepka from CNRS Meudon for the SIMS and Raman measurements. Thanks are also due to C. Mer and D. Tromson, for help with the B-NCD sample growth at CEA Saclay and the defect spectroscopy study and finally to P. Bergonzo for fruitful discussions. The authors acknowledge the partial support of the FP6 project DRAMS, Contract No.: STRP 033345. The work was also supported by the Czech Science Foundation, Contract No 202/06/0040. Furthermore, the research work at the Institute of Physics was supported by Institutional Research Plan No AV0Z10100521.

References

1 Gruen, D.M. (1999) *Annual Review of Materials Science*, **29**, 211.
2 Bhattacharyya, S., Auciello, O., Birrell, J., Carlisle, J.A., Curtiss, L.A., Goyette, A.N., Gruen, D.M., Krauss, A.R., Schlueter, J., Sumant, A. and Zapol, P. (2001) *Applied Physics Letters*, **79**, 1441.
3 Yang, W., Auciello, O., Butler, J.E., Cai, W., Carlisle, J.A., Gerbi, J.E., Gruen, D.M., Knickerbocker, T., Lasseter, T.L., Russell, J.N., Smith, L.M. and Hamers, R.J. (2002) *Nature Materials*, **1**, 253–7.
4 Härtl, A., Schmich, E., Garrido, J.A., Hernando, J., Catharino, S.C.R., Walter, S., Feulner, P., Kromka, A., Steinmüller, D. and Stutzmann, M. (2004) *Nature Materials*, **3**, 736–42.
5 Yang, W., Auciello, O., Butler, J.E., Cai, W., Carlisle, J.A., Gerbi, J.E., Gruen, D.M., Knickerbocker, T., Lasseter, T.L., Russell, J.N. Jr, Smith, L.M. and Hamers, R.J. (2003) *Materials Research Society Symposium Proceedings Volume 737: Quantum Confined Semiconductor Nanostructures* (eds J.M. Buriak, D.D.M. Wayner, F. Priolo, B. White, V. Klimov and L. Tsybeskov), paper F4.4.
6 Williams, O.A., Daenen, M., Haen, J.D., Haenen, K., Nesladek, M. and Gruen, M. (2006) *Diamond and Related Materials*, **15**, 654.
7 Williams, O.A., Curat, S., Gerbi, J., Gruen, D.M. and Jackman, R.B. (2004) *Applied Physics Letters*, **85**, 1680.
8 Mares, J.J., Hubík, P., Kritofik, J., Kindl, D., Fanta, M., Nesládek, M., Williams, O. and Gruen, D.M. (2006) *Applied Physics Letters*, **88**, 092107.
9 Nesladek, M., Tromson, D., Bergonzo, P., Hubik, P., Mares, J.J., Kristofik, J., Kindl, D., Williams, O.A. and Gruen, D. (2006) *Diamond and Related Materials* **15**, 607.
10 Nesladek, M., Tromson, D., Bergonzo, P., Hubik, P., Mares, J.J. and Kristofik, J. (2006) *Applied Physics Letters*, **88**, 232111.
11 Greiner, N.R., Philips, D.S., Johnson, J.D. and Volk, F. (1988) *Nature*, **333**, 440.
12 Williams, O.A., Douheret, O., Daenen, M., Haenen, K., Osawa, E. and Takahashi, M. (2007) *Chemical Physics Letters*, **445**, 255.
13 Gruen, D.M. (2001) *MRS Bulltein*, **26**, 71.
14 Ferrari, A.C. and Robertson, J. (2000) *Physical Review B*, **61**, 14095.
15 Birrell, J., Gerbi, J.E., Auciello, O., Gibson, J.M., Johnson, J. and Carlisle, J.A. (2005) *Diamond and Related Materials*, **14**, 86.
16 Bustarret, E., Kacmarik, J., Macenat, C., Gheeraert, E., Marcus, J. and Klein, T. (2004) *Applied Physics Letters*, **93**, 237005.
17 Nesládek, M. (2005) *Semiconductor Science and Technology*, **20**, R19.
18 Zapol, P., Sternberg, M., Curtiss, L.A., Frauenheim, T. and Gruen, D.M. (2001) *Physical Review B*, **65**, 045403.
19 Nesládek, M., Meykens, K., Stals, L.M., Vanecek, M. and Rosa, J. (1996) *Physical Review B*, **54** (8), 5552–61.
20 Mott, N.F. (1987) *Conduction in Non-crystalline Materials*, Oxford University Press, New York.
21 Bergmann, G. (1983) *Physical Review B*, **28**, 2914.
22 Mareš, J.J., Krištofik, J. and Hubík, P. (1997) *Solid State Communications*, **101**, 243.
23 Ekimov, E.A., Sidorov, V.A., Bauer, E.D., Melnik, N.N., Curro, N.J., Thompson, J.D. and Stishov, S.M. (2004) *Nature*, **428**, 542.
24 Takano, Y., Nagao, M., Sakaguchi, I. Tachiki, M., Hatano, T., Kobayashi, K., Umezawa, H. and Kawarada, H. (2004) *Applied Physics Letters*, **85**, 2851.
25 Yokoya, T. *et al.* (2005) *Nature*, **438**, 647–50.
26 Tshepe, T., Kasl, C., Prins, J.F. and Hoch, M.J.R. (2004) *Physical Review B*, **70**, 245107.

3
Chemical Vapor Deposition of Homoepitaxial Diamond Films

Tokuyuki Teraji
National Institute for Materials Science, 1-1 Namiki, Tsukuba, Ibaraki 305-0044, Japan

3.1
Introduction and Historical Background

3.1.1
Diamond – A Superior Semiconducting Material

Diamond will become an increasingly important future electronic device material with superior semiconducting properties. In addition to the large band gap corresponding to deep ultraviolet light region, the high breakdown electric field and high thermal conductivity of diamond are attractive for semiconductor use because these properties open up a new field of research. Fundamental values of diamond are listed in Table 3.1 with those of other common semiconductors. Because of the high physical/chemical stability of diamond, diamond devices can operate under harsh environments at high temperature, in high radiation environments, or in strong chemical conditions. Room-temperature electron mobility of diamond is rather high and hole mobility at room temperature is the highest among the applicable semiconductors. Then we can expect high-performance diamond-based optoelectronics such as ultraviolet light emitting diodes [1, 2] and ultraviolet photodetectors [3–5] using high-quality single crystalline diamond.

Because of sample availability, natural diamond crystals and single-crystalline diamond synthesized by the high-temperature/high-pressure (HPHT) method were mainly examined during initial phases of semiconducting diamond research [6]. Subsequently, after development of the low-pressure diamond chemical vapor deposition (CVD) method by Japanese researchers with the promise of realistic growth configuration [7, 8], CVD-grown diamond films came to be investigated because CVD diamond films offer advantages for electronic applications in crystal purification and doping for p-type and n-type conductivity.

Semiconductor research centering upon the use of diamond still faces various fundamental issues that must be resolved before advancing to practical applications. The most important subjects are the improvement of crystalline quality and

Physics and Applications of CVD Diamond. Satoshi Koizumi, Christoph Nebel, and Milos Nesladek

ISBN: 978-3-527-40801-6

Table 3.1 Fundamental values of semiconductors.

			Si	GaAs	SiC (6H)	SiC (4H)	GaN	Diamond
300K								
Bandgap	eV		1.12	1.42	2.86	3.36	3.39	5.47
Thermal conductivity	W/cmK		1.5	0.46	3.6	3.6	2	21.9
Carrier cobility	cm^2/Vs	Electron (Hall)	1450	8500	370	981	1245	660
		Electron (TOF)						4500
		Hole (Hall)	370	400	100	120	24	1650
		Hole (TOF)						3800
Dielectric constant			11.9	13.1	9.7	9.7	10.4	5.7
					10.03	10.03		
Saturation velocity	cm/s	Electron (calc.)	10^7	2×10^7	2×10^7	2.2×10^7	2.5×10^7	1.5×10^7
		Electron		1.9×10^7	2×10^7		2.5×10^7	2.5×10^7
		Hole	10^7					1.1×10^7
Breakdown field	MV/cm		0.3	0.4	3	2.7	2.7	10–20
Activation energy	meV	n-type	45 (P), 54 (As)	5.8 (Si), 5.9 (Se)	97_h, 141_c (N)	42_h, 84_c (N)	64, 30 (N_{vac})	570 (P)
						53_h, 93_c (P)		
						67_h, 127_c (As)		
		p-type	45 (B), 67 (Al)	31 (Zn), 35 (Cd)	270 (Al), 390 (B)	183 (Al), 285 (B)	250 (Mg)	360 (B)
500K								
Bandgap	eV		1.05	1.33	2.79			5.42
Thermal conductivity	W/cmK		0.8		1.7		1	11.2
Carrier mobility	cm^2/Vs	Electron (Hall)	400	2500	70	200	350	270
		Hole (Hall)	150					500
600K								
Bandgap	eV		0.86	1.26	2.74			5.38
Thermal conductivity	W/cmK		0.55		1.15			9.6
Carrier mobility	cm^2/Vs	Electron (Hall)	200		20	90		130
		Hole (Hall)	100					250

h: hexagonal site; c: cubic site; vac: vacancy.

control of electrical conductivity of diamond. For both of them, development of crystal growth techniques is a key to requirement. The study of diamond growth under low-pressure conditions started in the 1950s and has progressed to the present with the aim of growing high-quality diamond for the device applications. During the last 15 years, some breakthrough results have been reported from several research groups through advanced studies on homoepitaxial diamond growth. Excellent semiconducting properties that are comparable to or better than those of the common semiconductors have been obtained experimentally from high-quality homoepitaxial diamond grown by the method of plasma-assisted chemical vapor deposition. Synthesis of diamond large crystal is another important research issue at present. Heteroepitaxy is a different route for enlarging crystals [9], but the quality of the film remains far from that required for actual electronic device application at the present stage. Therefore, fundamental studies of semiconducting diamond including applicability to electronic devices are mainly making progress using homoepitaxially grown single-crystal diamond. Growth mechanisms of diamond crystal under low-pressure conditions have gradually been revealed through many studies on homoepitaxial diamond. In this chaper, recent studies on homoepitaxial diamond growth are reviewed. High-rate growth of homoepitaxial diamond will be discussed in later sections.

3.1.2
Low-Pressure Chemical Vapor Deposition

For most semiconductors, including silicon, chemical vapor deposition (CVD) is carried out under equilibrium conditions. Simple growth techniques such as thermal CVD are available for crystal growth. On the other hand, because graphite is the most stable state of carbon under normal or low-pressure conditions, low-pressure deposition of diamond must be conducted under nonequilibrium conditions. A difference from the case of the other IV-group materials such as Si or SiC, is that the reaction temperatures during diamond CVD are much lower than the Debye temperature that describes the degree of lattice vibration of a crystal. During diamond growth under low pressure, graphite phases are formed on the growth surface in a certain amount together with diamond phases. Consequently, the graphite must be removed effectively to continue the diamond CVD. Atomic hydrogen has the function of etching the graphite phases selectively while keeping the diamond phases stable. The carbon source gas must therefore be highly diluted with hydrogen, which is characteristic of diamond CVD under low pressure.

Several growth methods have been proposed for diamond film deposition under low-pressure conditions: hot-filament CVD, plasma-assisted CVD, combustion flame CVD, and so on. Because hot-filament CVD is a simple growth method [7] it has been widely accepted among diamond researchers in the early stages of research. This method has been applied for homoepitaxial diamond deposition. However, since the hot-filament CVD method presents problems of film contamination with filament materials, microwave plasma-assisted CVD, which has no discharging electrode inside the reaction chamber [8, 10], came to be used for high-quality homoepitaxial diamond growth. Direct-current plasma CVD [11] has been applied for heteroepitaxial diamond growth on foreign substrates [12, 13]. The flame combustion method is advantageous for its high growth rate; the method can provide thick diamond [14, 15]. Thickness and quality of homoepitaxial films grown by the flame combustion method are inhomogeneous in the film lateral direction [14, 15], which might result from the large spatial distribution of plasma in the fire flame. In addition, impurities tend to be doped unintentionally from ambient gas into the diamond film because no evacuation system is set up for this method, which is another problem from the viewpoint of electronic device application. Against this background, microwave plasma-assisted CVD is the major method for homoepitaxial growth at present. The following paragraph simply describes the configuration of the microwave plasma-assisted chemical vapor deposition (MPCVD) apparatus.

In a conventional MPCVD system, the 2.45-GHz microwave emitted from a generator propagates to a reaction chamber in TE_{10} (transverse electric) mode through an isolator, a tuner, and an applicator, which are all connected with waveguides. Both forward and reverse microwave powers are measured using power monitors set in the waveguide. A quartz tube is commonly used as a reaction chamber because it is transparent and is exchangeable. Moreover, it offers small dielectric power loss of microwaves. However, atomic hydrogen in plasma can etch the quartz wall and grown films are consequently contaminated with etched silicon. This is a considerable problem for synthesizing electronic device-grade diamond crystals. Recently, to suppress this contamination, researchers have begun to use stainless-steel reaction chambers whose inner diameter is sufficiently large to avoid the plasma coming into contact with it. Microwaves are injected into the chamber through the TE/TM mode converter which changes TE_{10} mode of microwave into TM_{01} (transverse magnetic) mode. In this case, we must establish these rather expensive CVD systems independently for each doping study because exchanging the stainless chamber is not an easy or routine task. Microwave generators, which are popularly used, have maximum power output of about 1 kW, although generators with higher output power have recently been employed for diamond growth. The MPCVD apparatus equipped with these high power generators, 5–10 kW for research purposes and 50–60 kW for development and production stages, effectively provide high-rate growth or large-area growth [16–18].

High-purity methane is widely used as a source gas for homoepitaxial diamond CVD. Purity of the methane gas is 6–7 N (99.9999–99.99999%), whereas that of hydrogen supplied from a gas cylinder is 7 N. A gas purifier using palladium is capable of purifying the hydrogen gas to more than 7 N. Hydrogen purification is effective to improve diamond purity, especially in the case of diamond deposition under very low methane concentration, because the impurities in hydrogen gas dominate residuals of the total feed gas. The substrate temperature and the total gas pressure are typically 700–1000 °C and 20–100 Torr, respectively. Higher temperatures and higher pressures are sometimes employed as process parameters to increase the diamond growth rate. The substrate temperature can be monitored using non-contact optical measurement with a pyrometer. The effect of the plasma radiation must be considered for precise evaluation and control of the substrate temperature. Boron and phosphorus are useful as p-type and n-type dopants. These doped homoepitaxial diamonds give superior electrical and optical properties, which originate from diamond electronic properties.

3.1.3 Homoepitaxial Diamond Films

Diamond growth mechanisms under low pressure condition have frequently been discussed based on polycrystalline diamond studies. The polycrystalline diamond film is composed of small crystallites with several small crystal planes, so-called facets. These crystallites with various facets are oriented randomly and are mutually attached to form one polycrystalline film. During low-pressure CVD, diamond grows simultaneously on each facet of the crystallites, whereas the effect of process parameters on diamond growth mode differs from facet to facet. In addition, the diamond crystallites knock together during diamond growth, which complicates the growth-mode analysis. For those reasons, it is difficult to understand fully the growth mechanism for aspects such as the growth rate, crystalline quality, and impurity incorporation from studies only of polycrystalline diamond. Semiconducting properties obtained from polycrystalline diamond are not as high as expected from their diamond characteristics because polycrystalline diamond has a certain amount of crystalline defect as crystal grains. Moreover, the defect density inside each grain of polycrystalline diamond is much higher than that for a single crystal. Therefore, homoepitaxial diamond has come to be studied intensively with a constant supply of HPHT substrates.

The first investigation of homoepitaxial diamond films using MPCVD was undertaken by Kamo *et al.* [19], following his demonstration of MPCVD for polycrystalline diamond growth. Following this study, homoepitaxial diamond has been investigated mainly in Japan, USA, and Europe. Natural or HPHT-synthesized single crystals, which contain nitrogen or boron as impurities, are available as substrate diamond. Mirror-polished HPHT type-Ib single crystals are commercially available at a reasonable price and have therefore been widely used as substrates for homoepitaxial growth. The type-Ib crystals contain substitutional nitrogen in concentrations of 10–100 ppm. Although the substitutional nitrogen in the substrate seems not to affect the homoepitaxial film growth much, the substrate-related features sometimes obscure characterization of electrical/optical properties of homoepitaxial films. We must therefore distinguish the properties of the homoepitaxial film from those of the substrate underneath the film. This problem can be resolved using a transparent type-IIa substrate containing a tiny amount of nitrogen. Typical dimensions of the substrate are 2×2–$4 \times 4\,\text{mm}^2$ in flat surface area and about 0.5 mm in thickness. A single crystal diamond whose diameter is circa 8 mm was grown using MPCVD [16], which indicates that the low-pressure growth method can provide a larger substrate. Crystal orientations that are commonly examined for homoepitaxial studies are (100), (111), and (110) faces.

In early research stages, identification of homoepitaxial diamond growth on the substrate was a fundamental issue. Therefore ^{13}C isotope was used in some cases as a source gas for diamond CVD [19–21]. Measurement of homoepitaxial film thickness is somewhat difficult. The thickness is commonly estimated from the weight change or selective growth of diamond by covering a part of the substrate surface [22, 23]. The nitrogen depth profile by secondary ion mass spectroscopy (SIMS) is a valid way to measure film thickness because the nitrogen concentration changes at the film/substrate interface when the substrate is a type-Ib crystal. The growth rate can be estimated simply from the homoepitaxial film thickness divided by the growth duration because we need not consider the diamond nucleation duration. Growth occurs more slowly than 1 μm h^{-1} under typical MPCVD conditions; the finally obtained film thickness is typically in the order of a few micrometers. Homoepitaxial diamond crystals which are thicker than 1 mm have recently been deposited using the high-rate growth mode.

Surface morphology can be characterized using a differential interference-contrast (Nomarski-mode) optical microscope (OM) or a scanning electron microscope (SEM). In contrast to polycrystalline diamond, homoepitaxial films grown under optimized conditions are morphologically smooth and electronically homogeneous. The SEM characterization is rather insensitive to the surface morphological features of homoepitaxial films such as hillocks or bunching steps. Nomarski-mode optical microscopy is frequently applied for macroscopic characterization of surface morphology because it clearly reveals the morphological features of homoepitaxial diamond films. Microscopic surface structures are characterized mainly by electron diffraction methods [24]. Near-field microscopy provides real-space information about atomic-scale surface structures. Furthermore, scanning tunneling microscopy (STM) is a suitable method for investigating bonding states of surface atoms because a high-resolution image is obtainable and the images reflect the electronic states of surface atoms, although STM requires electrical conductivity [25]. On the other hand, atomic force microscopy (AFM) is widely used because it yields information on the bunching step structure and surface roughness through comparably easy operations, in addition to the fact that AFM is applicable with insulators.

Crystallinity or purity of homoepitaxial diamond is frequently evaluated using Raman, cathodoluminescence, and photoluminescence spectroscopy, which are powerful but nondestructive methods. The transmission electron microscope and electron spin resonance are available to investigate the microscopic crystalline structure and electronic states of defects. Hall measurements are commonly used for electrical characterization, but the measurement is not easy because of the highly insulating property of diamond. AC magnetic-field-type Hall measurements are an effective means to characterize such highly insulating specimens as phosphorus-doped n-type diamond [26].

3.2 Effects of Process Parameters on Homoepitaxial Diamond Film Quality

This section presents a description of how each parameter of the growth conditions of MPCVD affects the diamond growth mode, based on the reported results from homoepitaxial diamond. Although some fundamental process parameters of growth condition are discussed below, it is noteworthy that these parameters are mutually related. For example, when the total gas pressure is changed, the substrate temperature is modified as well because the plasma radiation density at the substrate changes. Thus, the effects of each parameter on the diamond growth mode are hard to describe simply under MPCVD condition.

3.2.1 Methane Concentration

We refer to the fractional concentration of carbon source gas as *methane concentration* hereafter because methane gas is used in most cases. Methane concentration is defined as the flow rate ratio of methane gas to total gas, which is typically composed mainly of hydrogen. Because the hydrogen gas flow rate dominates in the total gas flow rate under typical growth conditions, the methane concentration is often defined as the flow rate ratio of methane gas to hydrogen gas (CH_4/H_2). Here, the effects of the methane concentration on the diamond growth mode are discussed.

The growth rate depends strongly on the methane concentration. In cases where homoepitaxial diamond films are grown on the (100) crystal plane under typical growth conditions of methane concentration of 0.1–1.0%, substrate temperature of 700–1000 °C, reaction pressure of 10–30 Torr, and microwave power of 300–1000 W, the growth rate of homoepitaxial diamond film is proportional to the methane concentration [22, 27–30], as shown in Figure 3.1.

The growth rate is described roughly according to the following simple liner relationship:

$$R_g = g_{(100)} \times C_{me} \tag{3.1}$$

In that equation, R_g denotes the growth rate, C_{me} is the methane concentration, and g is the rate constant. The value of $g_{(100)}$, which is g of the (100) crystal plane, is 9–300 μm h^{-1} under typical growth conditions and changes depending on other process parameters such as substrate temperature T_S and the reaction pressure. In a lower C_{me} range below 0.1%, R_g of the (100) crystal plane departs from the proportional relationship with C_{me}. This fact suggests diamond surface etching by atomic hydrogen becomes dominant with the increasing hydrogen dilution ratio of the source gas. Actually, when diamond is grown under the extremely low C_{me} condition of 0.05%, the surface morphology of the diamond film is reportedly similar to the case of the atomic hydrogen etching of a diamond surface [31]. In the higher C_{me} range above 1%, R_g tends to saturate in most cases, as seen in

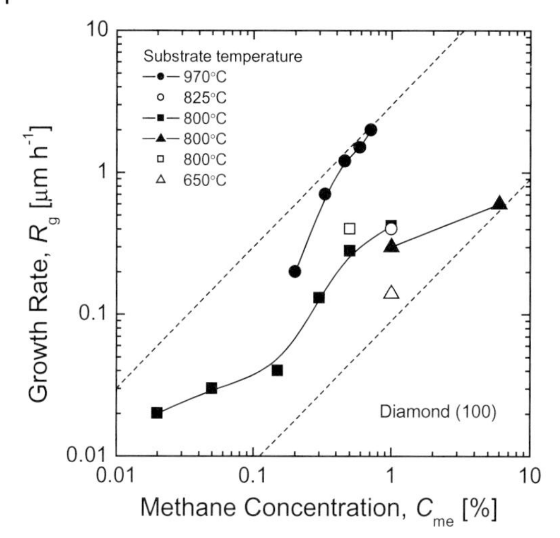

Figure 3.1 Growth rate of homoepitaxial diamond (100) films as a function of methane concentration. Different substrate temperatures are used by the authors. R_g were taken from several reports; • (Chu et al., 1992), ○ (Kasu et al., 2003), ■ (Takeuchi et al., 1999), ▲ (Tsuno et al., 1996), □ (Hayashi et al., 1996), and △ (Kasu et al., 2003).

Figure 3.1. Graphite comes to be incorporated in the diamond crystal with increasing C_{me} [32], which might be one reason to saturate R_g. Diamond growth is difficult to continue if the diamond growth surface is covered considerably with non-diamond phases. In addition, the surface excited sites are terminated significantly with methyl groups under the higher C_{me} condition. Subsequently, repulsion between these methyl groups lowers the surface reaction rate. These phenomena might result in saturation of the growth rate.

Figure 3.2 indicates that R_g increases in proportion to C_{me} for each crystal orientation of the substrate plane, but its proportional rate constant g varies depending on the crystal orientation. Chu *et al.* reported that R_g for the (111) crystal plane is independent of C_{me} in the range of 0.3–0.6% [27], in contrast to the (100) crystal plane results. In the C_{me} range below 0.3% the R_g of the (111) crystal plane is proportional to C_{me} [33] and $g_{(111)}$ is estimated as 300 μm h^{-1}. For the (110) plane, R_g is roughly proportional to C_{me}, while $g_{(110)}$ of circa 700 μm h^{-1} is the largest among the fundamental crystal planes [27]. The faster growth rate on the (110) crystal plane is consistent with the fact that {110} planes are not observed on spontaneously nucleated diamond particles of a certain size.

In the study of the growth mechanism using small diamond particles that are spontaneously nucleated, the growth parameter α, which is defined as the R_g ratio of (100) crystal plane ν_{100} to (111) crystal plane $\nu_{111}(=\sqrt{3}\,\nu_{100}\nu_{111}^{-1})$, is frequently used. At constant T_S, α is known to increase with increasing C_{me} [34]. The inset in Figure 3.2 shows the growth parameter α which was derived from homoepitaxial data [27]. The increase in α with C_{me} is consistent with that reported for polycrystalline diamond. It can be understood from the study of homoepitaxial

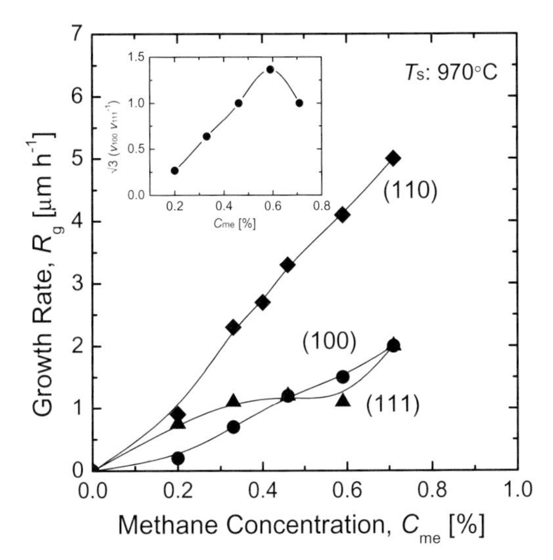

Figure 3.2 Crystal orientation dependence of growth rate of homoepitaxial diamond films (Chu et al., 1992). The inset shows the growth rate ratio of ν_{100} to ν_{111} corresponding to the growth parameter α.

growth that increase of α with C_{me} in the range of 0.3–0.6% originates from the constant ν_{111}. However, a comprehensive understanding of diamond growth mode from the viewpoint of the growth parameter α is difficult at present; α derived for homoepitaxial diamond using data from Chu *et al.* has a peak at C_{me} of 0.6% and then decreases with increasing C_{me}, whereas it maintains a weak increase in the case of polycrystalline diamond [34]. In addition, R_g of the (100) crystal plane reported by Chu *et al.* does not show R_g saturation at a higher C_{me} range, which is a different feature from the other reports [28, 29]. Systematic studies of R_g on both (100) and (111) planes are indispensable at each C_{me} and T_S for deeper understanding of the diamond growth mode.

We must be aware that the growth parameter α as a function of C_{me} and T_S is not universal but variable because R_g is affected by not only C_{me} and T_S but also other process parameters. Actually, when higher total gas pressure and higher microwave power were employed as growth parameters, α that gives a certain value (e.g. 1.5) at a certain C_{me} (e.g. 1.0%) moved to a higher T_S range [16, 36].

The configuration of methyl radicals and their concentrations near the reaction front of the substrate are difficult to investigate because the formation processes of radicals in plasma are rather complicated and the diffusion path of these radicals from the plasma to the substrate surface is not simple. Methyl radicals and acetylene are reportedly considered as the growth species of diamond. A report by D'Evelyn *et al.* indicates that R_g and g are affected by the kinds of the source gas and the crystal orientation of growth facets [37]. We must therefore consider numerous parameters for growth conditions to understand the diamond growth mechanism.

Surface morphology of homoepitaxial diamond films varies microscopically and macroscopically depending on the methane concentration. Intensive studies have been carried out using (100) diamond films because their crystalline quality is the best among homoepitaxial films grown on fundamental crystal planes of the available substrate. Under typical growth conditions in which C_{me} is 0.5–2%, T_S is 700–900 °C and with a misorientation angle θ_{mis} circa 3°, bunching steps are often observed on the homoepitaxial surface, whereas the 2×1 single-domain structure appears microscopically. The step-bunching surface comprises flat terraces with double-step height of 0.18 nm between two terraces. The average terrace width is consistent with θ_{mis} of the substrate examined. On the other hand, when C_{me} is higher than 2%, the surface has $2 \times 1/1 \times 2$ double-domain structures. These features are inferred to originate from higher growth rates of two-dimensional nucleation on terraces than the lateral growth rate from step edges [28, 38].

Unepitaxial crystallites and growth hillocks are often observed on the homoepitaxial diamond films as macroscopic anomalous growths. These anomalous features were already reported in the early stages of homoepitaxial growth study [19]. Especially when the film is substantially thick, the film surface tends to be covered with these anomalous-growth regions [39–41]. Two kinds of nucleation modes are reported for unepitaxial crystallites. One is nucleation on the substrate surface and the other is nucleation in the homoepitaxial film [29]. When the former mode dominates the anomalous nucleation, unepitaxial crystallites should have almost the same size. The number density of the unepitaxial crystallites is likely to be higher in the sample edge region than at its center [39], but their average number density depends on C_{me} and θ_{mis}. Tsuno *et al.* grew diamond on the substrates with several different misorientation angles and they found that the number density of growth hillocks decreases with increasing θ_{mis} under growth conditions in which C_{me} is 1% and T_S is 1000 °C [39]. Takeuchi *et al.* systematically investigated the effect of both C_{me} and θ_{mis} on the anomalous growth at a constant T_S of 800 °C [42]. When C_{me} is lower than 1% and θ_{mis} is greater than 3°, the nucleateation of unepitaxial crystallites is suppressed. On the other hand, when θ_{mis} is smaller than 3°, unepitaxial crystallites appear even if C_{me} is less than 1%. Because C_{me} is typically in the range of 0.3–2% for homoepitaxial diamond growth, appearance of anomalous growth in most cases is understood to depend on θ_{mis}. Moreover, when C_{me} is much less than 0.15%, no unepitaxial crystallites appear on the homoepitaxial films that are independent of θ_{mis} [31, 42]. These behaviors suggest that non-diamond phases are surely formed on terraces of the substrate surface but they are etched effectively by atomic hydrogen. Actually, the unepitaxial crystallites hardly appear when the density of atomic hydrogen increases with increasing microwave power density and reaction pressure, as discussed later. Most unepitaxial crystallites formed are oriented randomly, although some are twins of the single-crystal substrate [39]. The {111} facets of the twins have an angle of 15.8° with the (100) substrate surface. The crystalline quality of this region is poorer than that of the homoepitaxial region. The band A emissions are intensively observed from the regions in the crystallites where {112} $\Sigma 3$ grain boundaries or dislocations are formed dominantly [43]. The luminescence pattern of band A in the homoepitaxial film has a complimentary relationship with that of the free exciton recombination emission [29, 44]. Because these defects are considered to form non-radiative centers, anomalous growth must be suppressed for developing optoelectronic devices.

Band A emissions are also observed from flat regions of homoepitaxial layer other than the anomalous growth regions [44, 45]. It is deduced that this emission reflects dislocations that propagate from or locate near the substrate [45]. The nanosize unepitaxial crystallites that are buried in the homoepitaxial film are another possible cause of this emission [29]. Growth hillocks might grow faster than other homoepitaxial regions and bunching steps are formed beside the growth hillocks. Although the formation mechanism of the growth hillock remains unclear, some hillocks have defective regions in their center with unepitaxial crystallites [39, 45]. Structural defects of the substrate are one possible origion which for the hillocks. Etched pits appear sometimes depending on the growth conditions. The origin of the pits is inferred to be defects in the substrate [46, 47]. The anomalous growth described above results in the formation of a rough surface and serves to foster electronic defects such as the non-radiation centers. These defects have a small effect on the Hall mobility [48], but they increase the leakage current of Schottky contacts or degrade the breakdown voltage [49].

Interesting growth modes also appear in other methane concentration (C_{me}) ranges. When C_{me} is less than 0.15%, the diamond surface morphology resembles that of the case where diamond is exposed to the hydrogen plasma, which means that diamond growth under such low C_{me} obeys etching by atomic hydrogen [31]. Under extremely low C_{me} of 0.025%, nucleateation of unepitaxial crystallites is suppressed; consequently, atomically flat surfaces are formed throughout the crystal plane of the substrate [50, 51]. Because the growth rate is very slow, circa 20 nm h^{-1}, this growth condition cannot be used for practical purposes of homoepitaxial film deposition. This is useful for the initial growth process to etch or passivate subsurface defects in the substrate before beginning the main diamond deposition at a normal growth rate [52]. At a higher C_{me} range of 6–10%, a macroscopically flat surface sometimes appears [28, 53–55]. The surface is, however, microscopically rough because of substantial two-dimensional nucleation growths on terraces [28]. The electrical properties of boron-doped diamond films grown under these conditions are worse than those in the case of low C_{me} [55]. Those worse properties imply that the macroscopic flatness of homoepitaxial films is not always a good indicator of crystalline quality of homoepitaxial film.

3.2.2 Substrate Temperature

Similar to the methane concentration, the substrate temperature T_S during diamond deposition is also an important growth parameter that determines the growth rate R_g and crystalline quality of homoepitaxial diamond. Homoepitaxial diamond is typically grown in a T_S range of 600–1200 °C. Graphite is formed dominantly at T_S greater than 1200 °C, but R_g is very small at T_S below 500 °C [7, 8, 56]. These tendencies are theoretically reasonable [57, 58], in which R_g becomes maximum at a reaction temperature of 1000 °C.

The R_g of homoepitaxial films is known to depend on T_S according to the Arrhenius relationship. The activation energies E_a reported by Chu *et al.* were 8 ± 3 kcal mol^{-1} for (100) crystal plane, 18 ± 2 kcal mol^{-1} for the (110) crystal plane, and 12 ± 4 kcal mol^{-1} for the (111) crystal plane for T_S of 750–1000 °C [27]. Maeda *et al.* obtained 7–15 kcal mol^{-1} for the (100) crystal plane and 20 kcal mol^{-1} for the (111) crystal plane [59], whereas Sakaguchi *et al.* reported 8.5 kcal mol^{-1} for the (111) crystal plane[60]. These values are not identical with each other. T_S depends dominantly on the plasma radiation density at the sample, which is modified along with changes in the plasma conditions such as radical density. The modification of plasma condition affects E_a, which might cause the spreading of E_a. External heating of the sample holder allows T_S to be controlled without changing the plasma condition. The RF heating of graphite susceptor is reportedly used as external heating.

Previous reports on E_a have mostly been carried out using the CVD apparatus without external heating. Therefore the absolute values of the reported E_a must be carefully examined. E_a tends to be large for crystal orientations with a high growth rate. The average value of E_a for spontaneously nucleated diamond particles is reportedly 14 kcal mol^{-1} [56]. When the combustion flame method was applied for homoepitaxial diamond growth using acetylene as the source gas, E_a was reportedly 12 kcal mol^{-1} for both the (100) and the (111) crystal planes and 18 kcal mol^{-1} for the (110) plane, even though both T_S of 1000–1400 °C and the plasma density were much higher than for the case of typical MPCVD [14], indicating that E_a depends strongly on the diamond's crystal orientation.

Because a diamond surface is covered with hydrogen during chemical vapor deposition, the surface-terminating hydrogen must be first abstracted to form surface-excited sites; the second carbon radical adsorbs onto the surface-excited sites. Theoretical predictions of chemical reactions for diamond growth are described in detail by Spear *et al.* and Buttler *et al.* [58, 61]. Under typical low-pressure diamond CVD conditions, each reaction equation can be described as follows:

		Rate constant	
Hydrogen abstraction	$C_DH + H \rightarrow C_D^{\bullet} + H_2,$	(k_1)	(3.2)
Hydrogen adsorption	$C_D^{\bullet} + H \rightarrow C_DH,$	(k_2)	(3.3)
Diamond growth	$C_D^{\bullet} + C_xH_y \rightarrow C_D{-}C_xH_y,$	(k_3)	(3.4)
Graphite formation	$HC_D{-}C_D^{\bullet} \rightarrow C_{gr} = C_{gr} + H,$	(k_4)	(3.5)

$$\chi_{C_D^{\bullet}} = k_1[H]/(k_{-1}[H_2] + k_2[H] + k_3[C_xH_y] + k_4) \times \chi_{C_DH} \qquad (3.6)$$

Here, C_xH_y is a reaction precursor, C_D is carbon with a diamond structure, $C_D^{\cdot}$ is excited state of C_D, χ_{C_DH} is the number density of C_D–H sites, and $\chi_{C_D^{\cdot}}$ is the number density of $C_D^{\cdot}$ sites. Therefore, R_g increases with increasing $\chi_{C_D^{\cdot}}$. Under typical growth conditions, where methane is highly diluted and the reaction temperature is circa 900 °C, Equation 3.6 is approximated as [58]

$$\chi_{C_D^{\cdot}} \approx k_1/k_2 \times \chi_{C_DH}. \tag{3.7}$$

The reaction rate of hydrogen abstraction (k_1) has T_S dependence with activation energy of 5 kcal mol^{-1}, whereas that of hydrogen absorption (k_2) has no T_S dependence [61]. Therefore, the number density of surface active sites increases with increasing T_S. Decomposition of source methane gas into the methyl radicals cannot be expected on substrate surfaces in this T_S range. On the other hand, when T_S is greater than circa 1050 °C, graphite formation becomes considerable on the substrate [58]. Equation 3.7 also indicates that the number density of reaction site $\chi_{C_D^{\cdot}}$ does not depend on the atomic hydrogen density [58, 61]. Actual growth surfaces are not flat but have steps and terraces. The theoretical calculation indicates that desorption of hydrogen from the step edge is easier than that from the terrace. Therefore, carbon precursors are adsorbed preferentially at the step edges [62]. The activation energy E_a obtained experimentally is higher than that of $\chi_{C_D^{\cdot}}$, which suggests the diamond growth is limited by reaction processes other than the hydrogen abstraction.

On each crystal plane, several growth models of diamond are proposed theoretically as processes after hydrogen abstraction [61]. For example, after the adsorption of the methyl precursor on the diamond (100) surface, other hydrogen atoms must be abstracted both from substrate surface and the adsorbed methyl group to produce the second bond for the diamond structure. From this point of view, diamond is expected to grow from the step edges because of the considerable steric repulsion with the adjacent hydrogen atom [62]. On the (100) crystal plane, preferential growth from S_B-type step edges is observed experimentally [63], as discussed later. The largest activation energy among all reaction processes in this growth model is estimated theoretically as 6 kcal mol^{-1} [61]. The reaction process that limits growth rate is inferred to be different among the crystal planes upon which diamond film is grown. Fundamental experiments on diamond growth along with theoretical aspects are important to elucidate diamond growth mechanisms.

3.2.3
Total Gas Pressure

The total gas pressure P_T mainly influences the plasma gas temperature and the plasma volume. Under typical MPCVD conditions, the electrons in the plasma are accelerated using a high-frequency electric field; thereby, the total energy of the electron gas system increases. On the other hand, plasma ions cannot follow the microwave frequency because of their heavy mass. Therefore, the total energy of the ion gas system is much lower than that of the electron system unless interaction with the accelerated electrons becomes important. The ion gas temperature remains low under the lower P_T.

Because the ionization rate in plasma is not high under typical low-pressure MPCVD conditions for diamond growth, the small number of ionized electrons plays an important role in creating carbon radicals and atomic hydrogen, resulting in a low growth rate that is attributable to a low excitation density of these species. As P_T increases, the frequency of collisions among electrons and ions in the plasma is heightened. For that reason, the energy of the electron gas system is transported into the ion gas system, which increases the ion gas temperature. Configurations of both ionized species and radicals in plasma vary with changing P_T because the interaction between ions becomes stronger. In optical emission spectroscopy of MPCVD plasma, the hydrogen-related emission showing a purple color is observed dominantly from the plasma under P_T of 20–50 Torr with normal growth conditions. At P_T higher than 100 Torr with high methane concentration, the strong Swan band, which is the plasma emission band corresponding to C_2 molecules, appears [64].

Bombardment of diamond film with excited ions is also weakened when P_T is high due to the decreasing mean-free path of the ions. Increase of P_T improves the quality of diamond from this point of view. The mean-free path is approximately 1 μm at 100 Torr and is in inverse proportion to P_T.

The plasma radiation density on the substrate surface is enhanced with increasing P_T because the plasma density becomes high, which leads to an increase in the diamond growth rate. When the heat power source for plasma creation is weak and P_T is substantially high, recombination reactions among precursors become considerable. As a result, dependence of the growth rate on the methane concentration changes from linear to sublinear with increasing P_T [65]. This fact suggests that the plasma power source must be made larger to produce a sufficient quantity of radicals when the total pressure is increased.

3.2.4
Crystal Orientation

The (100), (111) and (110) crystal planes, the three fundamental crystal orientations with low Miller indices, are used widely as the substrate for homoepitaxial growth. As described above, optimal growth conditions for high crystalline quality differ according to the crystal orientations. In addition, the growth rate R_g for each crystal orientation has different dependence on growth conditions. The value of R_g for homoepitaxial diamond on (100) crystal plane is smaller than that for the other fundamental crystal planes under typical growth conditions [19, 27, 66], whereas the homoepitaxial (100) diamond films generally show the highest electrical/optical properties among the three fundamental planes. For example, Okushi *et al.* reportedly obtained strong free-exciton recombination emission at room temperature, with an atomically flat surface and high hole mobility, from high-quality homoepitaxial (100) films [67].

The as-grown homoepitaxial (100) diamond surface is reconstructed and terminated with hydrogen. Either the 2×1 single domain structure or the $2 \times 1/1 \times 2$ double domain structure is formed on the (100) surface [38, 68]. The domain structure depends on the misorientation angle of the substrate. Dimer rows were observed to extend beneath terraces in the direction perpendicular to the step edge lines. Steps on the (100) surface are, in general, categorized into the S_A-type step that is parallel to the dimer rows on a terrace or the S_B-type step that is perpendicular to the dimer rows. Experimental observations suggest diamond growth from the edge of the S_B-type step is dominant on the (100) crystal plane [63]. This growth mechanism is supported theoretically [61]. In this growth model, the dimer shift process from the edge of the S_B-type step is the largest endothermic process and its energy is reported to be 6 kcal mol^{-1}, as described before. On the other hand, some reports suggest that growth from the S_B-type step edge is unremarkable [69]. At this moment, the number of experimental studies on diamond growth mechanism from the microscopic point of view is limited. Approach of sequential diamond growth and atomic scale characterization will give information to elucidate how the surface atomic structure has changed.

The (111) crystal plane can provide a homoepitaxial diamond film whose surface is macroscopically flat. The (110) crystal plane has a rough surface and the quality of samples is typically low [54, 70]. Therefore, studies on homoepitaxial diamond growth have been conducted mostly for (100) and (111) crystal planes. Unlike those of the (100) plane, unepitaxial crystallites are only slightly nucleated on the (111) crystal plane. The grown surface is likely to be flat in the optical microscope image, but microscopic surface structures, such as the dimer row, are difficult to verify [20, 33, 71, 72], indicating formation of atomically rough surface. The impurity incorporation ratio for the (111) crystal plane is higher than that for the (100) crystal plane. For example, phosphorus, which has a larger atomic radius, can be incorporated readily into a substitutional site of the diamond crystal in the case of diamond growth using the (111) crystal plane [73]. Homoepitaxial diamond (111) film is reported to contain large stress. For that reason, cracks often appear in thick films grown on (111) substrate [72, 74]. A 1×1 surface structure is reportedly observed on the limited flat regions of the (111) film [39]. Similar to the case of the Si (111) surface, either a monohydride termination, in which each dandling bond on the growth surface is terminated with one hydrogen, or a trihydride termination, where methyl radicals cover the diamond surface, is considered as a possible structure. In STM observations, triangle structures with dislocations in their center were apparent, in which the step direction is <112> and bilayer terraces were formed [39]. Raman spectroscopy indicates the incorporation of graphite in some of the homoepitaxial diamond (111) film. Semiconducting properties of (111)-grown crystal are generally inferior to those of homoepitaxial film on (100) crystal plane, except for the case of phosphorus doping. These facts indicate that the diamond growth mode on the (111) crystal plane is considerably different from that on the (100) crystal plane.

3.2.5
Misorientation Angle

The misorientation angle θ_{mis} is defined as the declination angle of a flat plane of the polished substrate surface from the crystallographic plane. The angle θ_{mis} affects the growth mode of diamond crystal, as does the methane concentration C_{me} and the substrate temperature T_s. Therefore, growth conditions must be optimized considering θ_{mis} to improve homoepitaxial film quality. As θ_{mis} increases, the anomalous growth including formation of unepitaxial crystallites tends to be suppressed; then bunching steps appear over the entire sample surface with an average terrace width of circa 1 µm. Microscopically, the $2 \times 1/1 \times 2$ double domain structure is formed on the (100) crystal plane at lower θ_{mis}, whereas the surface comes to have the 2×1 single domain structure with increasing θ_{mis} [38, 68]. The relationship between the growth rate R_g and θ_{mis} was reported in detail by Tsuno *et al.* [28]. When C_{me} was as small as 1%, R_g increased in proportion to θ_{mis}. This dependence became remarkable with increasing T_s. Then, R_g saturated above a certain θ_{mis} (e.g. >5°). These authors mentioned this feature was explainable based on the model of the so-called step-flow growth mode. When the methane concentration was as high as 6%, R_g was constant, independent of θ_{mis}, and two-dimensional nucleation growth was observed morphologically. In the step-flow growth mode, R_g is proportional to the number of radicals that reach the step edges through the long terrace migration. With increasing θ_{mis}, the number of these radicals increases because the average terrace width is small, thereby increasing R_g. The saturation of R_g appears when the average terrace width determined by θ_{mis} is smaller than the average migration length of radicals. This growth model seems to explain well their experimental data. The possibility of surface migration of precursors on diamond surfaces under typical CVD conditions is theoretically reported [75]. This model predicts the average migration length of radicals to be 1–10 nm at a substrate temperature of 1000 K, which is consistent with experimental results.

The step-flow growth model is discussed frequently for metal deposition. We are aware that the growth condition of diamond under low-pressure CVD differs greatly from the case of metal. First, unlike the case of metal deposition, the diamond surface is covered with hydrogen, meaning that the number density of the surface reaction site depends on hydrogen abstraction and is very low because of stable hydrogen bonding with diamond. Second, the substrate temperature seems to be low for radical migration. Lattice vibration of the substrate promotes radical migration, and the degree of the lattice vibration is related to the Debye temperature of the substrate material. For example, the Debye temperature of silicon is circa 600 K, whereas the homoepitaxial deposition of crystalline silicon is carried out above this temperature. For diamond, however, the Debye temperature is circa 2000 K, meaning that the diamond substrate, of which the typical temperature is below 1000 K, is cold from the viewpoint of the lattice vibration under the diamond CVD condition. The step-flow growth model requires successive abstraction of terminating hydrogen on terraces in advance of the radical migration. Remarkable surface migration is unlikely to occur on the diamond surface under low temperature conditions, as suggested theoretically [76, 77].

Another model explaining the lateral growth of diamond was proposed by Harris *et al.* [77] and Tamura *et al.* [62]. They mentioned that adsorbed radicals are more stable at step edges than on terraces, and that growth occurs dominantly at step edges. In addition, because abstraction of terminating hydrogen from step edges is easier than from terraces, radicals are considered to adsorb at step edges in high probability. This growth model is called the trough model. Carbon radicals are inferred to diffuse in the gas phase near the substrate surface terminated with hydrogen. They are then absorbed at step edges that are excited by hydrogen abstraction. An important fact experimentally obtained is that the lateral growth from step edges governs the diamond growth mode when the misorientation angle is substantially large and the methane concentration is low, which dose not always mean step-flow growth in a strict sense. At present no direct experimental evidence that distinguishes the actual growth model has been reported.

3.2.6 Substrate Quality and Preparation Method

Crystalline defects in the subsurface region of the substrate have an unfavorable effect on the quality of homoepitaxial diamond films. Figure 3.3a shows polishing marks that are typically observed throughout the surface of mirror-polished single-crystalline diamond. These marks typically consist of large and small periodic patterns of circa 200 nm and circa 20 nm, respectively; their average height is about 4 nm [31]. When a homoepitaxial film is deposited on the substrate with polishing defects, the band A (BA) emissions, which evidence dislocations, are clearly apparent in CL images.

Figure 3.4a shows that the emission pattern corresponds to that of the polishing marks featured by line shape, as indicated by dashed line. The emission pattern of the free exciton (FE) recombination is complementary to the BA one, as evidenced by Figure 3.4b. In addition to the line-shape emission pattern, the point-like pattern of the BA emission, one of which is indicated by dashed circle, is

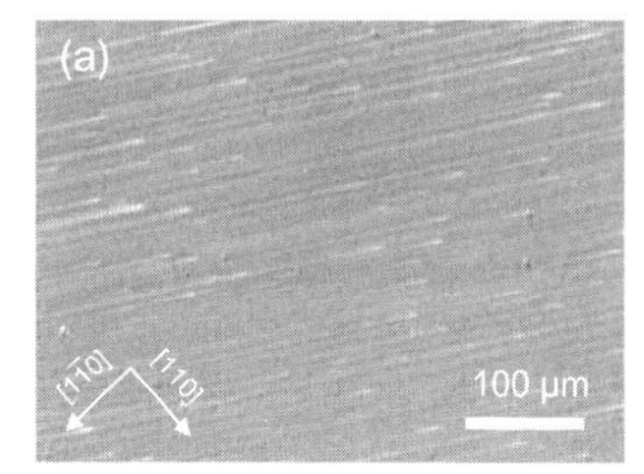

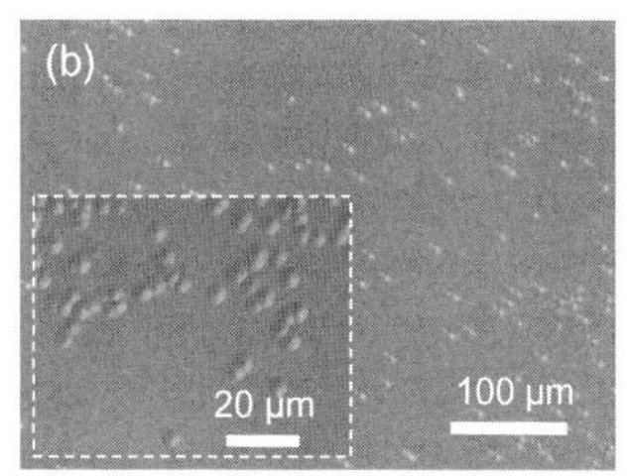

Figure 3.3 Optical microscope images of (a) an as-received HPHT substrate whose surface was mirror-polished, and (b) a HPHT substrate after oxygen quasi-ECR plasma etching, with an inset of a magnified image.

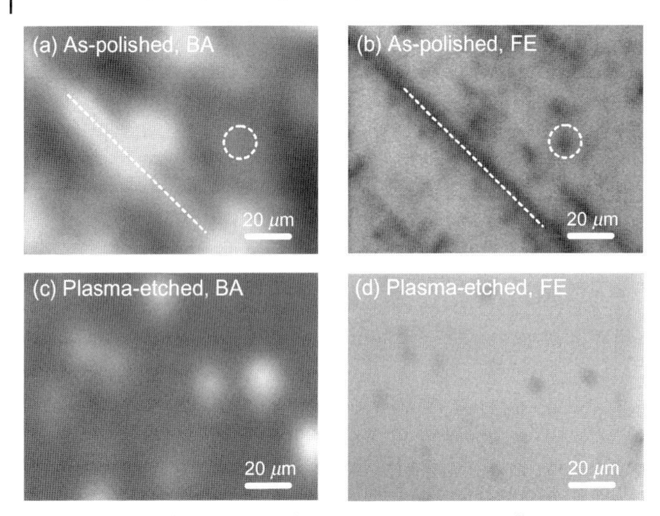

Figure 3.4 Typical CL images taken at room temperature from two homoepitaxial specimens deposited without (a and b) and with oxygen quasi-ECR plasma etching process for HPHT substrates (c and d). The measured wavelengths correspond to dislocation-related band-A (BA) emissions, being the maximum at 420 nm (a and c) or the free-exciton (FE) recombination emissions that have a peak at 235 nm (b and d).

confirmed with spreading in the flat homoepitaxial area. The origin of these dislocations might be in the substrate, but the distribution of the dislocation in the homoepitaxial film is unclear at the present stage. Analyses of CL images suggest that dislocations localize at the substrate/film interface and that the dislocations seem not to propagate in the film [45, 46]. Embedding of the unepitaxial crystallites in the homoepitaxial film is reported [29], which also produces a spreading pattern of the BA emission. These facts indicate defect formation depends on the quality of substrate and the diamond growth condition. Because these defects must be removed as low as possible for electronic device applications, and especially for optoelectronics, investigation of the dislocation distribution in the film and elucidation of the effects of substrate defects on the dislocation formation in the homoepitaxial film are important research subjects.

A pretreatment to remove the polishing induced defects is desirable because mechanical polishing of diamond surfaces is difficult to control precisely. In the case of silicon, a process of surface oxidation followed by etching of the oxidized layer is used to remove the surface damage layer. On the other hand, oxygen plasma or hydrogen plasma can be used for diamond to etch the surface damaged layer. Microwave plasma, quasi-ECR microwave plasma and RF plasma are utilized for this purpose. (The term quasi-ECR is used here because this process is not an ideal ECR plasma process in the strict sense of the word since the pressure employed is too high for electrons to complete a cyclotron orbit without scattering.)

The etching rate is typically 1 μm h^{-1} under optimized etching conditions; the etched depth of diamond is circa 1 μm in most cases [78]. A macroscopically flat surface can be formed by reactive ion etching using RF plasma [79], while, as shown in Figure 3.3b, the surface becomes rough with pits in the case of etching using microwave plasma [47, 78]. The homoepitaxial film deposited on the pre-etched substrate shows no BA emission corresponding to the polishing marks with line shape, as evidenced in Figure 3.4c [78, 79]. Figure 3.4d shows FE emission images taken for this sample, indicating that the high-quality area is rather homogeneous. The pits created by the plasma etching process are covered effectively by the following homoepitaxial growth dominated by lateral growth. The number density of point-like defects seems to be unchanged with etching depth by oxygen plasma. The origin of these point-like defects remains unclear at this stage, as mentioned above. Excellent electric properties such as higher carrier mobility are obtained with high reproducibility through the etching process [79, 80].

As a pretreatment procedure, chemical cleaning is another approach to improve the substrate surface quality. Suppression of etch-pits and growth hillock formation through chemical cleaning, including that by high-temperature hydrogen annealing, are reported [46]. This cleaning procedure is described in Table 3.2 [81]. We can infer that the metallic solvent that is utilized for HPHT as a catalyst, or

Table 3.2 Substrate cleaning procedure.

Step	Treatment	Parameters	Removal of
1	DI → Ethanol → Acetone → Ethanol → DI	Ultrasonic, 10 min. for each	Adsorbed organic matter
2	(5% NaOH : 36% H_2O_2) = (5 : 3)	60 °C, 20 min. (option)	Al oxides
3	(98% H_2SO_4 : 67% HNO_3 : $HClO_4$) = (3 : 4 : 1)	250 °C, 30 min.	Non-diamond phases
4	(33% HCl : 67% HNO_3) = (6 : 1)	80 °C, 30 min. (option)	Metals
5	(30% NH_4OH : 36% H_2O_2 : H_2O) = (1 : 1 : 5)	75~85 °C, 10 min.	Organic and metal powders
6	(50% HF : 67% HNO_3) = (1 : 1)	~20 °C, 10 min.	Si and SiOx
7	DI	100 °C, 10 min.	
8	Annealing in H_2 atmosphere	1190 °C, 2×10^{-4} Torr, 2 h	Impurities in bulk by diffusion
9	Repeat the sequence once more except steps 2 and 8.		

vacancies in crystals, which could be an origin of anomalous growth, are diffused out of the substrate bulk during high-temperature treatment and then they are removed by the successive acid treatment. This chemical etching procedure has, however, no function in removing the polishing-induced defects. These results indicate that suitable combinations of pre-treatments to lower the density of defects of substrate surface are required to obtain high-quality diamond growth. In addition, buffer layer deposition, which relaxes localization of substrate defects favorably, might be an effective way to suppress anomalous growth such as growth hillocks.

3.2.7
Impurity Doping into Homoepitaxial Diamond

For controlling electrical conductivity, boron and phosphorus are used respectively as p-type and n-type dopants of diamond. Their respective activation energies are 0.36 eV and 0.57 eV. Although other elements are investigated to examine their use as a donor or acceptor, reproducible results have not been obtained. Nitrogen is likely to form several types of complexes with vacancies and makes a deep donor level of circa 1.7 eV in diamond. Conductivity of nitrogen-doped diamond is negligibly small at room temperature. High purity diborane or phosphine gases are available for boron or phosphorus doping, respectively. Organometallic compound gases such as trimethylboron or tertiarybutylphosphineare also currently used because of their easy handling, although purity of these gases is not so high. Details on doping studies of diamond are described elsewhere [82, 83]. Unique p-type conductivity is known to appear on the as-grown CVD diamond surface. Here, electric conductivity arising in bulk homoepitaxial diamond is discussed.

Semiconducting properties depend strongly on the crystalline quality of diamond; the quality varies considerably among natural and HPHT diamonds. The crystalline quality of homoepitaxial diamond is highly reproducible using the MPCVD method. Hall mobility of boron-doped homoepitaxial diamond (100) films is almost identical to that of the best natural diamond [80, 84]. Boron doping during diamond CVD was first tried by Fujimori *et al.* [66, 85]. Since that time, studies for improving diamond quality have been carried out intensively, mainly in Japan and the USA. We can readily obtain samples showing hole mobility of greater than 1000 $cm^2\ V s^{-1}$ at room temperature from boron-doped homoepitaxial (100) films [55, 80, 84, 86, 87]. The highest Hall mobility for a hole so far is 1860 $cm^2\ V s^{-1}$ at 290 K [84]. On the other hand, mobility of boron-doped (111) crystal at room temperature is not as high as 1000 $cm^2\ V s^{-1}$ at present [33]. Because of the larger activation energy of boron, the electric conductivity of diamond is rather low, even in the case of high mobility. Room temperature conductivity is typically lower than 100 $mS\ cm^{-1}$. Electrical conductivity increase with increasing doping concentration of boron, whereas hole mobility decreases rapidly in such cases because of crystalline deterioration and impurity scattering. Heavy doping is one of the ways to increase the conductivity. The activation energy is circa 0.25 eV at concentrations of $5 \times 10^{19}\ cm^{-3}$. It becomes zero at concentrations greater than $3 \times 10^{20}\,\mathrm{cm}^{-3}$, where conductivity is greater than $10^{2}\,\mathrm{S\,cm}^{-1}$ [88]. Metallic conduction is dominant in this high concentration region and superconducting properties are reported from super heavily boron-doped samples [89, 90].

It is known that n-type doping is still a difficult task for diamond research. In the early stage of this study, incorporation of a certain amount of hydrogen with phosphorus posed a considerable problem because such hydrogen can passivate phosphorus donors. Koizumi *et al.* optimized growth conditions with selection of the (111) crystal plane as a substrate. They first obtained the homoepitaxial diamond films showing clear n-type conductivity at room temperature [26, 73, 83]. With promotion of optimization of growth conditions, they have recently achieved high electron mobility of $660\,\mathrm{cm}^{2}\,\mathrm{V\,s}^{-1}$ at room temperature [91]. In addition, n-type doping homoepitaxial growth is recently demonstrated on the (100) crystal plane [92]. With optimization of growth conditions, semiconducting properties of homoepitaxial diamond are approaching electronic device grade and device performance reflecting diamond innate properties have recently been observed from devices that have a simple structure [1].

3.3 Homoepitaxial Diamond Growth by High-Power Microwave-Assisted Chemical Vapor Deposition

As described above, the microwave plasma-assisted chemical-vapor-deposition (MPCVD) method has several advantages for growth of high-quality diamond. Homoepitaxial diamond films that can yield free-exciton recombination emissions with room-temperature cathode luminescence (CL) or photoluminescence (PL) spectra are obtained from homoepitaxial films grown by the MPCVD method. The room-temperature Hall and photo-excited drift mobilities reported for holes of the MPCVD-grown homoepitaxial films are as high as circa 2×10^{3} and $4 \times 10^{3}\,\mathrm{cm}^{2}\,\mathrm{V\,s}^{-1}$, respectively [84, 93].

For the MPCVD process, however, some critical issues must be resolved for industrial production and application of CVD diamond films. One of the most important issues among them is to increase the diamond growth rate while maintaining high crystalline quality. The growth rate of conventional MPCVD is typically less than $1\,\mu\mathrm{m\,h}^{-1}$, which arises partly from a low gas temperature of circa 3000 K [94, 95]. Diamond quality worsens because of an appreciable increase in the formation of secondary-nucleated unepitaxial crystallites and non-diamond phases that occurs if the methane concentration is increased to achieve a higher growth rate. The research target so far for homoepitaxial diamond research is to grow high-quality diamond films; reportedly, an effective method to achieve this aim is to decrease the methane concentration greatly, for example, below 0.01% [50, 51]. The resultant growth rate for the high-quality sample was therefore much lower than $0.1\,\mu\mathrm{m\,h}^{-1}$ in the conventional MPCVD cases, as described in the previous section.

In the last five years, several research groups have launched studies on the growth of thick homoepitaxial diamond (100) films at high growth rates of >20 μm h^{-1} by employing high microwave power and by applying special growth conditions that are much different from conventional ones [40, 81, 96–99]. In addition, high-quality homoepitaxial diamond films have been grown at reasonably high growth rates by optimizing high-power MPCVD conditions [23, 81]. This section includes descriptions of several advantages of high-power MPCVD for homoepitaxial diamond growth.

3.3.1 Growth Conditions

Typical growth conditions that are employed for high-power microwave plasma CVD using a research-grade growth system are as follows. The input microwave power is 3–6 kW, which is greater than that of conventional low-power MPCVD by 5–10 times. The reaction pressure is conventionally 20–50 Torr, whereas the pressure for high-power MPCVD is 100–200 Torr, which is 4–5 times larger. Because the volume of created plasma is visibly of the same order of magnitude in most cases, the plasma density is deduced to be larger by more than four times for the high-power MPCVD. The reported microwave power density is 30–100 W cm^{-3} for high-power MPCVD, but is circa 5 W cm^{-3} for the conventional case. These higher parameters are characteristic of high-power MPCVD. Typical methane concentration C_{me}, which is defined as $[CH_4]/([H_2] + [CH_4])$, and substrate temperature T_S are 1–20% and 900–1200 °C, respectively. These are also higher values than those of the conventional case. The plasma radiation toward the diamond substrate is so strong under high-power MPCVD conditions that the substrate holder must be water-cooled during diamond deposition, which contrasts markedly with the external heating of the substrate holder in the conventional case. The color of the resultant plasma is greenish, corresponding to the Swan band emissions of C_2 molecules [64]. The density ratio of precursors $[C_2H_2]/[CH_3]$is reported to increase with increasing substrate temperature or methane concentration [99]. Therefore, attention must be devoted not only to the density of the carbon radicals, but also to the kinds of carbon radicals playing as precursors. It does not always mean that molecular carbon plays an important role in increasing the growth rate. The C_2 molecules can be formed at the plasma boundary because atomic carbons that are decomposed in the plasma center are possibly cooled at the boundary where gas temperature is low. In another report, increase of C_2H is observed in addition to increase of C and C_2 when plasma density becomes high [100]. It is still not clear which radical is a dominant precursor for diamond growth, especially under the high-power MPCVD condition. Because the form of radical is changed in the plasma, no direct experimental evidence which shows the effect of each carbon radical on the diamond growth has been presented at this moment. Understanding of growth precursors is an indispensable research subject for finding effective diamond growth parameters under the non-equilibrium CVD conditions.

3.3.2
Growth Rate

Figure 3.5 shows the growth rate R_g of homoepitaxial (100) films deposited using the high-power MPCVD method, R_g^{HP}, as a function of C_{me}, both on logarithmic scales. The values of R_g^{HP} obtained for the two different T_s of 890 °C (open circles) and 1010 °C (filled circles) are plotted in this figure [81]. The total gas pressure and microwave power used were 120 Torr and 3.8 kW, respectively. The microwave power density under this condition is inferred as circa 40 W cm^{-3}. Two substrate temperatures of 890 and 1010 °C were employed. When C_{me} is less than 1.0%, R_g^{HP} increases sub-linearly with increasing C_{me} [81, 98]. This feature might be comparable with R_g of homoepitaxial diamond (100) films for the conventional lower power MPCVD, R_g^{LP} [29, 51]. In the latter case, as shown with filled triangles in Figure 3.5, R_g^{LP} is roughly proportional to $C_{me}^{1/2}$ when $C_{me} \leq 0.15\%$. Similar nonlinear behaviors are confirmed for the deposition of amorphous silicon by plasma-assisted CVD methods under highly hydrogen-diluted conditions. Because the number density of atomic hydrogen increases with increasing plasma density, the hydrogen etching effect on R_g^{HP} is not negligible in the low C_{me} range under high-density plasma conditions. Consequently, C_{me} dependence of R_g^{HP} is inferred to

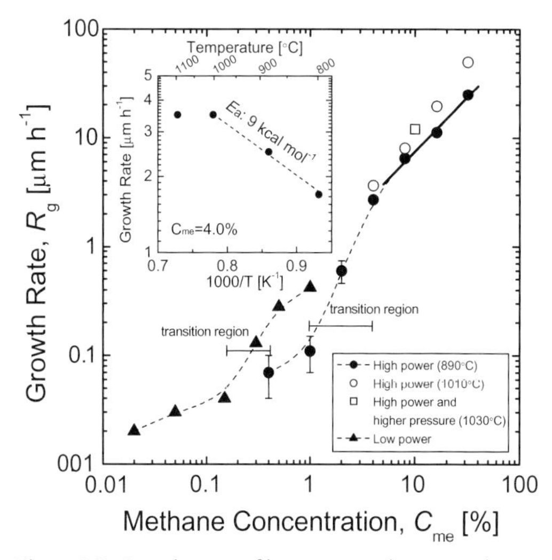

Figure 3.5 Growth rates of homoepitaxial diamond (100) films as a function of the methane concentration under various conditions: a 3.8-kW high power MPCVD at 890 °C (•) and at 1010 °C (○), a 4.7-kW high power MPCVD with higher total gas pressure of 160 Torr (□) (Teraji, T. and Ito, T., 2004) and conventional 0.75-kW MPCVD (Takeuchi, D. et al., 1999) (▲). Dashed lines are guides for the eye for linear dependencies. The inset shows the temperature dependence of the growth rate under high-power MPCVD conditions with methane concentration of 4.0%.

weaken. Actually, R_g^{HP} for high-power MPCVD is smaller than R_g^{LP} for low-power MPCVD at a certain C_{me} in the lower range (e.g. 0.5%). Therefore, the optimum C_{me} must shift to the higher C_{me} side when high-density plasma is applied for diamond deposition.

In the medium C_{me} range of 1.0–4.0%, R_g^{HP} increased super-linearly with increasing C_{me} [81, 98], suggesting that the growth mode changed in this C_{me} range. A similar transition was confirmed in R_g^{LP} as well. These transition ranges are indicated in Figure 3.5 as bars. The lower-side boundary C_{me}^{lsb} of the medium C_{me} range is circa 1.0% in the high-power MPCVD case whereas it is 0.1–0.2% in the low-power (conventional) MPCVD case. The rapid increase of R_g^{HP} means that the density of the concerned carbon precursors increased with increasing C_{me} whereas the growth rate and etching rate are comparable in the region below C_{me}^{lsb}. The value of C_{me}^{lsb} might be determined by densities of atomic hydrogen and hydrocarbon radicals as well as the substrate temperature. The total gas pressure and microwave power in the high-power MPCVD case were larger by circa 5 times than those in the conventional MPCVD case, although the plasma volume is inferred to be slightly (circa 1.5 times) larger in the former than in the latter. This means that the plasma density, which determines the density of atomic hydrogen, is 3–4 times larger in the former case than in the latter case. In addition, because the substrate temperature is slightly higher in the high-power case than in the low-power case, the difference between the two C_{me}^{lsb} must be further enlarged from the difference of the plasma density. Thus, the ratio of C_{me}^{lsb} between the high-power and the conventional low-power MPCVD methods, which can be estimated as 5–10 in Figure 3.5, is reasonable.

An approximately linear $C_{me} - R_g$ behavior was observed in the C_{me} range $\geq 4.0\%$ [81, 98, 101]. The linear dependence suggests that the growth model is relatively simple and rather similar to that of common thermal CVD. In this high C_{me} range, R_g^{HP} is recognized to increase with increasing substrate temperature, as shown in the inset of Figure 3.5. That is, R_g^{HP} is in the reaction-controlled growth mode [102]. In the reaction-controlled mode, the growth rate has activation energy, E_a, and follows the Arrhenius relationship, $\exp(-E_a/RT_s)$, with the substrate temperature T_s and the gas constant R. The E_a for the growth rate at a methane concentration of 4.0%, estimated in the substrate temperature range of 800–1000 °C, is 9 kcal mol^{-1}, which is consistent with the reported value for homoepitaxial (100) diamond in the case of low-power MPCVD [27]. We understand that the substrate temperature is an important process parameter controlling the growth rate. Saturation of the growth rate was observed in T_s at temperatures greater than 1000 °C. One can understand that the growth mode changes from reaction-controlled mode to a diffusion-controlled mode at this temperature. However, diamond might partly transform into graphite in this temperature range [58]. The diamond film quality grown at 1100 °C was actually not superior to that of crystal obtained at 1000 °C. The substrate temperature is demonstrably an important parameter that determines the growth rate and crystalline quality.

For homoepitaxial diamond growth using the low-power MPCVD method, the observed linear-dependence region is narrow in the C_{me} range; R_g^{LP} tends to be saturated at $C_{me} > 1.0\%$. A similar saturation behavior is confirmed for growth of polycrystalline diamond film [81, 103, 104]. In contrast, no saturation was observed in the $C_{me} - R_g^{HP}$ relation when the high-power MPCVD method was employed, suggesting that R_g^{LP} saturation arises from the low microwave power densities used. In other words, high-power MPCVD is appropriate for high-rate growth, whereas the conventional low-power MPCVD is not. A possible mechanism of this saturation phenomenon occurring for low-power MPCVD at such high C_{me} might be: (i) insufficient creation of both atomic hydrogen and effective carbon precursors attributable to low microwave power density; (ii) covering of the top surface with methyl group or non-diamond phases attributable to the low reaction rate, which leads to decreased nucleation-site density; or (iii) increase of the diamond-etching rate caused by substantial deterioration of diamond quality. We mention that another advantage of high-power MPCVD is that achieved from the viewpoint of total gas pressure P_T. As discussed in the previous section, P_T affects R_g, where α in the relationship $R_g \propto C_{me}^{\alpha}$ becomes small from unity to 0.5 with increasing P_T from 20 Torr to 200 Torr for the filament-assisted CVD condition [65]. The change was speculated to occur because of recombination reactions between diamond precursors, resulting in decrease of the precursors. Such a tendency was not observed even at high P_T of 120 Torr for high-power MPCVD, suggesting that carbon precursors were produced sufficiently. Consequently, R_g^{LP} were smaller than 1 μm h^{-1}, whereas R_g^{HP} were much larger than 1 μm h^{-1} under optimized growth conditions.

We might find optimal growth conditions that offer both high quality and a high growth rate through understanding the effect of each process parameter on R_g under high C_{me} conditions. The highest growth rate attained in this growth condition was 50 μm h^{-1} at a methane concentration of 32% and at substrate temperature of 1010 °C. The possible growth period under this condition was 10 min because of the considerable amount of graphite formed on a quartz window for microwave introduction when the growth duration lasts for longer than 10 min. The above discussion emphasizes that increased plasma density with higher methane concentration and higher substrate temperature is necessary to increase the diamond growth rate, although crystallinity of the homoepitaxial film tends to be poor when these parameters are reasonably high.

Furthermore, the microwave power density and total gas pressure are considered to be other key parameters to increase R_g while retaining good crystalline quality. Here, we examined them by increasing both the input microwave power and the total gas pressure. The open rectangle shown in Figure 3.5 represents R_g under higher power (4.7 kW) and higher pressure (160 Torr) conditions. The substrate temperature was 1030 °C. The power density, simply estimated under this condition, is 50 W cm^{-3}, although it is 40 W cm^{-3} for open circles representing the standard conditions (3.8 kW and 120 Torr) with a higher substrate temperature of 1010 °C. The R_g of these higher parameters locates almost on the line of the standard growth condition with almost the same T_s. This fact shows that R_g depends mainly on T_s and the methane concentration, that is, the growth mode is in the reaction-controlled one, as mentioned above. On the other hand, when the microwave power density increases considerably from $75\,W\,cm^{-3}$ to $130\,W\,cm^{-3}$, while keeping other process parameters constant and T_s is as low as 850 °C, R_g is reported to increase by a factor of 2–3 [98, 105]. The surface, however, has round morphology because of dominant island growth, indicating that the substrate temperature is lower than the optimized condition giving lateral growth. This fact suggests that temperature is the most important process parameter, along with a macroscopically flat surface, that must be optimized to grow high-quality diamond crystals.

The high-rate growth mechanism under high-power microwave plasma CVD is inferred as follows. The active-site density on the substrate depends on the substrate temperature, as mentioned in Equation 3.2 of the previous section. The site density is substantially higher under the higher substrate temperature condition of high-power MPCVD. Because of the higher methane concentration in the gas phase and its effective decomposition by high microwave power density, hydrocarbon radicals and atomic hydrogen are created in plasma at high density. Then, they adsorb on the active site of the substrate in high density. Consequently, the diamond growth rate increases. When the methane concentration is lower, because the abstraction of adsorbed hydrocarbon radicals by the high-density atomic hydrogen is considerable, the diamond growth rate under the high-power MPCVD becomes smaller than that in the low-power case. Diamond etching reportedly occurs at lower methane concentration below 1% when the higher T_s of 1200 °C and higher P_T of 150 Torr is employed [101]. The range of optimal growth conditions becomes narrower with increasing plasma density. Therefore, these process parameters should be controlled accurately to the optimal values.

3.3.3
Surface Morphology

Typical optical microscope images of diamond films grown at each C_{me} are shown in Figure 3.6. These optical microscope images were taken with a Nomarski filter. Growth duration and deposited film thickness d_F for these samples are listed in Table 3.3.

In low C_{me} cases, for example 1.0%, the sample surface had faceted regions whose ridges were extending to <110> directions with average length of a few-hundred micrometers, as indicated by the dotted ellipsoid in Figure 3.6a. The shape differs markedly from those of typical pyramidal hillocks. The regions were less than 0.5 μm high. Therefore, the larger two facets composing these regions were recognized to have a small tilt angle of less than 2° from the flat substrate plane into the <111> direction. These faceted regions do not come from twinning but are rather composed of a number of small facets with various tilt angles toward both (100) and {111} planes [81]. The reason for {111}-tilt facet formation might be described as follows. Under low C_{me} conditions in which collision frequencies between carbon precursors and the substrate surface are relatively low, the growing surface might be rearranged locally around surface structural defects for longer periods, compared with the high C_{me} case. As a result, several facets, some of

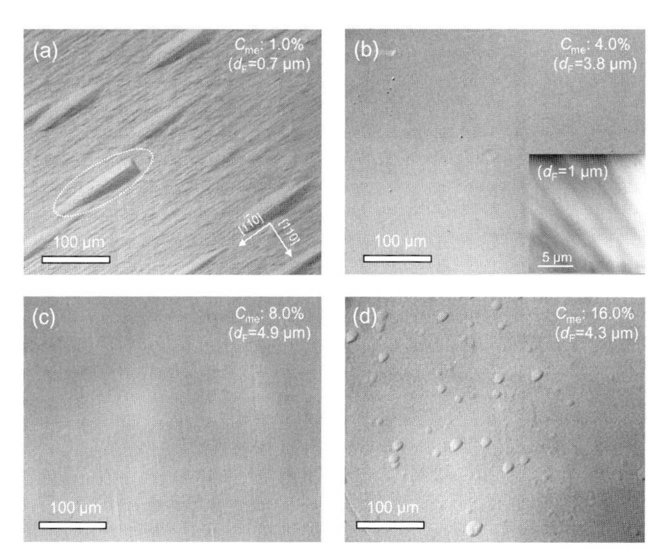

Figure 3.6 Typical optical microscope images of homoepitaxial diamond films grown at various methane concentrations of (a) 1.0%, (b) 4.0%, (c) 8.0%, and (d) 16.0%. Growth duration and film thickness d_F of these samples are listed in Table 3.3. The inset shows atomic force microscope images for the 4.0% sample at film thickness of 1 μm.

Table 3.3 Methane concentration C_{me}, growth time, film thickness, and growth rate R_g of homoepitaxial films deposited.

C_{me} (%)	Growth time (h)	Film thickness (μm)	R_g (μm h^{-1})
0.4	15	1.1	0.07
1	6	0.7	0.12
2	3	1.8	0.6
4	1.5	3.8	2.5
8	0.75	4.9	6.5
16	0.38	4.3	11.2
32	0.18	4.5	24.6

which give higher growth rate than that of (100) crystalline orientation, form near the defects and the diamond growth rate becomes higher at these sites. Diamond etching by atomic hydrogen, which simultaneously occurs with growth, is remarkable because of high-density creation of the atomic hydrogen under high-power microwave plasma condition. In this situation, the formation of facets tilting into the <111> direction seems to be reasonable from the viewpoint of lowered surface energy because the {111} crystal plan has the smallest dangling-bond density among the planes of the three primitive crystal orientations. Consequently, the stable facets tend to appear under the strong etching condition.

Bunched steps were observed macroscopically from remaining areas other than these faceted regions. These might also be composed of small {111}-tilt facets. In the case of conventional (low-power) MPCVD, macro-bunching steps are often observed on homoepitaxial diamond grown on substrates with misorientation angles, θ_{mis}, of >1° [42, 52], except for the extremely low C_{me} case. Ri *et al.* reported for the low C_{me} case that the surface roughness of a diamond film resembled that of an original substrate when diamond film was deposited on substrate with θ_{mis} >1° using the conventional MPCVD method. This fact indicates etching of surface defects by atomic hydrogen is insufficient to remove polishing marks on the original substrate surface [31]. They also found an atomically flat surface was formed on a diamond substrate with a small θ_{mis}. θ_{mis} of substrates examined here for the high-power MPCVD were ~3°. The boundary C_{me} that determined the macro-bunching step formation was circa 0.1% for the conventional MPCVD case [29, 51], which fell on the boundary between the nonlinear and linear growth regions of the $C_{me} - R_g^{LP}$ relationship in Figure 3.5. This means that the conditions when $C_{me} < 1.0\%$ for high-power MPCVD correspond to the extremely low C_{me} range for conventional low-power MPCVD, whereas the formation of faceted regions shown in Figure 3.6a is in great contrast to the conventional case. This fact means that surface morphology depends on the microwave power density for the lower C_{me} conditions. Contrary to a low etching rate for conventional MPCVD, etching of the substrate surface by atomic hydrogen is so strong in the high-power MPCVD case that the polishing marks can be removed morphologically. Then, the macro-bunching steps begin to form on the specimen surface, with the same mechanism as the {111}-tilt facet formation.

Both the films grown at C_{me} = 4.0% and at C_{me} = 8.0% had substantially flat surfaces, as evidenced in Figure 3.6b and c. Although a few pyramidal hillocks were observed on the film surfaces, no unepitaxial crystallites existed on the surfaces for all the growth conditions employed. These hillocks were difficult to confirm by SEM observation, which indicates that the grown films are rather uniform in terms of their electronic states. In SEM images, small dark spots, the positions of which were independent of those of hillocks, were observed from the grown sample. The change of images with the acceleration voltage of primary electrons suggested that the dark spots are located mainly near the Ib substrate region. The inset shows an atomic force microscope image of the film grown under C_{me} of 4.0%, but taken at $d_F \approx 1\,\mu m$, reflecting the initial growth stage. The characteristic step bunching structure, similar to the morphology of lower C_{me}, was observed in this image. We can understand that under high-power MPCVD conditions the surface first has a step-bunching structure and subsequently becomes macroscopically flat with increasing film thickness. Details are discussed below. Under higher C_{me} conditions such as 16.0%, a few dome-like hillocks as shown in Figure 3.6d, which differ from typical rectangular pyramidal hillocks

[29], were formed on the sample surface, suggesting a change of growth mode from lateral growth to island growth.

Figure 3.7a and b show typical surface morphologies of homoepitaxial films grown respectively at 890 °C and 1010 °C. The C_{me} was 4.0% and d_F was 10–11 µm for both samples. On one hand, as shown in Figure 3.7a, round hillocks were observed to spread over the whole surface of the lower temperature specimen. Magnified OM and SEM images taken from the respective square regions in the whole image are shown on the right-hand side of the main OM image. Hillocks with sizes of <50 µm were found to cover at least the entire magnified area. In contrast to the OM images, such growth hillocks are only slightly apparent in the corresponding SEM image, indicating that the hillocks resemble electronically those in the remaining flat regions. Such round hillocks grew in lateral size to 100 µm at $d_F = 20$ µm, whereas the surface morphology appearing in the Nomarski-mode OM images became macroscopically rough. This feature is contrasted with step-bunching or flat surfaces observed in less than 4-µm thick films grown under identical growth conditions. On the other hand, as shown in Figure 3.7b, the morphology of the higher temperature specimen was substantially flat at d_F =

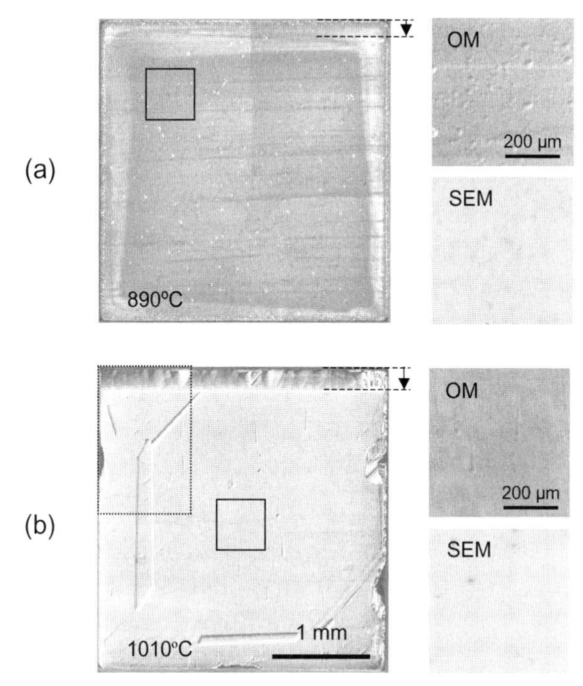

Figure 3.7 Optical microscope images of specimen surfaces of homoepitaxial diamond grown at substrate temperatures of (a) 890 °C and (b) 1010 °C. The images on the right in both panels were taken from respective square regions in the whole images. The upper image was taken using an optical microscope; the lower one was obtained by SEM. Surface morphological shift parallel to the sample edge is indicated with dashed lines. A magnified image of the rectangular region in the left-upper side is shown in Figure 3.8.

11 μm. Such surface flatness was maintained under the higher temperature condition even for $d_F = 20\,\mu m$. In this OM image, however, characteristic features originating from the substrate growth sectors were observed, suggesting that variation in R_g might be attributable to locally non-uniform crystalline quality and nitrogen impurities of the HPHT substrate used. A magnified image shown on the right-hand side indicates a limited number of small marks stretching in a certain direction, the origin of which is discussed below. Therefore, one can see that diamond growth at $T_s > 1000\,°C$ has a substantial effect in suppressing the round hillock formation.

Information on the growth mode can be inferred from a change of the morphological features in which lines parallel to a substrate edge were newly formed on both homoepitaxial film surfaces, as indicated by dashed lines in Figure 3.7a and b. The lines that are morphologically observed shifted from the substrate edge reflect misorientation angles of the substrate orientation from the (100) direction, as described below.

To investigate the growth mode under higher temperature conditions in more detail, we compared OM images taken before and after the homoepitaxial growth of an 11-μm-thick diamond film, which was carried out at C_{me} of 4.0% and at T_s of 1010 °C. The image shown in Figure 3.8b was taken from the upper left side area of the specimen whole image shown in Figure 3.7b, and Figure 3.8a is the image at the same area of the substrate before the homoepitaxial growth. After the diamond deposition, characteristic morphological features on the surface were found to shift by an identical distance in a certain direction, as indicated by vector **AB**. Here, vectors **P1–P3** have the same origin in each case (**A** in Figure 3.8a and

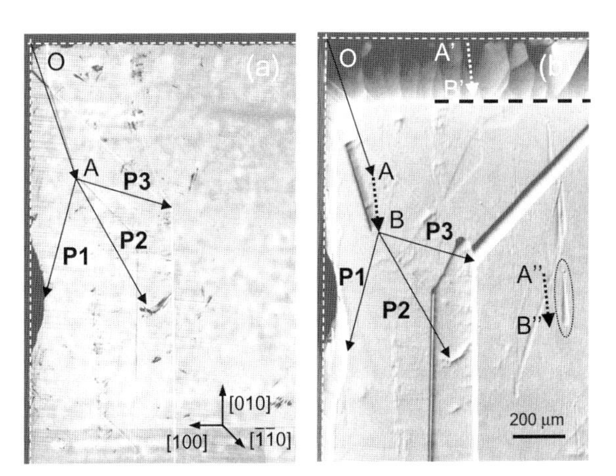

Figure 3.8 Magnified optical microscope images taken before (a) and after (b) the CVD growth of an 11-μm thick diamond film at the substrate temperature of 1010 °C using the CH_4/H_2 ratio of 4.0%. The images are taken from a rectangular area surrounded by dotted lines near the upper left corner of Figure 3.7b. Positions A, A′, A″, B, B′, B″ and vectors P1–P3 indicate morphologically characteristic regions.

B in Figure 3.8b), whereas the origin of vector **OA** is set at a corner of the specimen. The ending point **B′** of a vector **A′B′** starting at a point on the sample edge, **A′**, fell on the newly formed edge of the grown film parallel to one of the substrate edges, which is shown partly by a thick dotted line in Figure 3.8b, when **A′B′** = **AB**, meaning that the edge of the homoepitaxial film moved uniformly from the sample edge with the CVD growth. This serves as evidence that the growth mechanism of the present high-power MPCVD is dominated by apparent lateral growth. In addition, lobe-shaped features indicated by a dotted ellipsoidal circle were found to reflect the growth shift denoted by vector **A″B″** = **AB**, meaning that anomalous growth must originate from unrecognized crystalline defects existing on or near the substrate surface, although such morphological features were not observed clearly before film deposition. Results also showed that the growth rate changed near the crystalline defects. Lateral growth is inferred to be promoted with increasing T_s because of enhancement of the reaction rate at the steps, which results in an increase of R_g [106]. The misorientation angle from the exact <100> crystal orientation was estimated from measured values of d_F and |**AB**| to be 3.3°, which is consistent with the guaranteed misorientation angle for the type-Ib substrates employed here.

It is noteworthy that the hillock formation becomes a serious problem when the film thickness is greater than 1 mm because the film surface becomes covered with hillocks [97, 104]. The growth conditions should be optimized more strictly when thick diamond films are grown.

3.3.4 Optical Properties

In this section, crystalline qualities of homoepitaxial diamond grown utilizing high-power MPCVD are discussed from the viewpoint of the optical properties. In addition, dynamical behaviors of carriers excited in homoepitaxial diamond are described. Homoepitaxial films with a thickness of 20 μm were grown under various conditions. The methane concentration, the substrate temperature, the total gas pressure, and the microwave power were 4.0%, 1010 °C, 120 Torr and 3.8 kW, respectively. Their cathode luminescence (CL) spectra taken at room temperature are shown in Figure 3.9.

A readily visible emission band was observed at circa 520 nm from the sample grown on a type-Ib substrate, as shown by the spectrum (a). This emission band originates from the H3 center and its replica, which comes from a nitrogen-vacancy (NV) complex [107]. On the other hand, no NV emission appeared from the diamond film grown on a type-IIa substrate, as evidenced by the spectrum (b). Nitrogen incorporation into the homoepitaxial film is unlikely to occur substantially in the refined growth environment [81]. When nitrogen was fed intentionally into the growth chamber in the gas ratio (N/C) of 1 ppm, characteristic spectral features known as 533 and 575 centers [107] were observed from the film deposited on a type-IIa substrate, whereas the 520-nm band was negligibly small in intensity, as shown by spectrum (c). In addition, macro-bunching steps appear on the entire

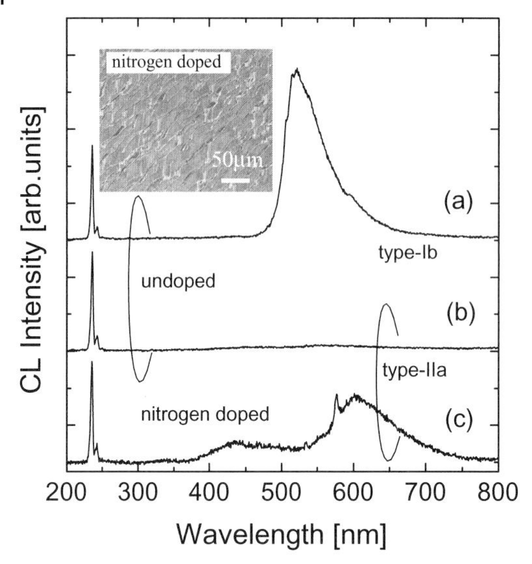

Figure 3.9 Cathodoluminescence spectra taken at room temperature from diamond films grown at methane concentration of 4.0% and substrate temperature of 1010 °C: (a) an undoped film grown on a type-Ib substrate, (b) an undoped film grown on a type-IIa substrate, and (c) a nitrogen-doped film grown on a type-IIa substrate.

surface area of grown film as shown by the OM image of the sample in the inset. This feature is common in the case of nitrogen doping [108]. For those reasons, NV emission of spectrum (a) is inferred not to occur dominantly in the homoepitaxial film. The possible mechanism inducing the emission is substantial long diffusion of excited carriers from the homoepitaxial film into the type-Ib substrate followed by recombination in the substrate through the NV emission path. Therefore, the NV emission intensities can provide information about the number of electron-beam-excited carriers that reached the substrate from the homoepitaxial film.

Figure 3.10 shows a series of typical OM and scanning room-temperature CL images of the homoepitaxial film taken by interrupting the CVD process at 1-μm steps at several film thicknesses from 1 μm to 6 μm [45]. The substrate temperature of the sample during the MPCVD was 890 °C. The CL images were taken at wavelengths of 235, 420 and 525 nm, which correspond to the free exciton (FE) recombination, the band A (BA) and nitrogen-vacancy complex (NV), respectively. The BA emissions are known to originate from dislocations. In OM images, the macro-bunching steps which extend to the <110> direction were observed at initial growth stages with film thickness $d_F \leq 2\,\mu m$. This feature disappeared gradually and the morphology became flat with increasing d_F except for the hillocks, which means that lateral growth dominates the growth mode under the optimized growth condition of high-power MPCVD. In addition, small hillocks, such as those marked by dotted circles in Figure 3.10, were also enlarged laterally with increasing d_F. The

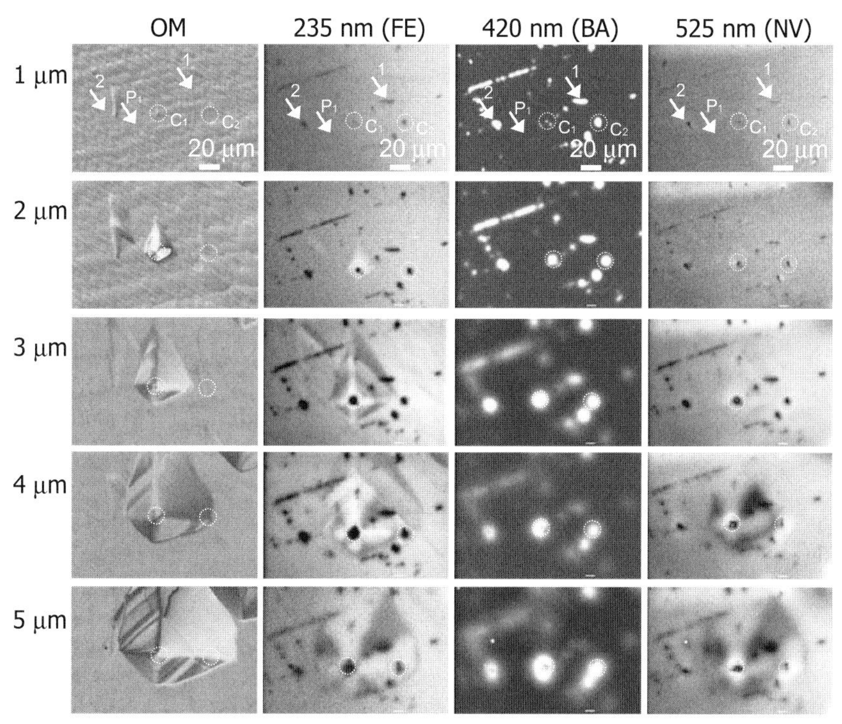

Figure 3.10 A series of optical microscope and scanning room-temperature CL images of the diamond film taken by interrupting homoepitaxial diamond growth several times. The wavelengths set for the CL images are 235 nm (FE), 420 nm (BA), and 525 nm (NV).

ratio of the growth rate in the lateral direction to that in the surface-vertical direction was greater than 20 at the hillock. Therefore, these hillocks finally had rather gentle slopes. The FE and NV emissions were observed from almost the entire specimen surface including the hillock regions, whereas the BA emission sites were localized in space as spot-like features on the specimen surface. The FE and NV emission images were complementary to the BA ones, as indicated by the arrows 1 and 2 in Figure 3.10. The SEM images (not shown) resemble the FE and NV emission images. For example, dark regions in the SEM image corresponded well to those in the FE and NV emission images. Characteristic features of surface morphology, such as hillocks and step bunching, were only slightly observable in the SEM images, compared to the corresponding OM images. Some BA emission sites corresponded to the morphologically protrusion region, as indicated by the circle C_1, although most emission sites were rather flat with no protrusion, as shown by the other circle C_2.

Here, let us discuss the NV emission feature of the film in greater detail. Figure 3.11a shows room-temperature CL spectra taken with V_{acc} of 15 kV at various d_F stages. The penetration depth of primary high-energy electrons from the specimen

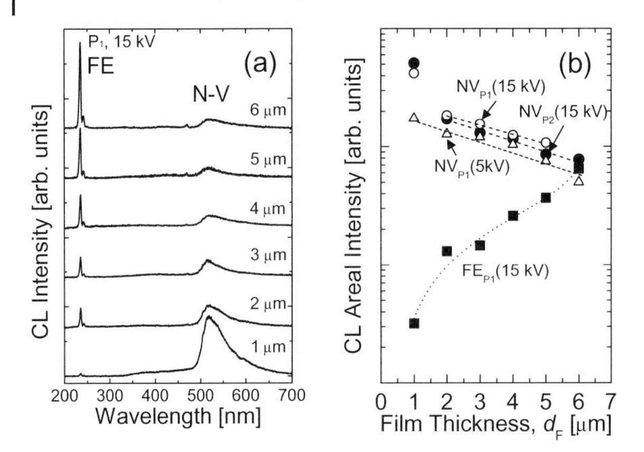

Figure 3.11 (a) Room-temperature CL spectra of the homoepitaxial diamond film taken at various film thicknesses (1–6 μm). The measured position is indicated by an arrow shown in Figure 3.10. (b) Dependencies of the integrated FE and NV emission intensities on the film thickness. Position P1 is a typical position denoted by the arrow in NV-center-emitting area without band A emission, whereas position P2 is another one that was selected arbitrarily on the specimen surface. Filled rectangles and filled circles are FE or NV emission taken at P1 with acceleration energy of 15 kV.

surface, P_e, which is roughly estimated by the projected electron range of $(0.6 \pm 0.2) \times R_e$, must better explain the observed d_F dependence of the CL intensity, compared with the electron range, R_e, itself. Reported values of R_e for diamond are 1.8–2.5 μm at 15 kV and 0.3 μm at 5 kV, respectively. The measured positions P_1, indicated by an arrow in Figure 3.10, are typical emission sites where the BA emission was negligibly small in intensity. (Note that the feature peaked at 470 nm is a replica of the FE emissions at double wavelengths and will not be discussed further.) The intensity of the NV emission, I_{NV}, decreased with increasing d_F, as shown in Figure 3.11b, whereas that of the FE emission, I_{FE}, increased with d_F. Here, the plotted CL intensities are the areal intensities integrated over the spectral full width of the corresponding emission peak. Filled circles were I_{NV} measured at position P_1, whereas open circles were taken at another typical position P_2. The measured I_{NV} is approximately represented by a single exponential function of d_F with a decay constant, C_D^N, to the film-growth direction in the d_F range of 2–5 μm. The C_D^Ns estimated for the positions of P_1 and P_2 at V_{acc} of 15 kV are 4.8 μm and 5.4 μm, respectively. The decay constant in the luminescence intensity reflects the density of non-radiative centers because the excited carriers are dominantly relaxed through non-radiative electronic transitions. From Figure 3.11b, the C_D^Ns are confirmed to be nearly independent of the depth and position, which means that the crystalline quality of the homoepitaxial film is three-dimensionally homogeneous, where the FE emissions were observed as a main feature. The C_D^N, estimated at V_{acc} of 5 kV, was circa 4.5 μm at position P_1. The value of I_{NV} at V_{acc} of 15 kV decreased rapidly for d_F smaller than 2 μm, whereas the turning point of the damping length corresponded roughly to the 15–keV electrons' P_e or R_e. Possible

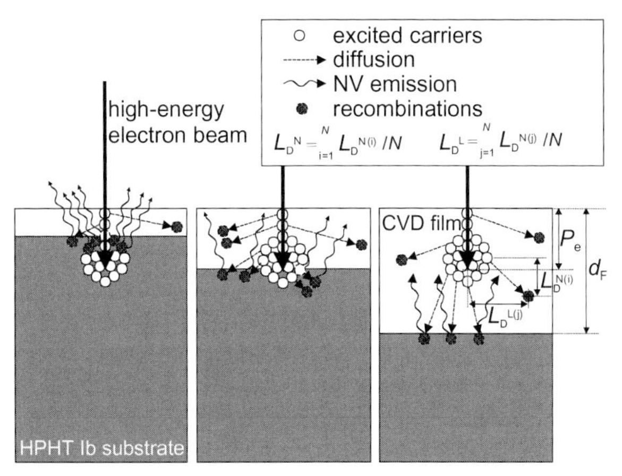

Figure 3.12 Schematic behaviors expected through carrier diffusions in a high-quality diamond layer homoepitaxially grown on a type-Ib HTHP substrate in three cases of $d_F < P_e$, $d_F \approx P_e$, and $d_F > P_e$, where d_F is the overlayer thickness and P_e the electron penetration depth.

major behaviors of the electron-beam-excited carriers in the specimen for different d_Fs are illustrated schematically in Figure 3.12.

In the case of $d_F < P_e$, electron–hole pairs are created mainly in the substrate, where the excited carriers are relaxed mainly through non-radiative recombination paths and slightly through radiative recombination via the NV and BA centers. In cases where $d_F > P_e$, most electron–hole pairs are created in the homoepitaxial film. For a high-quality film with small defect densities, the excited carriers are effectively prevented from relaxing through defects, so that the carrier diffusion length L_D is enlarged. In such a case, a part of the excited carriers diffusing in the homoepitaxial layer can penetrate into the substrate to yield NV emissions, as observed, although the number of such carriers decreases as an exponential function of d_F. When we applied positive bias voltage to the substrate from the film, the NV emission intensity increased with the voltage and vice versa, which means that diffusion carriers that contribute to the NV emission are mainly electrons.

Figure 3.11b shows the turning point of L_D^N, circa $2\,\mu m$, which corresponds roughly to P_e of the 15-keV electrons. Figure 3.12 also shows a situation that is expected when $d_F \approx P_e$, corresponding to the turning point in the d_F dependence of the substrate NV emissions. Furthermore, it is noteworthy that the estimated C_D^N of $5.1 \pm 0.3\,\mu m$ is an average one-dimensional diffusion length or a diffusion length projected to the specimen surface normal direction, L_D^N. The three-dimensional carrier diffusion length, L_D°, equals to $(\sqrt{2})^2 \times L_D^N \approx 10\,\mu m$ for isotropic diffusion. Details of carrier diffusion in the homoepitaxial film are discussed elsewhere [44].

Figure 3.13 shows typical room-temperature CL spectra of samples grown at various methane concentrations, C_{me}, of 4.0–16.0%. Strong FE emissions were observed from all samples. When the spectra were compared at the same C_{me} of 4.0%, total emission intensities were found to be higher for the specimen

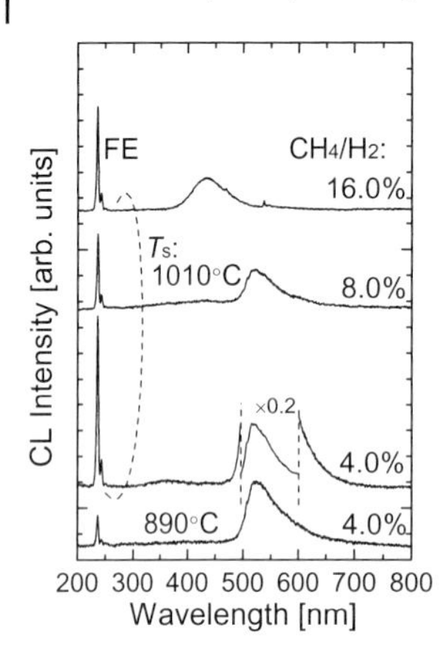

Figure 3.13 CL spectra taken at room temperature from homoepitaxial diamond films grown at different substrate temperatures using the high-power MPCVD method.

deposited at higher temperature than those for the lower-temperature specimen, indicating a decrease of a considerable amount of non-radiative centers in the specimen by increasing the substrate temperature. The FE emissions were still observed clearly from the sample grown under high C_{me} of 16.0%, for which a high growth rate of 20 μm h^{-1} was attained. These results indicate that an increase of substrate temperature to 1000 °C improves both the crystalline quality and the surface morphology of the homoepitaxial diamond (100) films under high-power MPCVD conditions. This advantage is favorable for practical device-grade diamond depositions because sufficiently thick, flat, and high-quality diamond films are useful for device fabrication without any post-deposition processes such as surface polishing. Figure 3.14a shows typical PL spectra of the diamond film grown on a type-Ib substrate and the standard HPHT type-IIa substrate. A Nd:YAG pulsed laser with 220-nm wavelength and ~30-ps width was used for PL measurements. The characterized homoepitaxial film is the same one that is shown in Figure 3.9 as CL spectrum (a). Both samples showed FE emission with substantial intensity. On the other hand, no FE emission was observed from the HPHT type-Ib substrate under the same measurement conditions as in the case of CL measurement. In contrast to the CL spectrum (a) taken from the diamond film grown on the type-Ib substrate, no broad NV emission was observed in the corresponding PL spectrum, indicating that the excitation condition affects the luminescence spectra strongly. (Note that the 470-nm peak is a replica of the FE emission.) Photoexcitation density during laser exposure is much larger than excitation density by high-energy electrons in CL measurements under the present experimental conditions. With decreasing laser power density, the FE intensity in PL spectra decreased for both samples (a) and (b), whereas NV emission appeared only from specimen (a) which was the homoepitaxial film grown on the type-Ib substrate. Consequently, PL spectra of both samples came to resemble those of corresponding CL spectra under low-density excitation conditions. We can understand these features as follows. When the laser power density is quite high, electron–hole pairs are created in diamond at high density. Therefore, these excited carriers tend to form excitons

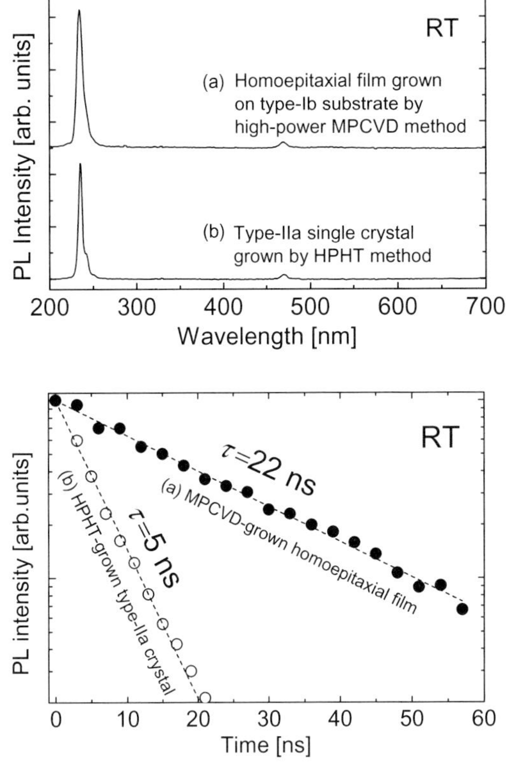

Figure 3.14 Photoluminescence spectra of (a) the MPCVD film and (b) the type-IIa HTHP substrate taken at room temperature and temporal changes of photoluminescence intensity corresponding to free-exciton recombination emission of these samples.

rather than a free carrier state. In the low-density excitation condition, most excited carriers might retain free carrier states because their mutual interaction is weaker as a result of relative remoteness. The inverse dependence of FE and NV emission intensities on the excitation density indicates that NV emission occurs dominantly through free carrier relaxation. As described above, a certain amount of the NV recombination center exists in the type-Ib substrate and electrons contribute to the NV emission as proved by CL results. Appearance of NV emission at a low-density excitation condition therefore indicates that the excited free carriers being possibly electrons diffused into the type-Ib substrate after passing through the thick homoepitaxial diamond film of 18 μm. We note that the long diffusion length reflects large carrier mobility and longer carrier lifetime of the homoepitaxial diamond films compared to the HPHT-grown diamond single crystal.

Temporal change of PL intensities of FE recombination emission, I_{FE}, is shown in Figure 3.14b. Filled circles denote the high-quality homoepitaxial diamond film deposited on a type-Ib substrate, whereas open circles signify the HPHT type-IIa substrate. The I_{FE} can be fitted well using a simple exponential decay function. The decay time τ_{FE} was 22 ns for the homoepitaxial film but 5 ns for the HPHT substrate. Because τ_{FE} is dominated by density of non-radiative center at room temperature, the homoepitaxial film is superior to that of the type-IIa HPHT diamond from the viewpoint of electronic quality.

3.3.5 Boron Doping

This section describes the electrical properties of homoepitaxial diamond film grown at high rate. Boron-doped homoepitaxial diamond (100) films were deposited on type Ib diamond (100) crystals using a high-power MPCVD apparatus. Growth conditions employed for boron doping are essentially the same as those of non-doped diamond. In this case, trimethylboron (TMB), $B(CH_3)_3$, was used as a doping gas. The growth rate was the same as the case without doping, $3.5\,\mu m\,h^{-1}$, which is circa 30 times faster than previously reported for the highest mobility sample grown by MPCVD [84]. The total film thickness examined was 25 µm; it comprised a B-doped layer and an undoped layer beneath the B-doped one. The resultant film surface was macroscopically flat; it had no unepitaxial crystallites in the entire sample surface area.

Electric properties were examined using Hall measurement methods under an AC magnetic field. The resistivity decreases with an increasing B/C ratio in the gas phase. The resistivity of the less-doped sample (i.e. 0.5 ppm) was at a minimum at circa 500 K, whereas that of the more-doped sample (i.e. 50 ppm) continued to decrease over the examined temperature range. The resistivity for the 50–ppm sample reached 0.05 Ω cm at 650 K, which corresponds to the resistivity of Si whose impurity concentration is circa $10^{18}\,cm^{-3}$ at the same temperature. Figure 3.15a shows the hole concentration in the Arrhenius coordinates. The hole concentration showed a thermally activated behavior, which is characteristic of doped semiconductors; the absolute value of the hole concentration increased with the increasing B/C ratio in the gas phase. No hopping conduction was observed for any sample in the examined temperature range, indicating a low density of electronic defects. All data fit well with the commonly used equation of carrier concentration considering partial compensation [87], as shown by the dashed curves. The fitting parameters are summarized in Table 3.4.

The acceptor densities estimated from the Hall measurement, N_A^{Hall}, increase with increasing B/C ratio in the gas phase. The boron acceptor activation energy E_A is 0.36 eV up to N_A^{Hall} of circa $10^{18}\,cm^{-3}$; above that level, E_A decreases gradually and becomes 0.29 eV at N_A^{Hall} of $10^{20}\,cm^{-3}$, which is consistent with previous reports [88]. The compensation ratios K are practically independent of N_A^{Hall} in the examined range; the lowest one, 0.45%, is comparable with the lowest one reported so far for CVD diamond with the highest mobility [84]. The certainty of the obtained N_A^{Hall} from Hall measurements was evaluated through comparison with those from C–V measurements N_A^{CV}. Those data agree well with an error of less than 30%, as shown for N_A^{CV} in Table 3.4, except for the case of B/C = 50 ppm. The disagree-

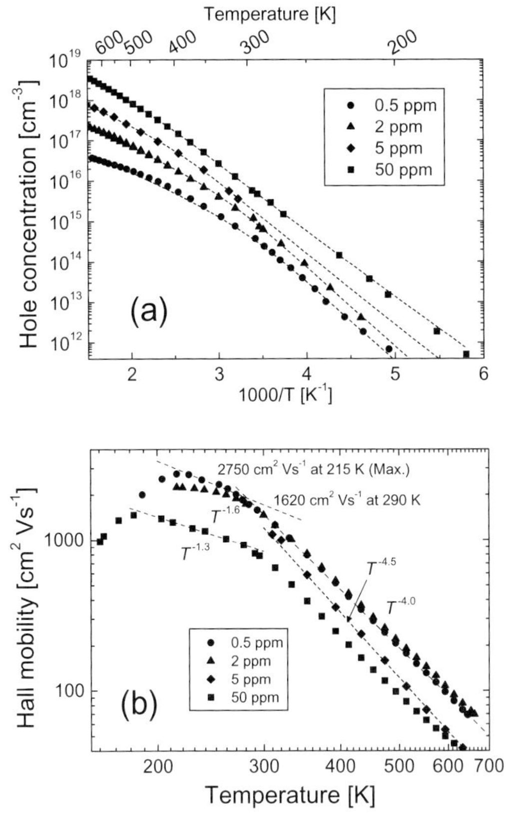

Figure 3.15 Temperature dependence of hole concentration and Hall mobility deduced for the B-doped diamond films grown using high-power MPCVD.

Table 3.4 The acceptor density estimated from Hall measurement N_A^{Hall}, the activation energy for ionization E_A, and the compensation ratio K estimated from Hall measurements. The acceptor density estimated from CV measurement N_A^{CV} is also listed.

B/C (ppm)	N_A^{Hall} (cm^{-3})	E_A (eV)	K (%)	N_A^{CV} (cm^{-3})
0.5	6.4×10^{16}	0.36	1.1	4.5×10^{16}
2	1.3×10^{18}	0.36	0.45	1.5×10^{18}
5	7.5×10^{18}	0.32	1.4	—
50	1×10^{20}	0.29	1.3	1.2×10^{19}

ment for the 50-ppm sample might be attributable to error in the *C–V* measurement: the leakage current for the Schottky contact is rather high for that sample because the tunneling effect in the Schottky barrier is considerable. Therefore, we inferred that N_A^{Hall} values are reliable.

Hall mobility is shown in Figure 3.15b as a function of temperature. High Hall mobilities of 1620 $cm^2 V s^{-1}$ at 290 K and 2750 $cm^2 V s^{-1}$ at 215 K were obtained from the 0.5-ppm sample [109]. These values are roughly equivalent to the highest previously reported mobility [84]. The Hall mobility μ decreased with T as $\mu \propto T^{-1.3}$–$T^{-1.6}$ for T of less than 270 K, above which μ has a dependence of $T^{-4.0}$–$T^{-4.5}$. The former is attributable to acoustic phonon scattering. Although the origin of the latter scattering factor remains unclear, strong temperature dependence is a common feature in diamond [84, 86, 87]. The possible mechanism arising in high-temperature range is inferred to the optical phonons scattering [117]. The acceptor concentration of $10^{20} cm^{-3}$ of the 50-ppm sample is particularly high, but higher Hall mobility of 830 $cm^2 V s^{-1}$ was achieved at 290 K [109]. This result indicates that high-power MPCVD is advantageous for deposition of high-quality and highly conductive diamond films.

Figure 3.16 shows the room-temperature Hall mobilities for the high-power MPCVD samples (filled circles) as a function of room-temperature hole concentration. A dashed line shown in Figure 3.16 was drawn by connecting the highest mobilities that were reported for the low-power MPCVD samples at each hole concentration [55, 84, 86, 87, 110, 111]. We found that Hall mobility decreases rapidly with increasing hole concentration in both cases, although the tendency is weak when the microwave power density is high. This result indicates that formation of defects that are created at rather high boron doping level is partly suppressed by increasing microwave power density. These results imply that the microwave power density plays an important role in both impurity incorporation and defect formation, although the the mechanisms are unclear at the present stage.

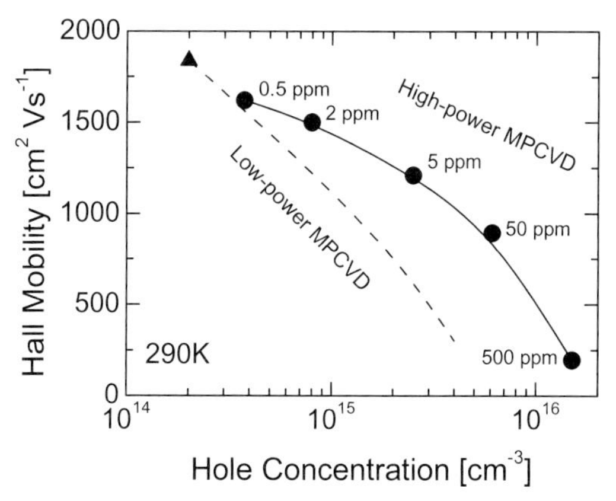

Figure 3.16 Hall mobility as a function of hole concentration: high-power MPCVD (filled circles) and an envelope of maximum mobility for low-power MPCVD (a dashed line). Both Hall mobility and hole concentration are room-temperature data

3.3.6 Nitrogen Doping

Nitrogen is known to increase the CVD diamond growth rate and tends to form (100) facets. Recently, rapid growth of homoepitaxial diamond has been investigated intensively using high-power MPCVD with nitrogen feeding, for which growth rates of greater than 100 $\mu m\,h^{-1}$ have been reported [96, 97]. Typical growth conditions for this super high-rate growth are 10% methane concentration, 1100–1200 °C substrate temperature and 100–200 Torr total pressure. In the case of non-doping, poor-quality diamond with a certain amount of graphite is likely to form under these conditions. The microwave power that is applied is 1.5–6 kW. Homoepitaxial growth was examined for N/C gas ratio below 10%.

The growth rate reportedly increases by a factor of circa 2 with nitrogen feeding during high-power MPCVD, whereas the rate increase is constant for the wide N/C ratio range of 500–100 000 ppm [97, 98]. A tiny amount of nitrogen is effective for stabilizing the diamond phase even under the higher growth conditions mentioned above [112, 113]. The nitrogen incorporation ratio from gas phase is in the range of 10^{-4}–10^{-3} [114, 115], which is not as high as that in the case of boron doping. Although the substrate temperature is in most cases set to be a higher than 1100 °C, transparent crystals can be grown as distinct from growth without nitrogen. The substrate temperature dependence of the growth rate is insignificant under this condition. On the other hand, the total pressure increases the growth rate [97]. Quality of the grown crystal, as characterized by X-ray diffraction, is of the same level as that of HPHT type-Ib substrate [114]. Surface morphology depends reportedly on the sample holder shape [97] and substrate pretreatment [47] as well as process parameters. Because a certain amount of diamond is deposited simultaneously on the sample holder during this high-rate growth, the holder must be cleaned by frequent halting of CVD growth. Otherwise, the diamond on the holder peels and contaminates the homoepitaxial film [116].

3.3.7 Large Area Deposition

Large area growth of single crystalline diamond is a current topic of homoepitaxial diamond study. Several approaches are proposed to increase the crystal surface area. The simplest is to deposit thick diamond crystal because diamond is grown not only perpendicular to the substrate crystal plane, but also parallel to it. A large crystal with surface diameter of circa 8 mm was grown [14], which implies that low-pressure CVD is a promising method for this purpose. However, diamond growth from the substrate side faces in the substrate lateral direction is slow and crystalline quality of the laterally grown layer is poor at the present stage. Another way to increase crystal size is through so-called mosaic growth; several substrates are placed regularly on the substrate holder and a diamond film is deposited on them. Diamond growth occurs simultaneously on each substrate. Then homoepitaxial films stick together side by side. Consequently, a rectangular diamond mosaic film is formed of $12 \times 12\,mm^2$ [14]. This film has defects in the junction regions in high density and is categorized as a highly oriented film because orientation of template substrates toward the {100} direction is imperfect. Because the growth rate is greater in the perpendicular direction than in the parallel direction at the substrate edges and because the quality is better for the former area, a complicated method is proposed as follows: first a diamond film thicker than 1 mm is deposited on the normal substrate; then the sample is rotated 90 degrees [116]. The crystal is snowballed by repeating this procedure. Improving crystalline quality of diamond is a fundamental issue that should be resolved in this method. At this stage, we have a view that heteroepitaxial growth is an effective method to grow larger area/medium quality diamond films, while high-rate homoepitaxial growth is a promising method to obtain medium size/higher quality diamond crystal.

3.4 Conclusions and Perspectives

Diamond is a semiconducting material that offers excellent physical and chemical properties. At this moment, the most important issue that must be resolved for realizing diamond-based electronic devices is the development of a growth technique for high-quality diamond at high growth rate with high reproducibility. The growth mechanism of diamond under low-pressure conditions remains unclear mainly because of the nonequilibrium growth condition, which limits the improvement of the crystalline quality and/or the increase of growth rate. The density profile of precursors and atomic hydrogen during the deposition process is still not understood well. From the viewpoint of elucidating the diamond growth mechanism, homoepitaxial growth study is beneficial because this growth system is simple. In addition, because the grown films have lower defect density and show excellent semiconducting properties of diamond, analyses of these properties are easier than for the case of polycrystalline diamond. In this chapter, recent research progress in homoepitaxial diamond growth has been reviewed from the viewpoint of growth conditions.

The growth parameters, including substrate conditions, are interrelated; the effects of each process parameter on the diamond growth mode must be understood well to optimize growth conditions. Sequential growth and characterization processes of diamond film, an example of which was described above, are effective for understanding diamond growth modes, not only in the flat epitaxial region, but also at defects. Microscopic characterization and *in situ* measurement of diamond growth features are powerful means for identifying the growth mode of homoepitaxial diamond; they lead to better understanding of diamond growth mechanisms under nonequilibrium conditions. In the case of microwave plasma-assisted chemical vapor deposition, plasma diagnostics is an effective approach to evaluate density and temperature of electrons, which helps to deduce the densities of carbon radicals or atomic hydrogen. Quantitative experimental analyses of reaction paths and reaction rates in diamond growth are important for a deeper understanding of diamond growth under nonequilibrium conditions. Advanced control of atomic hydrogen at the growth front enables *in situ* characterization. In addition, discovery of alternatives for atomic hydrogen might give us another kind of diamond growth condition, leading new deposition method instead of plasma-assisted chemical vapor deposition.

Control of electrical conductivity is necessary for electronic device applications. Impurities must be introduced at rather high concentrations for comparable conductivity with common semiconductors because of the large activation energy of dopant in diamond, whereas high impurity doping degrades the crystalline quality. In addition, the growth conditions and the substrate crystal plane are limited for each dopant. For that reason, diamond growth accompanied with impurity doping is much more complicated than in the case without doping. The search for a shallower dopant than boron is surely an important subject to reduce on-resistance during the electronic device operation.

Electrons often give superior performance in diamond-based devices, although electrons are unstable in the diamond conduction band. Therefore, the electron lifetime is regarded as an indicator of diamond crystalline quality. Enlarging crystals to obtain a single-crystalline wafer is a key issue for commercialization. Both super high-rate homoepitaxial growth and heteroepitaxial growth are candidates for this, but the knowledge gained from a basic study of homoepitaxial growth is valuable for success in these trials of crystal enlarging. At the moment, microwave plasma-assisted chemical vapor deposition is the only promising method to grow high-quality homoepitaxial diamond. Through a better understanding of single crystalline diamond growth under nonequilibrium conditions, advanced growth methods that have higher controllability than microwave-plasma chemical vapor deposition can be proposed in the future.

Acknowledgments

The author would like to acknowledge Professor Toshimichi Ito of Osaka University for fruitful discussion on the high rate growth of homoepitaxial diamond films. The author would also like to thank Dr Satoshi Koizumi of the National Institute for Materials Science for encouraging the preparation of this manuscript. This work was partly supported by a Grant-in-Aid for Scientific Research from the Japan Society for the Promotion of Science (No. 18760241 and 20360147).

References

1 Koizumi, S., Watanabe, K., Hasegawa, M. and Kanda, H. (2001) *Science*, **292**, 1899.

2 Makino, T., Tokuda, N., Kato, H., Ogura, M., Watanabe, H., -Gi Ri, S., Yamasaki, S. and Okushi, H (2006) *Japanese Journal of Applied Physics*, **45**, L1042.

3 Teraji, T., Yoshizaki, S., Wada, H., Hamada, M. and Ito, T. (2004) *Diamond and Related Materials*, **13**, 858.

4 Balducci, A., Marinelli, M., Milani, E., Morgada, E., Pucella, G., Tucciarone, A., Verona-Rinati, G., Angelone, M. and

Pillon, M. (2005) *Applied Physics Letters*, **86**, 213507.

5 Koide, Y., Liao, M.Y. and Alvarez, J. (2006) *Diamond and Related Materials*, **15**, 1962.

6 Collins, A.T. and Lightowlers, E.C. (1979) *The Properties of Diamond* (ed. Field, J.E.), Academic Press, London.

7 Matsumoto, S., Sato, Y., Tsutsumi, M. and Setaka, N. (1982) *Journal of Material Science*, **17**, 3106.

8 Kamo, M., Sato, Y., Matsumoto, S. and Setaka, N. (1983) *Journal of Crystal Growth*, **62**, 642.

9 Kobashi, K. (2005) *Diamond Films: Chemical Vapor Deposition for Oriented and Heteroepitaxial Growth*, Elsevier, Amsterdam.

10 Setaka, N. (1994) *Synthetic Diamond Emerging CVD Science and Technology* (eds Spear, K.E. and Dismukes, J.P.), John Wiley & Sons, New York.

11 Suzuki, K., Sawabe, A., Yasuda, H. and Inuzuka, T. (1987) *Applied Physics Letters*, **50**, 728.

12 Koizumi, S., Murakami, T., Inuzuka, T. and Suzuki, K. (1990) *Applied Physics Letters*, **57**, 563.

13 Ohtsuka, K., Fukuda, H., Suzuki, K. and Sawabe, A. (1997) *Japanese Journal of Applied Physics*, **36**, L1214.

14 Weimer, R.A., Thorpe, T.P., Snail, K.A. and Merzbacher, C.E. (1995) *Applied Physics Letters*, **7**, 1839.

15 Takeuchi, S., Murakawa, M. and Komaki, K. (2003) *Surface Coating Technology*, **169–70**, 277.

16 Kobashi, K., Nishibayashi, Y., Yokota, Y., Ando, Y., Tachibana, T., Kawakami, N., Hayashi, K., Inoue, K., Meguro, K., Imai, H., Furuta, H., Hirao, T., Oura, K., Gotoh, Y., Nakahara, H., Tsuji, H., Ishikawa, J., Koeck, F.A., Nemanich, R.J., Sakai, T., Sakuma, N. and Yoshida, H. (2003) *Diamond and Related Materials*, **12**, 233.

17 http://www.sekitech.biz/product/DiamondCVD/MicroCVD/index.html

18 www.aixtron.com

19 Kamo, M., Yurimoto, H. and Sato, Y. (1988) *Applied Surface Science*, **33–4**, 553.

20 Chu, C.J., Evelyn, M.P.D., Hauge, R.H. and Margrave, J.L. (1991) *Journal of Applied Physics*, **70**, 1695.

21 Behr, D., Wagner, J., Wild, C. and Koidl, P. (1993) *Applied Physics Letters*, **63**, 3005.

22 Hayashi, K., Yamanaka, S., Okushi, H. and Kajimura, K. (1996) *Applied Physics Letters*, **68**, 1220.

23 Teraji, T., Mitani, S., Wang, C.L. and Ito, T. (2002) *Journal of Crystal Growth*, **235**, 287.

24 Gamo, M.N., Low, K.P., Sakaguchi, I., Takami, T., Kusunoki, I. and Ando, T. (1999) *Journal of Vacuum Science and Technology A*, **17**, 2991.

25 Takami, T., Kusunoki, I., Gamo, M.N. and Ando, T. (2000) *Journal of Vacuum and Science Technology B*, **18**, 1198.

26 Koizumi, S., Teraji, T. and Kanda, H. (2000) *Diamond and Related Materials*, **9**, 935.

27 Chu, C.J., Hauge, R.H., Margrave, J.L. and D'Evelyn, M.P. (1992) *Applied Physics Letters*, **1**, 1393.

28 Tsuno, T., Shiomi, H., Kumazawa, Y., Shikata, S. and Akai, S. (1996) *Japanese Journal of Applied Physics*, **35**, 4724.

29 Takeuchi, D., Watanabe, H., Yamanaka, S., Okushi, H. and Kajimura, K. (1999) *Physica Status Solidi (a)*, **174**, 101.

30 Kasu, M. and Kobayashi, N. (2003) *Diamond and Related Materials*, **12**, 413.

31 Ri, S.G., Yoshida, H., Yamanaka, S., Watanabe, H., Takeuchi, D. and Okushi, H. (2002) *Journal of Crystal Growth*, **235**, 300.

32 Sato, Y. and Kamo, M. (1992) *The Properties of Natural and Synthetic Diamond* (ed. Field, J.E.), Academic, London.

33 Ri, S.G., Kato, H., Ogura, M., Watanabe, H., Makino, T., Yamasaki, S. and Okushi, H. (2005) *Diamond and Related Materials*, **14**, 1964.

34 Wild, C., Kohl, R., Herres, N., Müller-Sebert, W. and Koidl, P. (1994) *Diamond and Related Materials*, **3**, 373.

35 Ri, S.G., Nebel, C.E., Takeuchi, D., Rezek, B., Tokuda, N., Yamasaki, S. and Okushi, H. (2006) *Diamond and Related Materials*, **15**, 692.

36 Ando, Y., Tachibana, T. and Kobashi, K. (2001) *Diamond and Related Materials*, **10**, 312.

37 Evelyn, M.P.D., Graham, J.D. and Martin, L.R. (2001) *Diamond and Related Materials*, **10**, 1627.
38 Tsuno, T., Tomikawa, T., Shikata, S. and Fujimori, N. (1994) *Journal of Applied Physics*, **75**, 1526.
39 Tsuno, T., Imai, T. and Fujimori, N. (1994) *Japanese Journal of Applied Physics*, **33**, 4039.
40 Bauer, T., Schreck, M., Sternschulte, H. and Stritzker, B. (2005) *Diamond Related Material*, **14**, 266.
41 Hamada, M., Teraji, T. and Ito, T. (2005) *Japanese Journal of Applied Physics*. **44**, L216.
42 Takeuchi, D., Watanabe, H., Yamanaka, S., Okushi, H. and Kajimura, K. (2000) *Diamond and Related Materials*, **9**, 231.
43 Takeuchi, D., Watanabe, H., Yamanaka, S., Okushi, H., Sawada, H., Ichinose, H., Sekiguchi, T. and Kajimura, K. (2001) *Physical Review B*, **63**, 245328.
44 Teraji, T., Yoshizaki, S., Mitani, S., Watanabe, T. and Ito, T. (2004) *Journal of Applied Physics*, **96**, 7300.
45 Teraji, T., Mitani, S. and T. (2003) *physica status solidi (a)*, **198**, 395.
46 Wang, C.L., Irie, M. and Ito, T. (2000) *Diamond and Related Materials*, **9**, 1650.
47 Tallaire, A., Achard, J., Silva, F., Sussmann, R.S., Gicquel, A. and Rzepke, E. (2004) *physica status solidi (a)*, **201**, 2419.
48 Yamanaka, S., Takeuchi, D., Watanabe, H., Okushi, H. and Kajimura, K. (2000) *Diamond and Related Materials*, **9**, 956.
49 Yamanaka, S., Watanabe, H., Masai, S., Kawata, S., Hayashi, K., Takeuchi, D., Okushi, H. and Kajimura, K. (1998) *Journal of Applied Physics*, **84**, 6095.
50 Watanabe, H. and Okushi, H. (2000) *Japanese Journal of Applied Physics*, **39**, L835.
51 Watanabe, H., Takeuchi, D., Yamanaka, S., Okushi, H., Kajimura, K. and Sekiguchi, T. (1999) *Diamond and Related Materials*, **8**, 1272.
52 Takeuchi, D., Yamanaka, S., Watanabe, H., Sawada, S., Ichinose, H., Okushi, H. and Kajimura, K. (1999) *Diamond and Related Materials*, **8**, 1046.
53 Lee, N. and Badzian, A. (1995) *Applied Physics Letters*, **66**, 2203.
54 Shiomi, H., Tanabe, K., Nishibayashi, Y. and Fujimori, N. (1990) *Japanese Journal of Applied Physics*, **29**, 34.
55 Kiyota, H., Matsushima, E., Sato, K., Okushi, H., Ando, T., Tanaka, J., Kamo, M. and Sato, Y. (1997) *Diamond and Related Materials*, **6**, 1753.
56 Kamo, M., Ando, T., Sato, Y., Bando, M. and Ishikawa, J. (1992) *Diamond and Related Materials*, **1**, 104.
57 Frenklach, M. and Wang, H. (1991) *Physical Review B*, **43**, 1520.
58 Spear, K.E. and Frenklash, M. (1994) *Synthetic Diamond Emerging CVD Science and Technology* (eds Spear, K.E. and Dismukes, J.P.), John Wiley & Sons, New York.
59 Maeda, H., Ohtsubo, K., Irie, M., Ohya, N., Kusakabe, K. and Morooka, S. (1995) *Journal of Material Research*, **10**, 3115.
60 Sakaguchi, I., Gamo, M.N., Loh, K.P., Haneda, H. and Ando, T. (1999) *Journal of Applied Physics*, **86**, 1306.
61 Buttler, J.E. and Godwin, D.G. (2001) *Properties, Growth and Applications of Diamond* (eds Nazare, M.H. and Neves, A.J.), Inspec/IEE, London.
62 Tamura, H., Zhou, H., Hirano, Y., Takami, S., Kubo, M., Belosludov, R.V., Miyamoto, A., Imamura, A., Gamo, M.N. and Ando, T. (2000) *Physical Review B*, **62**, 16995.
63 Tsuno, T., Imai, T., Nishibayashi, Y., Hamada, K. and Fujimori, N. (1991) *Japanese Journal of Applied Physics*, **30**, 1063.
64 Sevillano, E. (1998) *Low Pressure Synthetic Diamond: Manufacturing and Applications* (eds Dischler, B. and Wild, C.), Springer-Verlag, Heidelberg.
65 Harris, S.J., Weiner, A.M. and Perry, T.A. (1991) *Journal of Applied Physics*, **70**, 1385.
66 Fujimori, N., Imai, T. and Doi, A. (1986) *Vacuum*, **36**, 99.
67 Okushi, H. (2001) *Diamond and Related Materials*, **10**, 281.
68 Takami, T., Suzuki, K., Kusunoki, I., Sakaguchi, I., Nishitani-Gamo, M. and Ando, T. (1999) *Diamond and Related Materials*, **8**, 701.
69 Kawarada, H., Sasaki, H. and Sato, A. (1995) *Physical Review B*, **52**, 11351.
70 Sutcu, L.F., Chu, C.J., Thompson, M.S., Hauge, R.H., Margrave, J.L. and

D'Evelyn, M.P. (1991) *Journal of Applied Physics*, **70**, 1695.

71 Rawles, R.E., Morris, W.G. and D'Evelyn, M.P. (1996) *Applied Physics Letters*, **69**, 4032.

72 Tajani, A., Mermoux, M., Marcus, B., Bustarret, E., Gheeraert, E. and Koizumi, S. (2003) *physica status solidi (a)*, **199**, 87.

73 Koizumi, S., Kamo, M., Sato, Y., Ozaki, H. and Inuzuka, T. (1997) *Applied Physics Letters*, **71**, 1065.

74 Sakaguchi, I., Gamo, M.N., Loh, K.P., Haneda, H. and Ando, T. (1999) *Diamond and Related Materials*, **8**, 1291.

75 Frenklach, M. and Skokov, S. (1997) *Journal of Physics Chemistry B*, **101**, 3025.

76 Mehandru, S.P. and Anderson, A.B. (1990) *Journal of Materials Research*, **5**, 2286.

77 Harris, S.J. and Goodwin, D.G. (1993) *Journal of Physical Chemistry*, **97**, 23.

78 Yamamoto, M., Teraji, T. and Ito, T. (2005) *Journal of Crystal Growth*, **285**, 130.

79 Tavares, C., Koizumi, S. and Kanda, H. (2005) *physica status solidi (a)*, **202**, 2129.

80 Teraji, T., Arima, K., Wada, H. and Ito, T. (2004) *Journal of Applied Physics*, **96**, 5906.

81 Teraji, T. and Ito, T. (2004) *Journal of Crystal Growth*, **271**, 409.

82 Deneuville, A. (2003) *"Thin-Film Diamond I" Semiconductors and Semimetals*, Vol. 76 (eds Nebel, C.E. and Ristein, J.), Elsevier Academic, Amsterdam.

83 Koizumi, S. (2003) *"Thin-Film Diamond I" Semiconductors and Semimetals*, Vol. 76 (eds C.E. Nebel and J. Ristein), Elsevier Academic, Amsterdam.

84 Yamanaka, S., Watanabe, H., Masai, S., Takeuchi, D., Okushi, H. and Kajimura, K. (1997) *Japanese Journal of Applied Physics*, **37**, L1129.

85 Fujimori, N., Nakahata, H. and Imai, T. (1990) *Japanese Journal of Applied Physics*, **29**, 824.

86 Hemley, R.J., Chen, Y.C. and Yan, C.S. (2005) *Elements*, **1**, 105.

87 Fox, B.A., Hartsell, M.L., Malta, D.M., Wynands, H.A., Kao, C.-T., Plano, L.S., Tessmer, G.J., Henard, R.B., Holmes, J.S., Tessmer, A.J. and Dreifus, D.L. (1995) *Diamond and Related Materials*, **4**, 622.

88 Lagrange, J.-P., Deneuville, A. and Gheeraert, E. (1998) *Diamond and Related Materials*, **7**, 1390.

89 Ekimov, E.A., Sidorov, V.A., Bauer, E.D., Mel'nik, N.N., Curro, N.J., Thompson, J. D. and Stishov, S.M. (2004) *Nature*, **428**, 542.

90 Bustarret, E., Kacmarcik, J., Marcenat, C., Gheeraert, E., Cytermann, C., Marcus, J. and Klein, T. (2004) *Physical Review Letters*, **93**, 237005.

91 Katagiri, M., Isoya, J., Koizumi, S. and Kanda, H. (2004) *Applied Physics Letters*, **85**, 6365.

92 Kato, H., Yamasaki, S. and Okushi, H. (2005) *Applied Physics Letters*, **86**, 222111.

93 Isberg, J., Hammersberg, J., Johansson, E., Wikstrom, T., Twitchen, D.J., Whitehead, A.J., Coe, S.E. and Scarsbrook, G.A. (2002) *Science*, **297**, 1670.

94 Moustakas, T.D. (1994) *Synthetic Diamond: Emerging CVD Science and Technology* (eds Spear, K.E. and Dismukes, J.P.), John Wiley & Sons, New York.

95 Leers, D. and Lydtin, H. (1991) *Diamond and Related Materials*, **1**, 1.

96 Yan, C.S., Vohra, Y.K., Mao, H.K. and Hemley, R.J. (2002) *Proceeding of National Academic Science*, **99**, 12523.

97 Chayahara, A., Mokuno, Y., Horino, Y., Takasu, Y., Kato, H., Yoshikawa, H. and Fujimori, N. (2004) *Diamond and Related Materials*, **13**, 1954.

98 Achard, J., Tallaire, A., Sussmann, R., Silva, F. and Gicquel, A. (2005) *Journal of Crystal Growth*, **284**, 396.

99 Zhang, J.Y., Wang, P.F., Ding, S.J., Zhang, D.W., Wang, J.T. and Liu, Z.J. (2000) *Thin Solid Films*, **368**, 266.

100 Deák, P., Kováts, A., Csíkváry, P., Maros, I. and Hárs, G. (2007) *Applied Physics Letters*, **90**, 051503.

101 Sternschulte, H., Bauer, T., Schreck, M. and Stritzker, B. (2006) *Diamond and Related Materials*, **15**, 542.

102 Teraji, T., Hamada, M., Wada, H., Yamamoto, M. and Ito, T. (2005) *Diamond and Related Materials*, **14**, 1747.

103 Fayette, L., Mermoux, M. and Marcus, B. (1994) *Diamond and Related Materials*, **3**, 480.
104 Park, Y.S., Kim, S.H., Jung, S.K., Shinn, M.N., Lee, J.–W., Hong, S.K. and Lee, J.Y. (1996) *Material Science of Engineering A*, **209**, 414.
105 Tallaire, A., Achard, J., Silva, F. and Gicquel, A. (2005) *Physica Status Solidi (a)*, **202**, 2059.
106 Teraji, T., Hamada, M., Wada, H., Yamamoto, M., Arima, K. and Ito, T. (2005) *Diamond and Related Materials*, **14**, 255.
107 Zaitsev, A.M. (2001) *Optical Properties of Diamond: A Data Handbook*, Springer, Berlin.
108 de Theije, F.K., Schermer, J.J. and van Enckevort, W.J.P. (2000) *Diamond and Related Materials*, **9**, 1439.
109 Teraji, T., Wada, H., Yamamoto, M., Arima, K. and Ito, T. (2006) *Diamond and Related Materials*, **15**, 602.
110 Takeuchi, D., Yamanaka, S., Watanabe, H. and Okushi, H. (2001) *Physica Status Solidi (a)*, **186**, 269.
111 Tsubota, T., Fukui, T., Saito, T., Kusakabe, K., Morooka, S. and Maeda, H. (2000) *Diamond and Related Materials*, **9**, 1362.
112 Tallaire, A., Achard, J., Silva, F., Sussmann, R.S. and Gicquel, A. (2005) *Diamond and Related Materials*, **14**, 1743.
113 Gryse, O.D. and Corte, K.D. (2005) Antwerp Facets, 14 November.
114 Mokuno, Y., Chayahara, A., Soda, Y., Yamada, H., Horino, Y. and Fujimori, N. (2006) *Diamond and Related Materials*, **15**, 455.
115 Tallaire, A., Collins, A.T., Charles, D., Achard, J., Sussmann, R., Gicquel, A., Newton, M.E., Edmonds, A.M. and Cruddace, R.J. (2006) *Diamond and Related Materials*, **15**, 1700.
116 Mokuno, Y., Chayahara, A., Soda, Y., Horino, Y. and Fujimori, N. (2005) *Diamond and Related Materials*, **14**, 1743.
117 Pernot, J., Contreras, A. and Camassel, J. (2005) *Journal of Applical Physics*, **98**, 023706.

4
Heteroepitaxy of Diamond

Yutaka Ando and Atsuhito Sawabe
Aoyama Gakuin Uni versity, Department of Electrical Engineering and Electronics, 5-10-1 Fuchinobe, Sagamihara, Kanagawa 229-8558, Japan

Diamond's extreme properties such as the highest thermal conductivity, hardness, wide band-gap semiconducting characteristics and so on have been reported by numerous researchers. However, realization of diamond devices with expected properties is not a simple task. People who would like to develop diamond devices need a high quality diamond substrate or film of large size which could be used in a lithography process at least and in an assembly line for device fabrication at most. In the 1970s and 1980s, publication methods for the synthetis of diamond from the gas phase in the low-pressure metastable conditions [1–3] attracted the people who needed large size diamonds. Polycrystalline diamond films or substrates with large sizes can be formed and are commercially available now. Some of these have a smooth polished surface which can be used in a lithography process and in a kind of assembly line for device fabrication. Several researchers in a company have been trying to fabricate a diamond device using the polycrystalline rather than the single crystal [4]. Such a compromise is not our intention. We would like to supply single-crystal diamond films or substrates with a large size which could open a new field of applied solid states physics.

An important key to the growth of single-crystal diamond films may be the choice of the substrate. Many kinds of materials such as c-BN [5–13], Si (β-SiC/Si) [14–28], Ni [29–36], Co. [37], Pt [38–48], Ir [49–64], and so on (Graphite [65], BeO [66], α-SiC [67], TiC [68], Ni_3Ge [69], Re [70]) have so far been chosen as candidate materials for the substrate for diamond heteroepitaxy. Details of each candidate will be given later.

Before consideration of heteroepitaxy of diamond, we must briefly discuss the best substrate for diamond epitaxy, which is diamond itself (detailed information about this is given in Chapter 3 on homoepitaxy). The merits of heteroepitaxial growth seem to be the possibility of size enlargement and its cost. That is why we have researched the heteroepitaxy of diamond. However, techniques for size enlargement of homoepitaxially grown diamond have been developed stage by stage (e.g. [71–73]). Recently, single-crystal diamonds with 15-mm diameter have become commercially available and the maximum size will become larger and

Physics and Applications of CVD Diamond. Satoshi Koizumi, Christoph Nebel, and Milos Nesladek

ISBN: 978-3-527-40801-6

larger over several years. Therefore, we may need to reconsider the merits of heteroepitaxy of diamond.

Before having such a discussion, we would like to look back at past results in the heteroepitaxy of diamond.

4.1 Cubic Boron Nitride

Diamond's extreme properties could also be demerits for the people who want to make large single crystals. For example, the surface free energy of diamond is so high that the two-dimensional growth of diamond on foreign substrates is very difficult. Furthermore, the covalent C—C bonds in diamond are so rigid that diamond nuclei, individually grown on a substrate, are very unlikely to coalesce without leaving grain boundaries between them, unless the diamond nuclei have been perfectly aligned with each other. From the viewpoint of materials which have similar properties and very good lattice match to diamond, the substrate which comes to mind is cubic boron nitride (c-BN). Cubic BN has properties which make it suitable as a substrate for heteroepitaxy of diamond: a zinc-blende structure (the two-element analog of diamond) with close lattice matching (0.3567-nm for diamond and 0.3615-nm for c-BN), a linear thermal-expansion coefficient similar to diamond and high surface energy.

Epitaxial growth of diamond on c-BN was reported by Koizumi *et al.* in 1990 [5]. Then, Inuzuka *et al.* summarized the epitaxial growth of diamond on c-BN as follows [9]: (i) on {100} facets of c-BN crystals, diamond thin films could be grown epitaxially with high crystal quality; (ii) on $\{111\}_B$ facets of c-BN crystals, diamond thin films grow epitaxially, but on $\{111\}_N$ facets, they do not because of the difference in binding energies of C—N, C—C, C—H ($E_{C-N} < E_{C-C} < E_{C-H}$) and B—C, B—H ($E_{B-H} < E_{B-C}$) [10]; (iii) appropriate growth conditions were different with the crystal plane: relatively high methane concentration (C_m) with lower temperature (T_s) for {100} and the lower C_m with higher T_s for {111} which is similar to the general approach to growing an oriented diamond by CVD [22, 71, 74, 75]. Maeda *et al.* also discussed the relation between the crystal orientation of c-BN substrate and appropriate epitaxial growth conditions of diamond [13]. They show the influence of C_m, T_s and also gas pressure (P_r). It is not easy to compare their results and the former because they actually varied four of the growth parameter (C_m, T_s, P_r and also the microwave power to control T_s.) and used a different growth system (microwave [13] and dc [5, 9, 10] plasma CVD), even so, we can note the similarity of these results. Higher C_m is needed for {100} growth and lower C_m is better for {111} plane. Although the growth temperature is not so high even for the growth on {111} plane, this contradiction can be explained by the shift of contour line of α parameters ($\sqrt{3}$*<100>/<111>growth rate ratio) on the C_m–T_s map with P_r [76]. At lower P_r, it is relatively easy to achieve the lower α parameter conditions which are appropriate for {111} growth, even not so high T_s. At higher P_r, they could find an appropriate condition for {100} growth even with the same C_m and T_s as for

{111} because the contour lines of the parameter αs are shifted toward the higher T_s and the lower C_m direction at higher P_r.

As described above, c-BN could be the best candidate for the heteroepitaxy of diamond. This suggestion, however, was soon denied because the preparation of the c-BN single-crystal is more difficult than for diamond. Conversely, a single-crystal diamond substrate can be a good candidate for the heteroepitaxy of c-BN [77, 78] which can give the interesting c-BN/diamond heterojunction.

4.2
Silicon and Silicon Carbide

Cubic BN as the substrate for the heteroepitaxy of diamond has the problem of size limitations. Accordingly, from the viewpoint of enlarged size as well as for its economical merit, silicon must be the best candidate substrate for diamond growth. Although the facts that silicon has a very large lattice mismatch with diamond (52%) and a much lower surface energy than diamond (1.5 J/m^2 for silicon (111) plane and 6 J/m^2 for diamond) should reduce interest in the use of silicon as a substrate for heteroepitaxy of diamond, some people have wisely overcome even such difficulties. Narayan *et al.* suggested matching the {111} or {200} lattice planes of diamond with the {022} planes of silicon and have shown <110> textured films of diamond grown on {100} silicon substrates [14]. This kind of concept is important for crystal growth on lattice-mismatched system. However, the heteroepitaxy of diamond on Si without interface layers has not actually been successful. Jeng *et al.* [16] reported a limited texturing of diamond locally on the *in situ* pretreated Si substrate. They explained that the role of the pretreatment is to form carbide materials as an interface layer and to reduce the untextured growth. Williams *et al.* [15] also found an epitaxial layer of β-SiC at the diamond–silicon interface. Then, Stoner *et al.* [17] used β-SiC(001) substrates for textured diamond growth. They have shown that approximately 50% of the diamond nuclei were textured with the C(001) planes parallel to the SiC(001) substrate and C[110] directions parallel to the SiC[110] within 3°. Even for β-SiC, the lattice mismatch with diamond is still large (0.3567 nm for diamond and 0.4359 nm for β-SiC). They remarked, however, that β-SiC grew epitaxially on Si despite a 24% lattice mismatch, thereby indicating that such a mismatch did not prohibit heteroepitaxy. Key techniques for the heteroepitaxy of diamond on such a large mismatched substrate are the bias enhanced nucleation (BEN) method [79] and also the following textured growth process [74, 75]. BEN, where the sample was negatively biased in the initial phase of the film growth, was originally just a method to enhance the density of diamond nuclei. However, it attracted further attention because it was found to be useful for generating oriented diamond nuclei on β-SiC. The textured growth process is also important because the surface of bias-treated samples was composed of both epitaxially and nonepitaxially nucleated diamond crystals. Even not used in the initial phase of growth, the process could play an important role in restriction of secondary nucleation as well as defect reduction. Stoner *et al.*

[20] demonstrated that highly oriented diamonds (HOD) in which close to 100% of the grains were in epitaxial alignment with the silicon substrate could be grown via a three-step, carburization, BEN and textured-growth, process. Hall measurements were also made for the HOD grown on Si as a measurement of the crystal quality. Hall mobility of B-doped diamond layers deposited on HOD films are found to be 165 cm^2/Vs at room temperature (278–135 cm^2/Vs at 180–440 K), which is a marked improvement over the values of B-doped polycrystalline diamond films which are less than about 50 cm^2/Vs [21]. Kawarada *et al.* [22] and also Tachibana *et al.* [23] refined the nucleation and growth process, and successfully fabricated HOD films having a smooth surface and high quality which can be used for some kinds of electrical or optical devices. For example, durable ultraviolet sensors which are of practical use were successfully fabricated by using HOD films [80].

4.3 Nickel and Cobalt

Heteroepitaxy of diamond on Si or β-SiC has met with some success as a route to highly oriented diamond. As a next step, we would like to find a substrate which has a close lattice match to diamond and so the possibility of size enlargement. Nickel and cobalt have very good lattice matches to diamond (Ni: 0.3524 nm, Co: 0.3545 nm, diamond: 0.3567 nm) and so the possibility of size enlargement (e.g. [81]). Preliminary studies of the heteroepitaxy of diamond on nickel and cobalt were presented by Sato *et al.* in ref. [29]. They have performed diamond deposition on Ni and Co under various conditions (500–1000 °C substrate temperatures, 40–100 torr of total gas pressure and 0.2–5% methane concentration) and epitaxially grown diamonds with low nucleation density have been observed only in conditions up to 0.9% methane concentration. They explained that the formation of graphite in other conditions prevented diamond growth as a result of the catalytic action of these materials. Detailed information about the difficulty of heteroepitaxy of diamond on Ni has been reported by Belton *et al.* [82]. Their results showed that carbon is deposited on the Ni(100) surface in an ordered c(2 × 2) array during the initial stage of growth, then a graphite layer with poor azimuthal order grows in islands on top of the c(2 × 2) deposit, and finally diamonds nucleate and grow on the grassy carbon surface. In order to realize heteroepitaxy of diamond on Ni, therefore, the diamonds have to nucleate on the c(2 × 2) carbon layer before the disordered graphite grows. However, their data also showed that diamond, graphite, and glassy carbon can all be formed under identical growth conditions.

Yang *et al.* [31] reported a novel approach to overcome the adverse effects of Ni on diamond formation. They applied a seeding and multistep deposition process to nucleate and grow diamond films directly on Ni substrates. As a result, highly oriented diamond nuclei have been deposited without graphite codeposition on both <100> oriented single-crystal Ni and polycrystalline Ni substrates depending upon the underlying substrate orientation. They postulated the existence of a liquid Ni—C—H layer during diamond growth and the reorientation of seeded diamond particles to align with the Ni substrate. Zhu *et al.* [30] improved the deposition process by refining the process parameters such as the degree of surface seeding, the annealing temperature and duration, the substrate temperature and the methane concentration. They found that heavy seeding followed by annealing resulted in high densities of both <100> and <111> oriented diamond nuclei on a single crystal Ni surface. They mentioned the importance of the annealing time which must be long enough to allow for sufficient reactions between the nickel, seeded carbon, and hydrogen to form Ni—C—H intermediate states, but must not be over long because the diamond seeds will then completely dissolve and diffuse into the bulk of the Ni lattice. Then, Yang *et al.* [32] found that non-diamond carbons such as graphite powders, fullerene (60) powders, and gaseous carbon species, were also effective for the encouragement of oriented nuclei of diamond. Therefore, the seeds for oriented diamond nuclei on Ni are not necessarily reoriented diamond particles. A real-time, in-situ laser reflectometry system was developed to monitor the changes in surface morphology observed during the high-temperature annealing stage [33, 34]. Using this technique, oriented nucleation and growth of diamond on Ni was reproducibly achieved. Analytical [35] and theoretical [36] work on the Ni surface during diamond growth were also reported.

A similar multistep deposition process which involves seeding, annealing, nucleation and growth has been applied with cobalt substrate [37]. As a result, <111> oriented diamond particles were obtained on Co(0001) oriented single crystal substrates. A mechanistic model which emphasized the suppression of graphite codeposition and oriented diamond nucleation through formation of the molten Co-C—H surface layer was proposed in a similar way to the heteroepitaxy of diamond on Ni.

4.4 Platinum

Although Ni and Co have very good lattice match to diamond and could be good candidates for heteroepitaxy of diamond, special surface treatments are necessary to achieve a high degree of orientation and a high density of diamond nuclei because of their properties such as high solubility for carbon and strong catalytic effects on hydrocarbon decomposition as mentioned above. Platinum is also a strong catalyst for hydrogen and hydrocarbons and has a relatively large lattice mismatch (~10%) to diamond. Thus, it was expected that research on Pt as a substrate for the heteroepitaxy of diamond would face more difficulties than for Ni or Co. However, in the results of ref. [83], Pt, which has no carbon atom interaction, is found to generate high density diamond nucleation, whereas Ni and Co have a negative energy of carbide formation and generate low density nucleation. Therefore, novel results were expected for the trial of heteroepitaxy of diamond on Pt. Shintani [38] has shown that a highly <111>-oriented, highly coalesced diamond film (not isolated grains) can be grown on the (111) domains in thinning platinum foil. Furthermore, the surface treatment for the diamond nucleation was a conventional method in which the surface of the Pt was ultrasonically scratched in ethanol mixed with diamond powder for 10–30 minutes. This result has attracted the attention of many related researchers since large areas of monocrystalline Pt substrate are available. Tachibana *et al.* [39, 40] tried a similar experiments but using a single crystal Pt(111) substrate and successfully found that diamond films with (111) facets were grown heteroepitaxially on the Pt(111). The diamond grown on Pt substrate was found by a high-resolution TEM to be directly connected to the substrate without any intermediate layer [41, 42]. The Eeitaxial relationships of these are $(111)_{diamond}$ // $(111)_{Pt}$ and $<1\bar{1}0>_{diamond}$// $<1\bar{1}0>_{Pt}$ at some local interface areas. However, the interface structure is complicated, including a graphite layer as an unavoidable film. Further information was reported in ref. [44]. The existence of an intricate structure including a graphite layer at the interface of Pt–diamond seems to be a demerit of Pt. Even if this is the case, a large merit of Pt is that diamond films can be grown on Pt{111} heteroepitaxially by conventional diamond nucleation methods such as a ultrasonic abrasion with diamond powder suspended in ethanol. Thus, there is in principle no size limit for the nucleation process. Tachibana *et al.* [46] have shown that heteroepitaxial diamond films can be successfully grown on Pt{111} films deposited on sapphire{0001} substrates which are commercially available in diameters up to several inches while bulk single crystal Pt is expensive and limited in size. Although the problem with graphite formation between Pt and sapphire was also found in their results, they overcame it soon by using the Pt/Ir/Pt films deposited on sapphire to prevent the diffusion of hydrocarbons through Pt during the diamond growth process [48]. The deposited diamond crystals were highly <111>-oriented with a FWHM value of 1.1° using an XRD rocking curve of diamond{111} diffraction, azimuthally oriented. While the substrate size used in the experiment was $10 \times 10\ mm^2$, the possibility of the depositing a larger area of heteroepitaxial diamond film has been shown.

4.5 Iridium

Heteroepitaxy of diamond on Pt seems to be successful but only on the {111} plane. From the viewpoint of diamond growth with higher quality, <100> growth is more desirable than <111> growth in general. One of the authors of this chapter (Y.A.) and others have found {100} epitaxial growth of diamond on a domain of Pt polycrystalline foil and presented this at a domestic conference. However, we could not reproduce the sample at that time.

Epitaxial growth of diamond grains on iridium thin films was performed by Ohtsuka *et al.* [49]. The Ir film was deposited epitaxially on a (001) cleaved surface of an MgO single crystal. Then the diamond was grown heteroepitaxially on the pretreated Ir surface by direct-current plasma chemical vapor deposition. The epitaxial relationship of diamond on iridium (001) was written as $(001)_{diamond}$ // $(001)_{iridium}$ and $[001]_{diamond}$ // $[001]_{iridium}$. In the pretreatment process, positively charged ion species impinge on the surface of the Ir substrate by applying a negative dc voltage to the substrate in a plasma of CH_4/H_2 gas mixture. Without the pretreatment process, diamond is not heteroepitaxially grown on Ir but nonepitaxial diamond grains are grown only on some unintentionally roughened spots on the Ir surface, unfortunately. The epitaxially grown diamond islands coalesced and became a smooth thin film over a longer time period [50]. Even with a film of 1.5-μm thickness, the surface was already smooth with an average roughness (R_a) of approximately 1 nm. Suzuki *et al.* have described why they have chosen Ir as a substrate for the heteroepitaxy of diamond [51, 54]. Saito *et al.* have shown similar heteroepitaxy of diamond on Ir/MgO(001) by using a microwave plasma-assisted chemical vapor deposition (MPCVD) system [52]. Fujisaki *et al.* have also shown diamond heteroepitaxy on Ir/MgO(001) substrate by using antenna-edge-type MPCVD for the pretreatment (bias-enhanced nucleation) process [53]. The removal process for the substrate (Ir/MgO) to fabricate a free-standing diamond is described in ref. [54].

Schreck *et al.* [55] have used a Ir/$SrTiO_3$(001) substrate instead of Ir/MgO(001) and improved the FWHM value of diamond (004) rocking curve and (311) azimuthal scan by X-ray diffraction measurements. The values of these were 0.34° and 0.65° for an 8-μm thick film [55], and furthermore, 0.17° and 0.38° for a 34-μm thick film after thinning from the growth side down to 10-μm [56], respectively. This mosaicity reduction could be achieved by an appropriate textured growth technique. However, the mechanism is not as simple as in the case of the highly oriented diamond on silicon which could be explained by the principle of evolutionary selection [22, 74, 75]. The mosaicity reduction during textured growth on Ir includes tilt as well as twist which is one of the advantages of Ir. Detailed information about the mosaicity reduction has been discussed by Schreck *et al.* in ref. [57]. Since he and his co-workers have done much respected work in this research area, I would like to recommend referring to their works [55–60]. Bednarski *et al.* [61] have shown epitaxial growth of Ir on an A-plane α-Al_2O_3 as well as $SrTiO_3$ to obtain a large scale heteroepitaxial diamond. They explained that Al_2O_3 possesses superior interfacial stability at high temperatures in vacuum or in hydrogen plasma with a better thermal expansivity match to diamond compared with that of $SrTiO_3$. Lee *et al.* [62] have shown that high quality epitaxial Ir films can be grown on Si by using CaF_2 as an intermediate layer. They have demonstrated epitaxial growth of diamond on Ir/CaF_2/Si substrate. The results suggested the possibility of integrated Si and diamond devices. Gsell *et al.* [63] also offered Ir/YSZ (yttria-stabilized zirconia)/Si as a substrate for heteroepitaxy of diamond from the viewpoint of thermal expansion mismatch between diamond and substrates. For example, the values of thermal stress for dia–MgO, dia–$SrTiO_3$, dia–Al_2O_3 and dia–Si are −8.30, −6.44, −4.05 and −0.68 GPa, respectively, when diamond growth temperature is 700 °C [63]. Thus, no delamination occurred for the diamond films grown on the Ir/YSZ/Si substrates. Similar Ir/oxide/Si systems such as Ir/$SrTiO_3$/Si, Ir/CeO_2/YSZ/Si, Ir/$SrTiO_3$/CeO_2/YSZ/Si have been demonstrated as substrates in ref. [64]. Although it is not clear which substrate is most suitable for high quality diamond growth, the results have shown that a large-area Ir film was producible. Recently, preparation of 4-inch Ir/YSZ/Si(001) substrate has been presented at the "18th European Confernce on Diamond, Diamond like materials, carbon nanotubes and nitrides (2007)" conference.

4.6 Recent Progress in Heteroepitaxy of Diamond on Iridium

Since epitaxial Ir films became available with a large size and commercial CVD systems have capacity for large size deposition of diamond, the most important task for us may be the development of a nucleation technique for epitaxial diamond. As mentioned above, diamond is not heteroepitaxially grown on Ir without a pretreatment process but only unepitaxial grains grow on some roughened spots on an Ir surface. It is not easy to nucleate heteroepitaxial diamond uniformly on the whole Ir surface when it is large. Even for highly oriented diamonds on Si, the largest size of substrates with uniform quality I have ever heard of are 1-inch in diameter at most, probably because of the difficulty of uniform nucleation.

In our laboratory, diamond nucleation on Ir surface has been done by applying a negative dc voltage to the substrate in a plasma of CH_4/H_2 gas mixture. The system we used is a simple planar diode type with a Mo anode and a cathode of the substrate itself on Mo folder. From the viewpoint of treatment time t_B, our dc plasma system ($t_B = 60$–90 seconds) is better than a bias enhanced nucleation using microwave plasma (typically $t_B = 60$ min). However, there are many things to be adjusted such as cathode–anode spacing and sizes, substrate temperature, discharge current density, contact resistance between substrate and folder, and so on. In principle for larger substrates, absolute values of the discharge current should be large to keep the discharge current density at a suitable level. It is not easy, however, to increase the current not only for economical reason but also because of the difficulty of keeping a constant and uniform value of substrate temperature.

Figure 4.1 shows heteroepitaxially grown diamond films on Ir/MgO(001) substrates with a size of 10×10 mm^2 (a) and 1-inch in diameter (b). The diamond nucleation process for the 1-inch substrates were accomplished by a geometrical enlargement of the planar diode system, although the discharge current density was lowered because of a limited power supply. Thus, the FWHM value of diamond (004) rocking curve or Raman 1332.5 cm^{-1} peak for the 1-inch diamonds are still larger than these for the 10×10 mm samples, nevertheless device performance of p^+-i-p^+ FETs fabricated on the 1-inch diamond are found to be comparable with those fabricated on a monocrystal diamond substrate which was categorized as natural type-IIa [84]. Although this result has raised our motivation in developing this technique for the heteroepitaxy of diamond, it may be necessary for further enlargement of the substrate to invent a novel nucleation technique or a new substrate which does not need nucleation. Fabrication of several alloys for use as substrates has been tried but has not been successful to date.

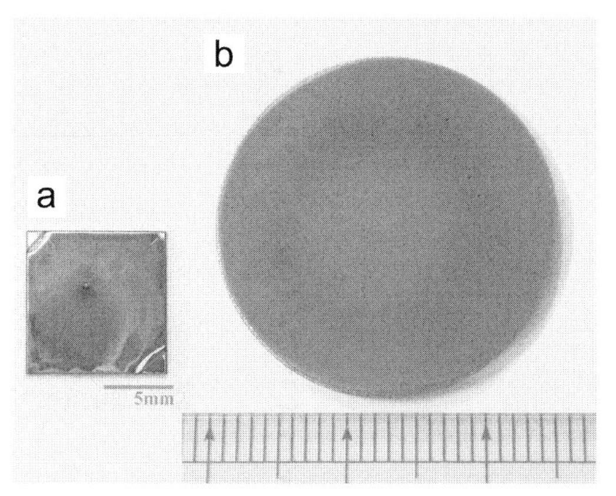

Figure 4.1 Heteroepitaxial diamond films grown on Ir/MgO(001) substrates, size 10 × 10 mm (a) and 1-inch in diameter (b).

Figure 4.2 Heteroepitaxial diamonds grown on Ir(001) surface by a patterned nucleation and growth method [85]. Growth time is 2 min. Black and white areas indicate iridium surface (no growth areas) and heteroepitaxial diamond, respectively.

Although the need for a nucleation process for Ir surfaces could be a demerit for size enlargement of the substrate, it is possible to view it as a merit. Patterned nucleation and growth [85] (PNG) is the method using a property of Ir in which diamond is grown without a pretreatment process. Figure 4.2 shows the patterned heteroepitaxial diamond on an Ir surface after 2 minutes of growth. Black and white areas indicate the Ir surface (no growth areas) and the heteroepitaxial diamonds, respectively. It was found that diamond grew only on the areas patterned by a lithography process. Since the diamond is heteroepitaxially grown, diamond facets in alignment with crystal orientation appear after longer growth. Figure 4.3 shows heteroepitaxial diamond after 15-min growth using the PNG method. Note that the "diamond shape" grown in diamond was drawn with lines that were initially the same width. The diamond lines after 15-min growth, however, have different widths depending on line direction. Figure 4.4 shows a typical example of patterned nucleated diamond after short and long growth times. The diamond

Figure 4.3 Heteroepitaxial diamond grown on patterned areas of Ir(001) surface. Growth time is 15 min.

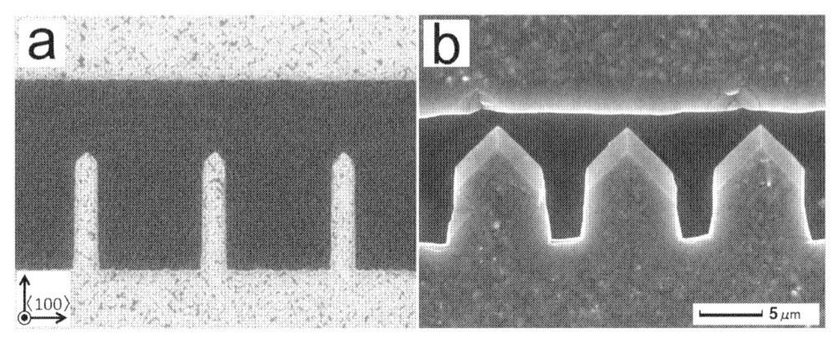

Figure 4.4 Heteroepitaxial diamonds on Ir(001) surface after 2 min (a) and 15 min (b) growth.

grows in accordance with the pattern after 2minutes (Figure 4.4a). When the diamond is grown for 15minutes (Figure 4.4b), diamond {100} and {111} facets appear depending on crystal orientation and growth conditions. Since the growth rate ratio <100>/<111> is larger for the sample shown in Figure 4.4b, a tip shape composed of {111} facets is observed along the <100> direction. Figure 4.5 shows an example of epitaxial lateral overgrowth (ELO) of diamond grown on an Ir(001) surface [86]. The nucleation site of the diamond shown in the center of SEM image is a very narrow line with 3-μm width and 200-μm length. The growth rates vertically (thickness) and laterally for the diamond are approximately 5-μm/h and 25-μm/h, respectively. This result was obtained by a suitable combination of the growth conditions and the line directions.

Not only the PNG method but also the nucleation technique itself is being improved day by day. Figure 4.6 shows a dot array pattern of heteroepitaxial diamond grown on Ir by the PNG method. The insets show magnified views of 0.8 and 6.0-μmϕ dots. Note that the domain structures which are usually found on the Ir surface during the nucleation process for heteroepitaxial diamond [87]

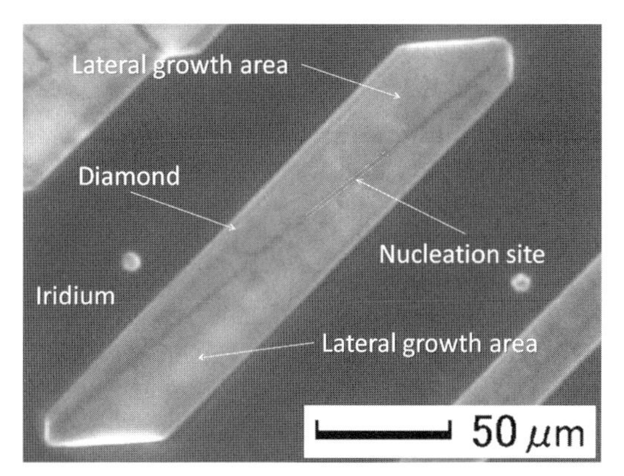

Figure 4.5 Epitaxial lateral overgrowth of diamond on Ir(001) surface. Initial nucleation site is a narrow line with 3 μm width and 200 μm length. Growth time is 1 h.

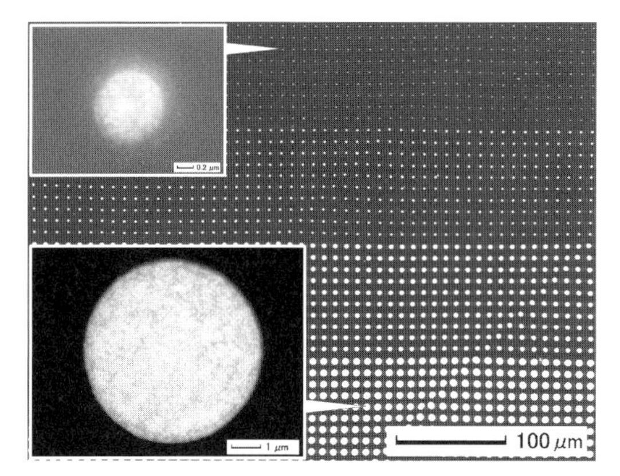

Figure 4.6 Dot array pattern of heteroepitaxial diamond on Ir(001) surface with higher nucleation density. Growth time is 30 s.

were not observed even after 30 seconds growth (cf. Figure 4.4a). Thus no lack of dots is observed even in the 0.8-μmϕ dot array. Such high nucleation density of heteroepitaxial diamond with no domain structure makes it possible for us to even fabricate a 100-nm linear shaped heteroepitaxial diamond as shown in Figure 4.7.

4.7 Other Trials for Heteroepitaxy of Diamond

Although we have obtained several successful results related to heteroepitaxy of diamond by using epitaxial Ir films, there are still areas for improvement in the

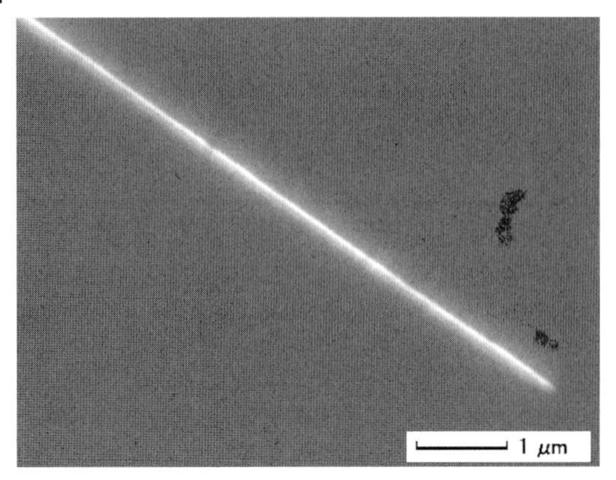

Figure 4.7 Linear shaped diamond of approximately 100-nm width heteroepitaxially grown on Ir(001) by a patterned nucleation and growth method.

diamond/Ir system such as the need for a special pretreatment process, relatively large lattice mismatch between them, and so on. Thus, we are looking for a new or an improved substrate for diamond heteroepitaxy even now. In order to consider new substrates, we would like to reviw various past attempts to grow heteroepitaxial diamond other than by the above-mentioned methods.

Graphite: Carbon atoms in the graphite (0001) basal plane have a similar geometrical arrangement to those in diamond (111) plane. Both planes are comprised of hexagonal rings although the hexagons are corrugated in the (111) plane of diamond. The C—C bond length of diamond is 0.154 nm but its projected length on the (111) plane is 0.145 nm which is only 2% greater than that for the (0001) plane of graphite (0.142 nm). The projected hexagon side-to-side distance for diamond (111) and the (0001) plane of graphite are 0.252-nm and 0.246-nm, respectively. Li *et al.* [65] have found that graphite and diamond have a preferential orientation relationship in which the diamond (111) plane is parallel to the graphite (0001) plane, and the diamond $[1\bar{1}0]$ direction is parallel to the graphite $[11\bar{2}0]$ direction.

BeO: Theoretical studies [88] show that beryllium oxide is a potentially good substrate for heteroepitaxy of diamond. Furthermore, since large crystals of BeO can be grown, this material might be a good candidate for a substrate to grow diamond single crystals. The Be—O bond length is 0.165 nm, 7% greater than the C—C bond length of 0.154-nm in diamond. Argoitia *et al.* [66] show that diamond particles were grown heteroepitaxially on the basal plane of BeO with an epitaxial relationship $\{111\}_{diamond}$ // $\{0001\}_{BeO}$ and $<1\bar{1}0>_{diamond}$ rotated by less than 6° with respect to $<1120>_{BeO}$. No interfacial phases between diamond and BeO were formed during the deposition of diamond on BeO. On the other hand, no epitaxial relationship has been found when diamond is grown on $(11\bar{2}0)_{BeO}$. Small particles of Be_2C have been found when diamond is grown on the $(11\bar{2}0)$ prism plane of BeO.

α-SiC: Diamond deposition on a (0001) plane of an α-silicon carbide single crystal is examined by Suzuki *et al.* [67] based on similar concepts of diamond growth on graphite or BeO. In this experiment, the α-SiC substrate surface was cleaned by pretreatment with hydrogen gas at 1200 °C. After a 12-h deposition, cubo-octahedral diamond grains with a crystallographic relationship of $(0001)_{SiC}$ // $(111)_{diamond}$, $<1120>_{SiC}$ // $<1\bar{1}0>_{diamond}$ were obtained.

TiC: In the case of Si substrate, the formation of a β-SiC interlayer was important to grow the highly oriented diamond with a smooth surface and high quality. From the viewpoint of carbide formation, TiC(111)was chosen as a candidate for the substrate [68]. Highly oriented diamond particles have been grown on TiC (111) substrates by BEN despite a 21% lattice mismatch, although the percentage of diamond particles oriented to each other across the substrate ranged from 10–15% for a density of 1.5×10^8 cm^{-2} of diamond particles.

Ni_3Ge: R. Haubner *et al.* [69] showed that cubic phase Ni_3Ge substrate is an interesting potential material for the heteroepitaxy of diamond because of its lattice parameter matching within <1% that of diamond and its coexistence with carbon up to its melting point. As a result of diamond growth in a microwave plasma system, epitaxial nucleation and growth of isolated diamond crystals on Ni_3Ge substrates was found to be possible. However, diamond films could not be grown because the diamond nucleation density on the Ni_3Ge was too low to form films. Additionally the single crystalline areas of the Ni_3Ge substrates were not large enough to allow the production of large single crystalline diamond layers. However, it is interesting that the results were obtained without any surface treatment for nucleation enhancement such as BEN, diamond scratching, or other seeding treatment. A similar attempt was made with a polycrystalline Ni_3Al (lattice constant 0.357 nm) substrate by Chen *et al.* [89], although heteroepitaxy of diamond was not observed.

Re: Rhenium films epitaxially deposited on c-sapphire surfaces were studied as a substrate for the heteroepitaxial growth of (111)-oriented diamond films by T. Bauer *et al.* [70]. They have reported that Re did not form any stable carbide as iridium does and also that Re has thermal (extremely high melting point of 3180 °C) and economic merits compared to the metals of the platinum group. On epitaxial Re films deposited by electron-beam evaporation on c-sapphire substrates, diamond nucleation can be performed by BEN under conditions similar to the standard conditions for nucleation on Ir. As a result, (111)-oriented epitaxial diamond was formed with two equivalent orientations due to the sixfold symmetry of the Re(0001) surface. Full width at half maximum for tilt and twist were 4.8° and 3.0°, respectively. The epitaxial relation was Re(0001)[100] // dia$(1\bar{1}1)$[110] and Re(0001)[010] // dia(111)$[1\bar{1}0]$. Please see ref. [70] for detailed information.

4.8 Summary

We have reviewed various attempts to grow heteroepitaxial diamond, and also reported recent results of diamond heteroepitaxy using Ir.

The quality of heteroepitaxial diamond on Ir is already sufficiently good for certain kinds of device application. However, it does not yet reach the quality standards of homoepitaxial diamond. When using Ir as a substrate for the heteroepitaxy of diamond, an important key for size enlargement is the development of nucleation techniques. In order to accomplish the two goals of quality improvement and size enlargement, a new or improved substrate and a new nucleation technique need to be developed.

Acknowledgments

The authors are grateful to Mr S. Maeda and S. Ichihara for the experimental work on 1-inch heteroepitaxial diamond growth and bias-enhanced nucleation with higher nucleation density.

References

1 Deryagin, B.V., Spitsyn, B.V., Builov, L.L., Klochkov, A.A., Gorodetskii, A.E. and Smol'yaninov, A.V. (1976) *Soviet Physics – Doklady*, **21**, 676.

2 Spitsyn, B.V., Bouilov, L.L. and Derjaguin, M.A. (1981) *Journal of Crystal Growth*, **52**, 219.

3 Matsumoto, S., Sato, Y., Tsutsumi, M. and Setaka, N. (1982) *Journal of Material Science*, **17**, 3106.

4 Kasu, M., Ueda, K., Yamauchi, Y., Tallaire, A. and Makimoto, T. (2007) *Diamond and Related Materials*, **16**, 1010.

5 Koizumi, S., Murakami, T., Inuzuka, T. and Suzuki, K. (1990) *Applied Physics Letters*, **57** (*6*), 563.

6 Yoshikawa, M., Ishida, H., Ishitani, A., Murakami, T., Koizumi, S. and Inuzuka, T. (1990) *Applied Physics Letters*, **57** (*5*), 428.

7 Braun, W.H.M., Kong, H.S., Glass, J.T. and Davis, R.F. (1991) *Journal of Applied Physics*, **69** (*4*), 2679.

8 Yoshikawa, M., Ishida, H., Ishitani, A., Koizumi, S. and Inuzuka, T. (1991) *Applied Physics Letters*, **58** (*13*), 1387.

9 Inuzuka, T., Koizumi, S. and Suzuki, K. (1992) *Diamond and Related Materials*, **1**, 175.

10 Koizumi, S. and Inuzuka, T. (1993) *Japanese Journal of Applied Physics*, **32**, 3920.

11 Wang, L., Pirouz, P., Argoitia, A., Ma, J.S. and Angus, J.C. (1993) *Applied Physics Letters*, **63** (*10*), 1336.

12 Tomikawa, T. and Shikata, S. (1993) *Japanese Journal of Applied Physics*, **32**, 3938.

13 Maeda, H., Masuda, S., Kusakabe, K. and Morooka, S. (1994) *Diamond and Related Materials*, **3**, 398.

14 Narayan, J., Srivatsa, A.R., Peters, M., Yokota, S. and Ravi, K.V. (1988) *Applied Physics Letters*, **53** (*19*), 1823.

15 Williams, B.E. and Glass, J.T. (1989) *Journal of Materials Research*, **4** (*2*), 373.

16 Jeng, D.G., Tuan, H.S., Salat, R.F. and Fricano, G.J. (1990) *Applied Physics Letters*, **56** (*20*), 1968.

17 Stoner, B.R. and Glass, J.T. (1992) *Applied Physics Letters*, **60**, 698.

18 Jiang, X. and Klages, C.-P. (1993) *Diamond and Related Materials*, **2**, 1112.

19 Wolter, S.D., Stoner, B.R., Glass, J.T., Ellis, P.J., Buhaenko, D.S., Jenkins, C.E. and Southworth, P. (1993) *Applied Physics Letters*, **62** (*11*), 1215.

20 Stoner, B.R., Sahaida, S.R., Bade, J.P., Southworth, P. and Ellis, P.J. (1993) *Journal of Materials Research*, **8** (*6*), 1334.

21 Stoner, B.R., Kao, C., Malta, D.M. and Glass, R.C. (1993) *Applied Physics Letters*, **62** (*19*), 2347.

22 Kawarada, H., Suesada, T. and Nagasawa, H. (1995) *Applied Physics Letters*, **66** (5), 583.

23 Tachibana, T., Hayashi, K. and Kobashi, K. (1996) *Applied Physics Letters*, **68** (*11*), 1491.

24 Jiang, X. and Jia, C.L. (1996) *Applied Physics Letters*, **69** (*25*), 3902.

25 Schreck, M. and Stritzker, B. (1996) *physica status solidi (a)*, **154**, 197.

26 Thürer, K.-H., Schreck, M. and Stritzker, B. (1998) *Physical Review B*, **57** (*24*), 15454.

27 Kono, S., Goto, T., Abukawa, T., Wild, C., Koidl, P. and Kawarada, H. (2000) *Japanese Journal of Applied Physics*, **39**, 4372.

28 Lee, S.T., Peng, H.Y., Zhou, X.T., Wang, N., Lee, C.S., Bello, I. and Lifshitz, Y. (2000) *Science*, **287**, 104.

29 Sato, Y., Yashima, I., Fujita, H., Ando, T. and Kamo, M. (1991) Proceedings of 2nd International Conference of New Diamond Science and Technology (eds R.J. Messier, J.T. Glass, E. Butler and R. Roy), ••, p. 371.

30 Zhu, W., Yang, P.C. and Glass, J.T. (1993) *Applied Physics Letters*, **63** (*12*), 1640.

31 Yang, P.C., Zhu, W. and Glass, J.T. (1993) *Journal of Materials Research*, **8** (*8*), 1773.

32 Yang, P.C., Zhu, W. and Glass, J.T. (1994) *Journal of Materials Research*, **9** (*5*), 1063.

33 Yang, P.C., Schlesser, R., Wolden, C.A., Liu, W., Davis, R.F. and Sitar, Z. (1997) *Applied Physics Letters*, **70** (*22*), 2960.

34 Sitar, Z., Liu, W., Yang, P.C., Wolden, C.A., Schlesser, R. and Prater, J.T. (1998) *Diamond and Related Materials*, **7**, 276.

35 Yang, P.C., Liu, W., Schlesser, R., Wolden, C.A., Davis, R.F., Prater, J.T. and Sitar, Z. (1998) *Journal of Crystal Growth*, **187**, 81.

36 Yang, H. and Whitten, J.L. (1992) *Journal of Chemistry Physics*, **96** (*7*), 5529.

37 Liu, W., Tucker, D.A., Yang, P. and Glass, J.T. (1995) *Journal of Applied Physics*, **78** (*2*), 1291.

38 Shintani, Y. (1996) *Journal of Materials Research*, **11** (*12*), 2955.

39 Tachibana, T., Yokota, Y., Nishimura, K., Miyata, K., Kobashi, K. and Shintani, Y. (1996) *Diamond and Related Materials*, **5**, 197.

40 Tachibana, T., Yokota, Y., Miyata, K., Kobashi, K. and Shintani, Y. (1997) *Diamond and Related Materials*, **6**, 266.

41 Tarutani, M., Zhou, G., Takai, Y., Shimizu, R., Tachibana, T., Kobashi, K. and Shintani, Y. (1997) *Diamond and Related Materials*, **6**, 272.

42 Zhou, G., Tarutani, M., Takai, Y. and Shimizu, R. (1997) *Japanese Journal of Applied Physics*, **36**, 2298.

43 Tachibana, T., Yokota, Y., Kobashi, K. and Shintani, Y. (1997) *Journal of Applied Physics*, **82** (9), 4327.

44 Tachibana, T., Yokota, Y., Miyata, K., Onishi, T., Kobashi, K., Tarutani, M., Takai, Y., Shimizu, R. and Shintani, Y. (1997) *Physical Review B*, **56** (*24*), 15967.

45 Zhou, G., Takai, Y. and Shimizu, R. (1998) *Japanese Journal of Applied Physics*, **37**, L752.

46 Tachibana, T., Yokota, Y., Kobashi, K. and Yoshimoto, M. (1999) *Journal of Crystal Growth*, **205**, 163.

47 Tachibana, T., Yokota, Y., Hayashi, K., Miyata, K., Kobashi, K. and Shintani, Y. (2000) *Diamond and Related Materials*, **9**, 251.

48 Tachibana, T., Yokota, Y., Hayashi, K. and Kobashi, K. (2001) *Diamond and Related Materials*, **10**, 1633.

49 Ohtsuka, K., Suzuki, K., Sawabe, A. and Inuzuka, T. (1996) *Japanese Journal of Applied Physics*, **35**, L1072.

50 Ohtsuka, K., Fukuda, H., Suzuki, K. and Sawabe, A. (1997) *Japanese Journal of Applied Physics*, **36**, L1214.

51 Suzuki, K., Fukuda, H. and Sawabe, A. (1998) Proceedings of ISAM '98, the 5th NIRIM International Symposium on Advanced Materials, p. 89.

52 Saito, T., Tsuruga, S., Ohya, N., Kusakabe, K., Morooka, S., Maeda, H., Sawabe, A. and Suzuki, K. (1998) *Diamond and Related Materials*, **7**, 1381.

53 Fujisaki, T., Tachiki, M., Taniyama, N., Kudo, M. and Kawarada, H. (2002) *Diamond and Related Materials*, **11**, 478.

54 Suzuki, K., Fukuda, H., Yamada, T. and Sawabe, A. (2001) *Diamond and Related Materials*, **10**, 2153.

55 Schreck, M., Roll, H. and Stritzker, B. (1999) *Applied Physics Letters*, **74** (5), 650.

56 Schreck, M., Hörmann, F., Roll, H., Linder, J.K.N. and Stritzker, B. (2001) *Applied Physics Letters*, **78** (*2*), 192.
57 Schreck, M., Schury, A., Hörmann, F., Roll, H. and Stritzker, B. (2002) *Journal of Applied Physics*, **91** (*2*), 676.
58 Schreck, M., Roll, H., Michler, J., Blank, E. and Stritzker, B. (2000) *Journal of Applied Physics*, **88** (*5*), 2456.
59 Hörmann, F., Roll, H., Schreck, M. and Stritzker, B. (2000) *Diamond and Related Materials*, **9**, 256.
60 Hörmann, F., Schreck, M. and Stritzker, B. (2001) *Diamond and Related Materials*, **10**, 1617.
61 Bednarski, C., Dai, Z., Li, A.-P. and Golding, B. (2003) *Diamond and Related Materials*, **12**, 241.
62 Lee, C.H., Qi, J., Lee, S.T. and Hung, L.S. (2003) *Diamond and Related Materials*, **12**, 1335.
63 Gsell, S., Bauer, T., Goldfuß, J., Schreck, M. and Strizker, B. (2004) *Applied Physics Letters*, **84** (*22*), 4541.
64 Bauer, T., Gsell, S., Schreck, M., Goldfuß, J., Lettiri, J., Schlom, D.G. and Stritzker, B. (2005) *Diamond and Related Materials*, **14**, 314.
65 Li, Z., Wang, L., Suzuki, T., Argoitia, A., Pirouz, P. and Angus, J.C. (1993) *Journal of Applied Physics*, **73** (*2*), 711.
66 Argoitia, A., Angus, J.C., Wang, L., Ning, X.I. and Pirouz, P. (1993) *Journal of Applied Physics*, **73** (*9*), 4305.
67 Suzuki, T., Yagi, M. and Shibuki, K. (1994) *Applied Physics Letters*, **64** (*5*), 557.
68 Wolter, S.D., McClure, M.T., Glass, J.T. and Stoner, B.R. (1995) *Applied Physics Letters.*, **66** (*21*), 2810.
69 Haubner, R., Lux, B., Gruber, U. and Schuster, J.C. (1999) *Diamond and Related Materials*, **8**, 246.
70 Bauer, T., Schreck, M., Gsell, S., Hörmann, F. and Stritzker, B. (2003) *physica status solidi (a)*, **199** (*1*), 19.
71 Meguro, K., Ando, Y., Nishibayashi, Y., Ishibashi, K., Yamamoto, Y. and Imai, T. (2006) *New Diamond and Frontier Carbon Technology*, **16**, 71.
72 Mokuno, Y., Chayahara, A., Soda, Y., Horino, Y. and Fujimori, N. (2005) *Diamond and Related Materials*, **14**, 1743.
73 Mokuno, Y., Chayahara, A., Soda, Y., Yamada, H., Horino, Y. and Fujimori, N. (2006) *Diamond and Related Materials*, **15**, 455.
74 Wild, C., Herres, N. and Koidl, P. (1990) *Journal of Applied Physics*, **68** (*3*), 973.
75 Wild, C., Kohl, R., Herres, N., Müller-Sebert, W. and Koidl, P. (1994) *Diamond and Related Materials*, **3**, 373.
76 Ando, Y., Yokota, Y., Tachibana, T., Watanabe, A., Nishibayashi, Y., Kobashi, K., Hirao, T. and Oura, K. (2002) *Diamond and Related Materials*, **11**, 596.
77 Zhang, X.W., Boyen, H.-G., Deyneka, N., Ziemann, P., Banhart, F. and Schreck, M. (2003) *Nature Materials*, **2**, 312.
78 Zhang, X.W., Boyen, H.-G., Ziemann, P. and Banhart, F. (2005) *Applied Physics A*, **80**, 735.
79 Yugo, S., Kanai, T., Kimura, T. and Muto, T. (1991) *Applied Physics Letters*, **58** (*10*), 1036.
80 Hayashi, K., Tachibana, T., Kawakami, N., Yokota, Y., Kobashi, K., Ishihara, H., Uchida, K., Nippashi, K. and Matsuoka, M. (2006) *Diamond and Related Materials*, **15**, 792.
81 Ogawa, S., Ino, S., Kato, T. and Ota, H. (1956) *Journal of the Physical Society of Japan*, **21** (*10*), 1963.
82 Belton, D.N. and Schmieg, S.J. (1989) *Journal of Applied Physics*, **66** (*9*), 4223.
83 Kawarada, M., Kurihara, K. and Sasaki, K. (1993) *Diamond and Related Materials*, **2**, 1083.
84 Yokota, Y., Kawakami, N., Maeda, S., Ando, Y., Tachibana, T., Kobashi K. and Sawabe, A. (2008) *New Diamond and Frontier Carbon Technology* (in press).
85 Ando, Y., Kuwabara, J., Suzuki, K. and Sawabe, A. (2004) *Diamond and Related Materials*, **13**, 1975.
86 Ando, Y., Kamano, T., Suzuki, K. and Sawabe, A. (2008) *Diamond and Related Materials* (in press).
87 Gsell, S., Schreck, M., Benstetter, G., Lodermeier, E. and Stritzker, B. (2007) *Diamond and Related Materials*, **16**, 665.
88 Lambrecht, W.R.L. and Segall, B. (1992) *Journal of Materials Research*, **7**, 696.
89 Chen, H.-G. and Chang, L. (2005) *Diamond and Related Materials*, **14**, 183.

5
Electrochemical Properties of Undoped Diamond

Christoph E. Nebel, Bohuslav Rezek and Dongchan Shin
Diamond Research Center, National Institute for Advanced Industrial Science and Technology, Tsukuba 305-8568, Japan

5.1
Introduction

Due to increasing demands in areas like environmental protection, health care and bionanotechnology, biosensors and chemical sensors based on semiconductors attract increasing attention because they can be miniaturized and integrated into small sized electronic equipment for real-time detection with high sensitivity. In 1990 Bergveld introduced the first ion-sensitive field effect transistor (ISFET) to measure pH or ions in solutions [1]. It was a silicon-based metal-oxide-semiconductor field effect transistor (MOSFET) where the metal gate electrode has been replaced by an electrolyte and a reference electrode. After 15 years of continuous progress, ISFETs can be provided with chemical and/or biological sensitivity by modifying the gate insulator with specific molecular layers [2]. Many applications in biological, environmental, and medical science involve chemically or biologically modified surfaces that must be stable in contact with electrolytes for long periods of time to be used on a continuous basis. The limited stability of silicon and other established semiconductor materials in comparatively harsh environments, characteristic of physiological conditions, has greatly hampered their practical utility [3–6]. While the unusual mechanical, thermal and electrical properties of diamond have been studied over years in detail [7], it is only recently that the outstanding chemical properties of diamond have been discovered which makes it a perfect interface to electrolytes, biological molecules and biological systems [8–15]. In addition, diamond is considered to be biocompatible [16] and can be grown with manifold structural properties like single-, poly- or nanocrystalline, either by homo- or heteroepitaxy [7].

The diamond surface can be terminated by a variety of elements or molecules, giving rise to new phenomena and applications. For oxygen termination, insulating properties are achieved with increased electron affinities while hydrogen gives rise to a negative electron affinity due to the formation of carbon–hydrogen dipoles [17]. In vacuum such surfaces are insulating too. However, if hydrogen-terminated

Physics and Applications of CVD Diamond. Satoshi Koizumi, Christoph Nebel, and Milos Nesladek

ISBN: 978-3-527-40801-6

diamond films are exposed to air, surface conductivity can be detected [18, 19]. Hall effect experiments revealed the p-type nature of surface conductivity [20, 21], with typical hole sheet-densities in the range 10^{10}–$10^{13}\,cm^{-2}$ [21, 22], and Hall mobilities between 1 and $100\,cm^2/Vs$ [22, 23]. The confusing mass of data about hole densities and mobilities as published in the literature is summarized in Figure 5.1 [24–30] (and C.E. Nebel, unpublished data). Most papers have applied different hydrogen termination parameters or even techniques (see Table 5.1 [27, 31, 32]), with the result that a coherent picture is not obvious. There is an

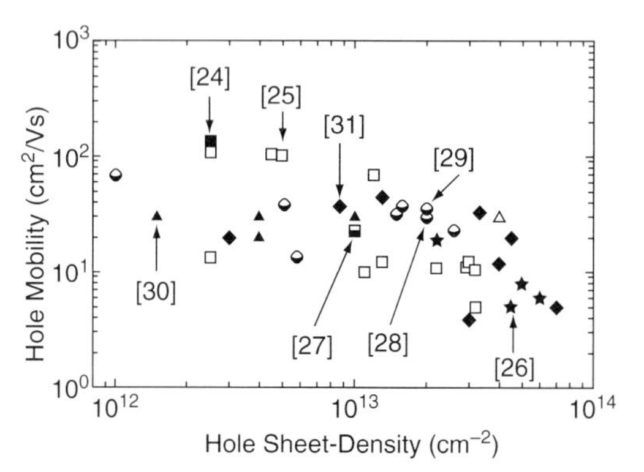

Figure 5.1 Summary of hole mobilities and hole sheet-densities as detected at room temperature on a variety of films. Data are from [24–30] and C.E. Nebel, unpublished data.

Table 5.1 Parameters for hydrogen termination of diamond.

	University College of London (UK) [27]	**LIST (France) [32]**	**Diamond Center, AIST (Japan) [31]**	**Institute for Technical Physics, Univ. Erlangen, Germany [32]**
Technique	Microwave plasma (2.45 GHz)	Microwave plasma (2.45 GHz)	Microwave plasma (2.45 GHz)	Hot filament
Power (W)	800	900	750	2200 °C tungsten temp.
Substrate temperature (°C)	500	820	800	690
Pressure (Torr)	40	38	25	38
Duration (minutes)	5	60	5	10
H_2-flow (sccm)	100	120	400	100

overall trend towards lower mobilities with increasing hole sheet-densities, but the scatter in the data is large.

A variety of experiments have shown that the conductivity is related to hydrogen termination in combination with adsorbate coverage, which led to the "transfer doping model" as the most plausible explanation [18, 19, 33].

In this model, valence band electrons tunnel into electronic empty states of an adjacent adsorbate layer as shown schematically in Figure 5.2. In order to act as a sink for electrons, the adsorbate layer must have its lowest unoccupied electronic level below the valence band maximum (VBM) of diamond. Maier *et al.* [18] proposed that for standard atmospheric conditions, the pH value of water is about 6 due to CO_2 content or other ionic contamination. They calculate the chemical potential μ_e for such an aqueous wetting layer to be about −4.26 eV below the vacuum. To calculate the valence band maximum, E_{VBM}, of H-terminated diamond with respect to the vacuum level, E_{VAC}, we take into account the band gap of diamond, for example, of 5.47 eV, and the negative electron affinity χ of −1.1 to −1.3 eV [18, 34]. This results in a gap between vacuum level and valence band maximum of −4.17 to −4.37 eV if we refer to the vacuum level as zero. The Fermi-level E_F at the water/diamond interface is therefore either 90 meV below or 110 meV above E_{VBM}. A generalized summary of chemical potentials with respect to hydrogen- and oxygen-terminated diamond is given in Figure 5.3 [35]. It shows that for well-defined pH liquids the chemical potential μ(pH) may indeed be below E_{VBM}. In such a case, a hole accumulation layer at the surface of diamond is generated by transfer doping. However, as the pH value of the adsorbate layers cannot be detected experimentally, a discussion based on assumptions will not elucidate the real features.

To characterize surface electronic properties of H-terminated diamond, first we performed several experiments on diamond/adsorbate layers and then on diamond/electrolyte combinations with well-defined redox couples and pH levels. Experiments to investigate Hall effect, contact potential difference ("Kelvin force"), Schottky junction characterizations, cyclic voltammetry, and field effect

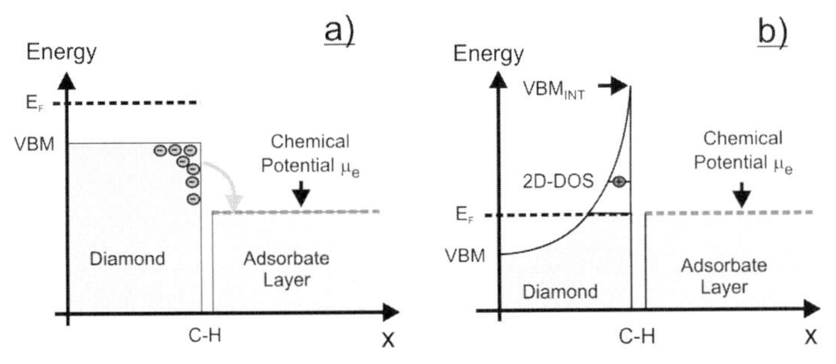

Figure 5.2 Schematic description of the diamond/adsorbate heterojunction (a) non-equilibrated and (b) equilibrated. Electrons from the valence-band tunnel into empty electronic states of the adsorbate layer as long as the chemical potential μ_e is lower than the Fermi energy E_F.

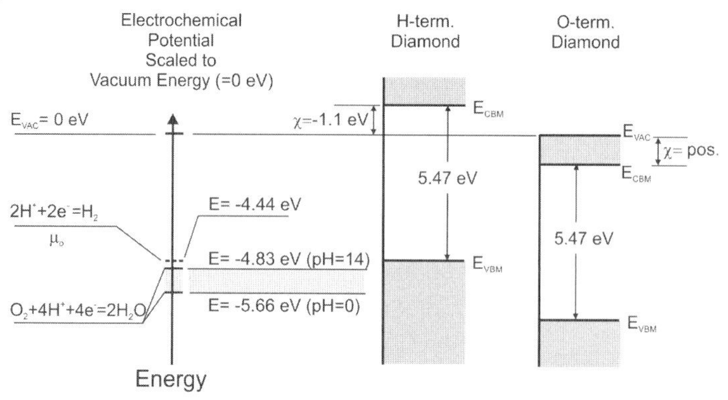

Figure 5.3 Chemical potentials referred to the standard hydrogen electrode and semiconductor energies referred to the vacuum level. The electron energies for the couple O_2 + $4H^+ + 4e^-$ = H_2O at pH 0 and 14 are shown along with the band edges for hydrogen-terminated and oxidized diamond. The negative electron affinity has been assumed to be −1.1 eV on H-terminated diamond [34].

measurements on ion-sensitive field effect transistors (ISFET) fabricated from diamond have been made. We give some coherence to the experimental results of Hall effect investigations by discussing temperature dependent mobility and hole density variations in the regime from 100 K to 500 K, detected on a single crystalline chemical vapor deposited (CVD) diamond film with H-termination.

We use one-dimensional numerical solutions of the Schrödinger and Poisson equations to show that the density-of-state distribution at the surface of diamond in contact with electrolyte solutions will be two-dimensional (2D), if the interface is not disturbed by disorder and ionic scattering [36–38]. Such Fermi-level shifts into the valence band take place only if the sub-surface (bulk) and the surface of diamond is defect-free, which requires high-quality growth and perfect H-termination.

One of the cornerstones of the transfer doping model is the requirement for chemical equilibrium between diamond and the electrolyte solutions in contact with diamond. A pH-dependent variation of the surface conductivity is predicted with variations following the Nernst relation. On polycrystalline diamond such experiments did result in no pH sensitivity (Kawarada *et al.* [39]) or in a weak and opposite pH sensitivity (Garrido *et al.* [40]). Polycrystalline diamond contains, however, grain boundaries with a continuous density of sp^2 states distributed throughout the gap of diamond [41–43]. This pins the surface Fermi level and causes a reduced or even zero response to electrolytes.

For hydrogen-terminated, undoped single crystalline CVD diamond, which is discussed in this paper, we detect an insulator–metal transition if the diamond is immersed in a redox–electrolyte solution with a chemical potential below the valence-band maximum of diamond. Metallic properties of diamond are generated by tunneling of valence-band electrons into the redox–electrolyte solution as predicted by the transfer doping model. To characterize this phenomenon, we have applied cyclic voltammetry experiments [44–46] using different redox systems with different chemical potentials and discuss the oxidizing and reduction currents in detail. In addition, the properties of ion sensitive field effect transistors (ISFET) are introduced and discussed: these show a pH sensitivity around 66 mV/pH, in agreement with the prediction of the Nernst relationand the transfer doping model [47, 48].

5.2 Surface Electronic Properties of Diamond Covered with Adsorbates

5.2.1 Contact Potential Difference (CPD) Experiments

For contact potential difference (CPD) experiments a CVD diamond film has been grown on (100)-oriented Ib high-pressure high-temperature (HPHT) diamond by use of hot-filament chemical vapor deposition. Please note that before homoepitaxial growth or hydrogen termination, generally all samples were cleaned and oxidized by boiling in aqua regia (a mixture of hydrochloric acid and nitric acid) to remove metals, followed by wet chemical etching in a CrO_3/H_2SO_4 solution at 180 °C for one hour to removed graphitic parts from the surface and to oxidize the surface. The CVD growth parameters were: 1.2% CH_4 in H_2, tungsten filament temperature 2200 °C, sample temperature 690 °C, filament sample distance 6 mm, pressure 38 Torrr, gas flow 100 standard cubic centimeter, growth time 60 minutes. To achieve hydrogen termination the CH_4 admixture had been stopped so that the surface was exposed to a pure hydrogen gas for 10 minutes with otherwise identical parameters.

Hydrogen-terminated patterns on the surface of diamond have been realized by oxygen plasma treatment through photolithographic masks. Typical plasma parameters were: DC plasma power 300 W, 1.3 mbar O_2 pressure, 3 minutes duration. For ohmic contacts on H-terminated diamond, we thermally evaporated 200 nm thick Au. Aluminum is used for Schottky contacts on H-terminated diamond with 600 nm thickness. Square-shaped contact geometry was used with (i) $50 \times 50\,\mu m$, (ii) $100 \times 100\,\mu m$, and (iii) $250 \times 250\,\mu m$ size. Please note that surface conductivity of diamond is generated by a "non-defined" adsorbate layer (= electrolyte), which deposits on the surface by exposure to air. To achieve reproducibility, samples were exposed to air at least 1 day before experiments.

In contact potential difference experiments we used Au as reference (for details see [49]) as Au is chemically inert towards oxidation or other chemical reactions. As the work function of Au differs significantly and textbook data may reflect only ideal properties, we have measured χ_{Au} of our gold layer by secondary photoelectron emission experiments. This results in $\chi_{Au} = 4.3\ (\pm 0.1)$ eV. It is smaller than values reported in the literature for polycrystalline Au of 4.9 to 5.1 eV [50].

A typical scanning electron microscopy image (SEM) and a related CPD result as detected on Au (region A), hydrogen-terminated diamond surface (region B), and on a surface which has been exposed to oxygen plasma (regions C) are shown in Figure 5.4a and b. In Figure 5.4c a line scan in units of potential (mV) (white line in Figure 5.4b) is shown which shows no difference between Au and H-terminated diamond. The oxidized area is dark in SEM, which is a result of the lower electron back scattering from the positive electron affinity surface. The hydrogen-terminated surface appears bright due to its negative electron affinity. Figure 5.5 summarizes the result schematically, taking into account a negative electron affinity of −1.2 eV (±0.1 eV) [37, 51] and surface energies as detected with respect to the work-function χ_{Au} of Au.

CPD measurements on Al contacts on H-terminated diamond result in a surface potential difference of +588 mV. The work function of Al, χ_{Al}, is therefore about 588 meV smaller than the work function of Au, which results in $\chi_{Al} = 3.7$ eV. We assume that this result is governed by partial oxidation of the Al surface as χ_{Al} reported in the literature is 4.3 eV [50]. Figure 5.5 summarizes the results schematically. Based on CPD data, the Fermi level of H-terminated diamond with an adsorbate layer is slightly in the valence band (in this case, about 30 meV). Such

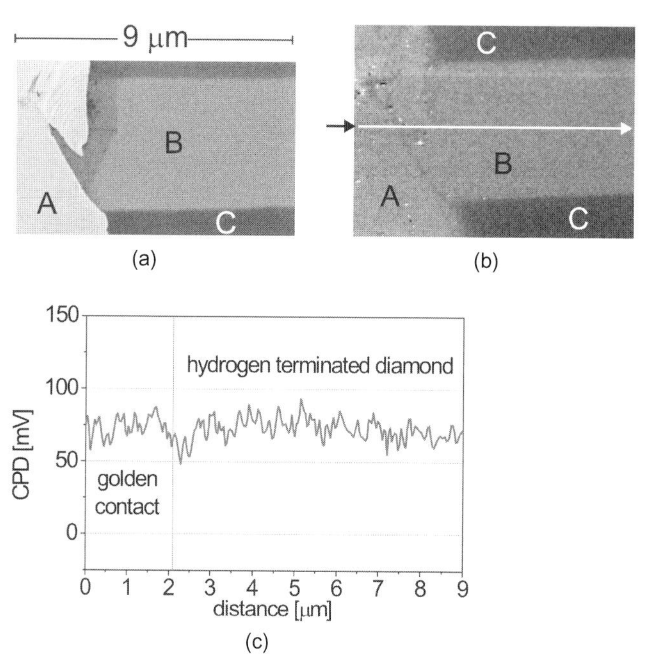

Figure 5.4 (a) Scanning electron microscopy image of diamond which has been partially covered with Au (A), hydrogen-terminated (B) and oxidized (C). (b) Two-dimensional contact potential measurement (CPD) on the same sample as shown in (a). (c) The white line in (b) indicates the scan position of the spatial CPD profile shown here. Note, within experimental accuracy, no contact potential difference between Au and H-terminated diamond can be detected.

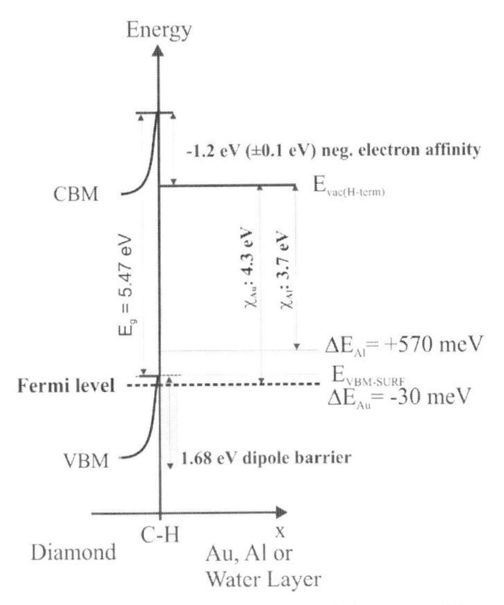

Figure 5.5 Schematic energy band diagram of the interface of H-terminated diamond covered with Au or Al. The ΔE_{Al} and ΔE_{Au} refer to the valence band maximum at the surface, $E_{VBM\text{-}SURF}$.

a shift of Fermi level is only possible if the Fermi level is not pinned by defects at the surface. Obviously, the optimized H-termination generates such a defect-free diamond surface, which is a perfect interface for electrochemical applications.

The ohmic properties of Au are confirmed, as the Fermi level of Au is perfectly aligned with H-terminated diamond, and the Schottky properties of Al are also confirmed as the Fermi-level is about 570 meV above the VBM. The in-plane electronic properties of Al and Au on hydrogen-terminated diamond are summarized in Figure 5.6.

5.2.2
Current–Voltage (IV) Properties

The current–voltage characteristics of Al on H-terminated diamond measured at 300 K are shown in Figure 5.7 (Al size: 250 × 250 μm) measured on two different junctions. Nearly perfect Schottky properties are detected with an ideality factor *n* of 1. Applying negative voltages of more than 0.6 V (threshold voltage) to the Al contacts gives rise to an exponential increase of current of over seven orders of magnitude. The threshold voltage of 600 meV is reasonably well in agreement with the detected energy barrier of 570 meV measured by CPD experiments. Positive voltages result in minor current variations (reverse currents) in the range of 10^{-13} A. Please note that the current has not been normalized by contact areas, as it is an in-plane current, flowing at the surface of diamond between Al and Au.

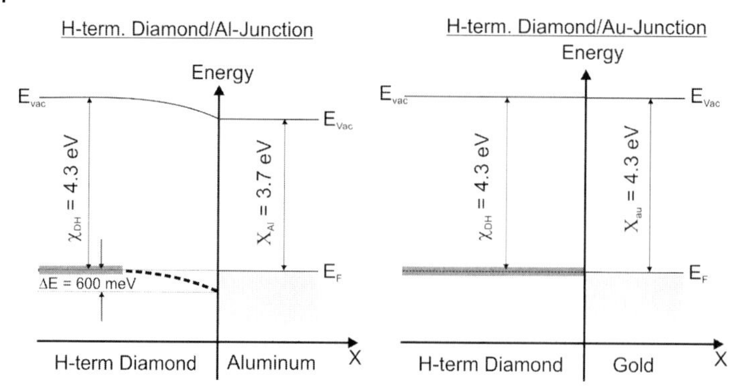

Figure 5.6 Schematic surface energy diagrams of H-terminated diamond covered with an adsorbate layer which is in contact with aluminum and with gold. The data was calculated using contact potential difference experiments and assuming a negative electron affinity of −1.1 eV.

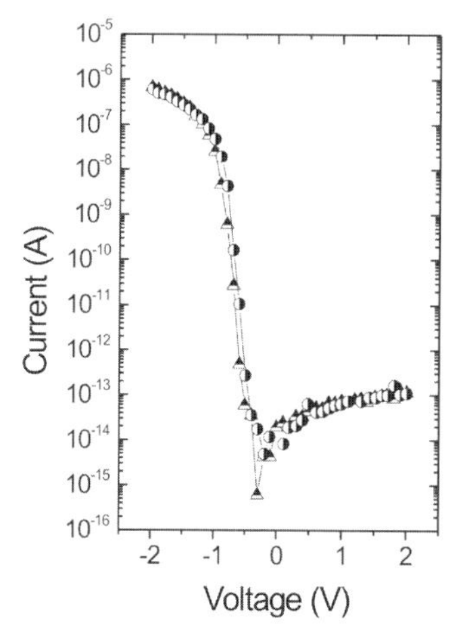

Figure 5.7 Two IV characteristics measured on Al/H-terminated diamond Schottky-junctions in air at $T = 300\,\mathrm{K}$ with $250 \times 250\,\mu\mathrm{m}$ contact size.

5.2.3
Capacitance-Voltage (CV) Experiments

Capacitance–voltage (CV) experiments on these contacts were carried out at $T = 300\,\mathrm{K}$ using a Boonton 7200 capacitance meter in the reverse regime of the diodes. Typical CV-data are shown in Figure 5.8 for contacts $50 \times 50\,\mu\mathrm{m}$, $100 \times 100\,\mu\mathrm{m}$ and $250 \times 250\,\mu\mathrm{m}$ [52]. It is important to note that the capacitance is very small with values $C \leq 1\,\mathrm{pF}$ in the regime $-3\,\mathrm{V} \leq V \leq 0\,\mathrm{V}$. If we assume a three-dimensional parallel-plate capacitor, the capacitance C can be calculated by:

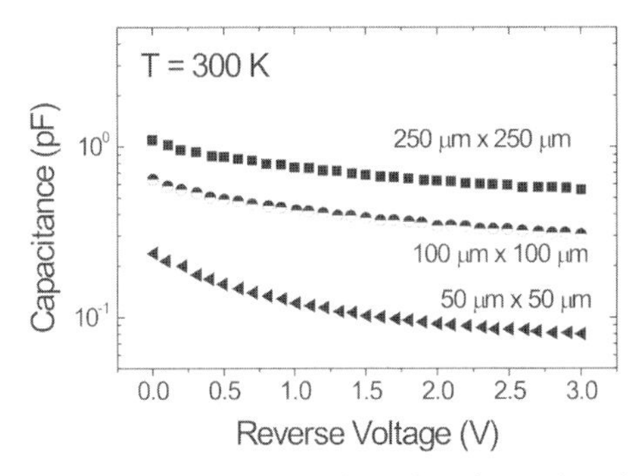

Figure 5.8 Capacitance–voltage data detected on three contact configurations with Al areas $250 \times 250\,\mu m$, $100 \times 100\,\mu m$ and $50 \times 50\,\mu m$.

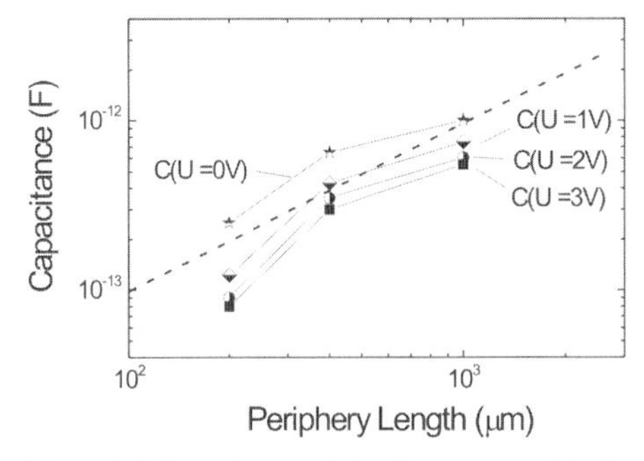

Figure 5.9 Capacitance of the Al contacts with sizes $50 \times 50\,\mu m$, $100 \times 100\,\mu m$ and $250 \times 250\,\mu m$ plotted for $U =$ 0, 1, 2 and 3 V as a function of periphery length (dashed line indicates linear dependence).

$$C = \varepsilon_o \varepsilon_r (A/d)$$

where ε_o is the dielectric constant, ε_r is the relative dielectric constant of diamond ($\varepsilon_r = 5.7$), A is the area of the Al contact, and d is the depletion layer width or the distance between the Al top contact and the doping layer. Considering an area of $250 \times 250\,\mu m$ and a distance of 10 nm between the Al contact and a H-induced doping layer, the capacitance would be about 310 pF. We detect, however, much smaller values of typically ≤ 1 pF on all contacts realized. In addition, the capacitance follows an approximate linear relationship with the length of the periphery of the Al contact as shown in Figure 5.9. We therefore conclude that the charge below Al contacts is negligible. The diode characteristics are governed by in-plane properties, where a p-type channel exists at the surface of hydrogen-terminated diamond which connects Al with Au. This channel is not present or depleted below Al. In case of work function differences between metals and H-terminated diamond, the energy gap generates in-plane Schottky properties which are not distinguishable from 3D Schottky junctions as discussed in detail by Petrosyan *et al.* [53] and by Gelmont *et al.* [54]. Before we address this problem, we want to discuss the electronic properties at the surface of a perfectly hydrogen-terminated diamond surface, with no defects in the bulk, in contact with adsorbates.

5.2.4 Two Dimensional Properties of a Perfectly H-Terminated Diamond Surface

Figure 5.2 shows a schematic view of the electronic properties at the surface of H-terminated diamond, where valence-band electrons can tunnel into the empty electronic states of an adsorbate layer. Tunneling gives rise to band-bending which decreases in diamond with increasing distance to the surface. To calculate the width and bending of the valence band, one has to take into account light hole (LL), heavy hole (HH) and split-off (SO) bands. The band structure of diamond is described by Luttinger parameters and has been discussed by Willatzen, Cardona and Christensen [55]. They derive $\gamma_1 = 2.54$, $\gamma_2 = -0.1$, and $\gamma_3 = 0.63$.

To calculate the band-bending in the vicinity of the surface of diamond we used a numerical approach which has been developed to solve the Schrödinger and Poisson equations simultaneously to calculate the energy levels in narrow GaAs/$Ga_{1-x}Al_xAs$ heterojunctions. Details can be found in [56]. In the case of band-bending over a distance which is shorter than the De Broglie wavelength of about 100 Å for holes (in diamond), the three-dimensional (3D) density-of-states (DOS) changes to a two-dimensional (2D) DOS as shown schematically in Figure 5.2b. For this calculatios the heterojunction effect is modeled using a graded interface in which the barrier height, as well as the effective mass is assumed to change smoothly in a transition layer whose thickness is specified. Holes move in an effective potential given by:

$$V(x) = e\phi(x) + V_{\mathrm{h}}(x)$$

where $\phi(x)$ is the electrostatic potential, $V_{\mathrm{h}}(x)$ is the effective potential energy associated with the heterojunction discontinuity, which we assume to be 1.68 eV [57]. The normalized envelope function $\zeta_i(x)$ for hole sub-band i is given by the Schrödinger equation of the BenDaniel–Duke form:

$$-\frac{\hbar^2}{2}\frac{d}{dx}\frac{1}{m_n(z)}\frac{d\zeta_i(x)}{dx} + V(x)\zeta_i(x) = E_i\zeta_i(x)$$

where $m_n(x)$ is the position-dependent effective mass (n stands for: HH, LH, SO) and E_i is the energy of the bottom of the i-th sub-band. The Poisson equation for the electrostatic potential takes the form:

$$\frac{d}{dz}\varepsilon_0\varepsilon_r(x)\frac{d\phi(x)}{dx} = e\sum N_i\zeta_i^2(x) - \rho_t(x)$$

$$N_i = \frac{m_n kT}{\pi\hbar^2}\ln\left[1+\exp\left(\frac{E_i - E_F}{kT}\right)\right]$$

where $\varepsilon_r(x)$ is the position-dependent dielectric constant, assumed to be constant in diamond ($\varepsilon_r = 5.7$). For the adsorbate layer we varied ε_r between 1 and 5.7. The calculations show that ε_r does not affect the energy levels in the quantum well but does affect the width of the wave function of holes extending out of the diamond into the water layer. In the following we show results deduced for $\varepsilon_r = 5.7$. N_i is the number of holes per unit area in the sub-band i, E_F is the Fermi-energy and m_n represents the mass of holes (HH, LH, SO). As a first order approximation we neglect impurities ($\rho_i(x) = 0$). Calculations have been performed for hole sheet-densities in the range $5 \times 10^{12}\,\text{cm}^{-2}$ to $5 \times 10^{13}\,\text{cm}^{-2}$.

A typical result is shown in Figure 5.10. In diamond at the interface to the H-terminated surface, three discrete energy levels for holes govern the electronic properties, namely the first sub-bands of the LH-, HH- and SO-holes. For a hole sheet-density of $5 \times 10^{12}\,\text{cm}^{-2}$, levels at 221 meV (HH), 228 meV (SO) and 231 meV (LH) below the valence band maximum at the surface (VBM_S) are deduced. The Fermi-energy is 237 meV below VBM_S. Also shown are normalized hole wave-functions labeled LH, HH and SO. In thermodynamic equilibrium the chemical potential of the adsorbate layer and the Fermi-level of diamond are in equilibrium. Our calculations reveal an energy gap between electrons in the adsorbate layer and holes in the quantum well. In the case of $5 \times 10^{12}\,\text{cm}^{-2}$, holes must overcome 6 meV or more to recombine with electrons. The wave-function of holes extends about 5 Å into the adsorbate layer calculated for a dielectric constant ε_r of 5.7. For a hole sheet-density of $5 \times 10^{13}\,\text{cm}^{-2}$, three discrete energy levels are calculated, namely

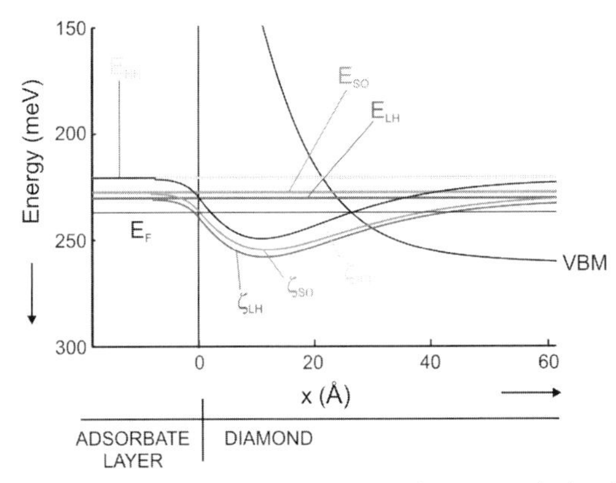

Figure 5.10 Energy band diagrams at the interface of hydrogen-terminated diamond and an adsorbate layer calculated for a hole sheet-density of $5 \times 10^{12}\,\text{cm}^{-2}$. The energies refer to the valence band maximum at the interface (VBM_{INT}). The figure shows the calculated energy levels of the first sub-bands of the light hole (LH), heavy hole (HH) and split-off band (SO). Also shown are the normalized wave functions ζ of holes (LH, SO, HH).

770 meV (HH), 786 meV (SO) and 791 meV (LH) below the VBM_S. The first three sublevels are occupied by holes and recombination is prevented by a gap of 90 meV.

Such two-dimensional properties can be expected only theoretically as the real surface is governed by several additional properties like: (i) Surface and bulk defects which may pin the surface Fermi level. (ii) Surface roughness. (iii) Ions in the Helmholtz layer of the adsorbate film in close vicinity to the hole channel. These parameters may cause broadening of the 2D levels so that a continuous, semimetallic DOS may be a better description [58, 59].

5.2.5
In-Plane Capacitance–Voltage Properties of Al on H-Terminated Diamond

The capacitance–voltage data shown in Figures 5.8 and 5.9 indicate that a peripheral depletion layer is present between Al and H-terminated diamond that is covered with an adsorbate layer to generate transfer doping. This is schematically displayed in Figure 5.6. Towards the Al contact, the accumulation layer is depleted as the work function of Al relative to H-terminated diamond is misaligned. This depletion region generates an in-plane capacitance which can be detected experimentally and which scales approximately with the length of the periphery (Figure 5.9). As contacts were squares, some deviations from a perfect fit can be expected. In the future, such experiments should be performed using contacts with rotational symmetry. The data indicate that in-plane Schottky-junction (2D) properties of Al on H-terminated diamond dominate the electronic properties.

General features of a junction between a two-dimensional electron gas and a metal contact with Schottky properties have been discussed by S.G. Petrosyan and A.Y. Shik 1989 [53] and by B. Gelmont and M. Shur 1992 [54]. Following their arguments the capacitance of a metal in contact with a 2D gas is well described by:

$$C = \frac{\varepsilon_o \varepsilon_r L}{\pi} \ln \left\{ \frac{(d_{AI}^2 + x_{dep}^2)^{0.5} + d_{AI}}{(d_{AI}^2 + x_{dep}^2)^{0.5} - d_{AI}} \right\}$$

where L is the length of the metal periphery, d_{Al} is the thickness of Al and x_{dep} is the width of the space charge region. Taking into account the thickness of Al as 600 nm, $\varepsilon_r = 5.7$, and the detected variation of the capacitance, the variation of the depletion width in our experiments is in the range of 10 to 300 nm (see Figure 5.11). The thinness of the depletion layer is a result of the high hole sheet-carrier density, which is in the range 10^{12} to 10^{14} cm^{-2}. Petrosyan and Shik [53] calculated an inverse proportional relationship between the width of the depletion layer and the sheet carrier density, given by:

$$x_{dep} \propto \frac{\varepsilon_o \varepsilon_r}{2\pi e} \frac{V}{p_{sh}}$$

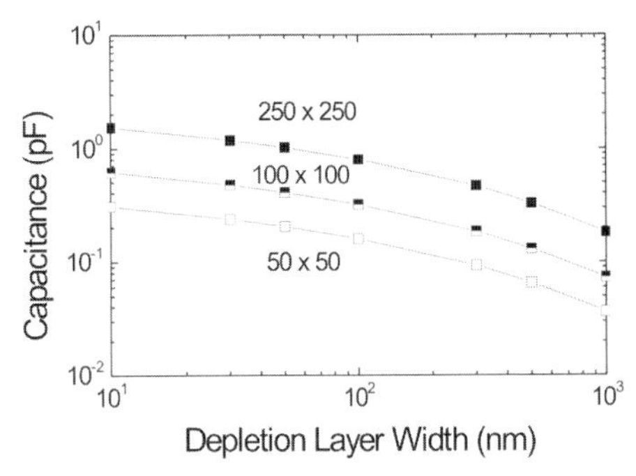

Figure 5.11 Calculated capacitance variations as a function of width of the depletion layer. Experimentally the capacitance is in the range 0.06 to 1 pF, which shows that the depletion layer width varies between 10 and 300 nm.

where V is the applied voltage, e is the elementary charge, and p_{sh} the (sheet) hole density. It is interesting to note that such an in-plane (2D) junction also follows the exponential law given by [53]:

$$j \approx j_{rev}\left[\exp\left(\frac{eU}{kT}\right) - 1\right]$$

where j_{rev} is the contact specific reverse current (for details see [53]). To summarize these results briefly: The detected absolute values of the capacitance are orders of magnitude too small to be discussed using a parallel plate model. The most reasonable model is an in-plane Schottky model. Unfortunately, such junctions show comparable characteristics to three-dimensional (conventional) Schottky junctions and are therefore not distinguishable if only IV-experiments are conducted. Only a combination of CV and IV helps to elucidate the detailed properties.

5.2.6
Hole Carrier Propagation and Scattering in the Surface Layer

To elucidate the transport properties of holes in the surface channel of diamond without dominant scattering by grain boundaries and surface roughness, we have grown several undoped single crystalline CVD diamond films of 200 nm thickness homoepitaxially on 3×3 mm (100) oriented synthetic Ib substrates, using microwave plasma chemical vapor deposition (CVD). Growth parameters were: substrate temperature 800 °C, microwave power 750 W, total gas pressure 25 Torr, total gas flow 400 standard cubic centimeter with 0.025 % CH_4 in H_2. To achieve H-termination after growth, the CH_4 is switch off and diamond is exposed to a pure hydrogen plasma for 5 minutes with otherwise identical conditions. Samples are then cooled down to room temperature in H_2 atmosphere. Temperature ramping,

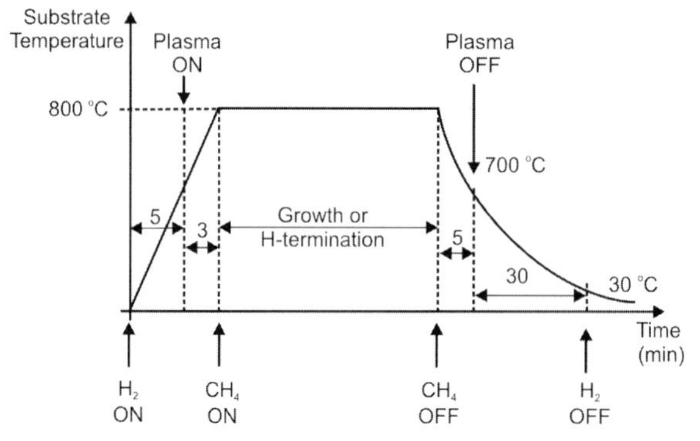

Figure 5.12 Temperature and gas flow variations, plasma switching cycles for growth or H-termination of CVD diamond.

gas flow, and plasma switching are shown as functions of time schematically in Figure 5.12. A detailed discussion of sample growth properties is given in [60–63].

Free-exciton emission in cathode-luminescence at 5.27 eV and 5.12 eV, measured at room temperature, indicates a high quality of the CVD layer and AFM shows atomically flat surface properties.

To characterize the propagation of holes in the accumulation layer, we have performed temperature-dependent Hall-effect experiments on these films. In the following we focus on results deduced for the same H-terminated diamond sample. To vary the electronic properties, the sample has been treated as follows: (a) Exposure for 3 days to humid air, insertion into the Hall set-up (base pressure 1 Torr), heating to 400 K and keeping it at 400 K for 60 minutes before start of the Hall experiment. (b) After the first experimental cycle (a), Hall experiments were carried out again in which the sample was first heated to 400 K and held at 400 K for 30 minutes before start of the Hall experiments. (c) The sample was cleaned mechanically to remove a graphitic layer which had been detected by contact mode AFM, electron microscopy and wetting angle experiments, and then measured as described for case (a) [49].

The results are summarized in Figures 5.13 and 5.14, where hole sheet-densities and mobilities are plotted as a function of temperature. Data, plotted as full squares, are detected in the first Hall measurement cycle (a) after exposing the surface for an extended period of time to air (more than 3 days) without mechanical cleaning. In this case the hole sheet-density is weakly temperature dependent, showing a decrease from $2 \times 10^{12}\,cm^{-2}$ at 400 K to 1.2×10^{12} at 100 K. The second Hall experiment (case b) was undertaken after cycle (a) without breaking the 1 Torr vacuum, but after annealing the layer for about 30 minutes at 400 K. These data are shown as open squares. The thermal annealing at 400 K gives rise to a decrease in the hole sheet-density which varies between $6 \times 10^{11}\,cm^{-2}$ at 400 K and $10^{11}\,cm^{-2}$ at 100 K. The density is slightly activated at about 13 meV at low temperatures, rising to 37 meV at higher temperatures (see solid lines in Figure 5.13). Exposing

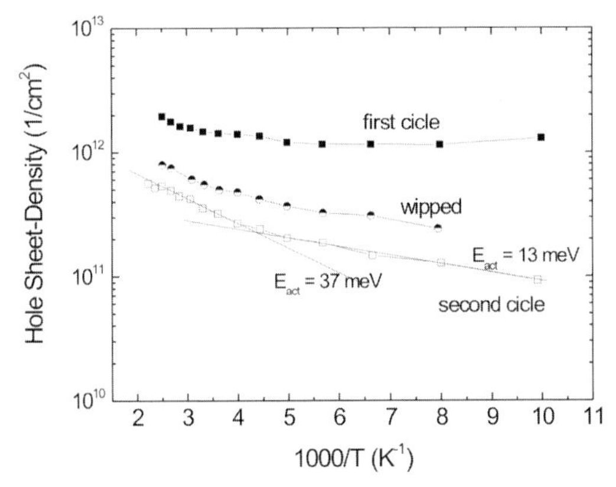

Figure 5.13 Hole sheet-carrier densities measured for two different conditions on the same sample. The straight lines are fits to the data to calculate the activation energies, which are 37 and 13 meV.

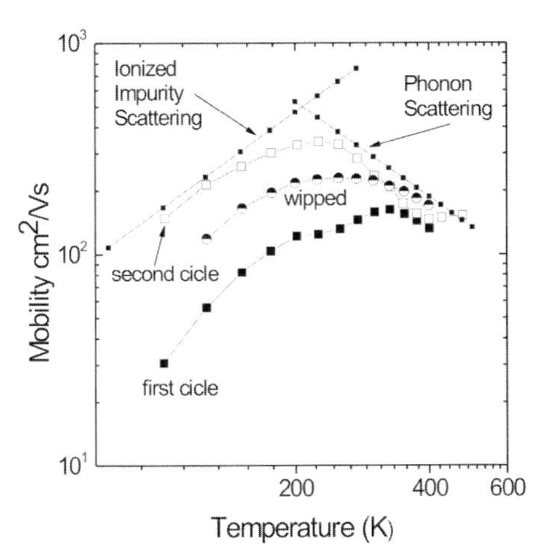

Figure 5.14 Hole mobilities measured for two different conditions on the same sample. The data were detected on the same layer as shown in Figure 5.13. Straight lines indicate the rise expected by ionized impurity scattering or the decrease due to phonon scattering.

the film to air for an extended period of time recovers the initial properties (detected in cycle a) or heating the sample for several hours at 550 K gives rise to insulating properties which can be converted to conductive properties by exposing the sample to air. The data support results of other groups who demonstrated that the application of annealing treatments gives rise to conductivity variations as described previously.

When switching off the hydrogen plasma, due to afterglow, a thin non-diamond carbon layer condenses onto the diamond surface which we detect on all samples by AFM and electron microscopy. This layer can be removed by mechanical cleaning. After such a treatment we performed Hall effect experiments again as in cycle (a) and found that the hole sheet-density was affected by surface cleaning. The holesheet-density and the mobility are somewhere between the results of the first (cycle a) and second (cycle b) Hall-effect measurements. Obviously, the interaction of the clean hydrogen-terminated diamond surface with the adsorbate layer changes. Further experiments are, however, needed to elucidate the details of this process.

These results support the transfer-doping model. In this model, the activation energy of holes is not due to thermal activation from acceptor states (classical doping) but the energy represents the trapping of holes in localized electronic states which are present at the surface of H-terminated diamond [58, 59]. The origin of these traps may be attributed to surface roughness, non-perfect H-termination and ion-induced potential modulation where ions in the Helmholtz layer of the adsorbate give rise to Coulomb scattering in the p-type channel.

In the following we discuss the variation of the hole mobility as a function of temperature which is shown in Figure 5.14. Generally, an increase of mobility is detected with decreasing hole sheet-density. This is in agreement with data summarized in Figure 5.1. For the first time, however, a "semiconductor" type of temperature dependence was detected with a maximum mobility around 230 K of 340 cm^2/Vs. Toward lower or higher temperatures the mobility decreases. Also shown are variations expected as a result of ionized impurity scattering which follows a $\mu \approx T^{-1.5}$ dependence toward lower temperatures and phonon scattering, which gives rise to a $\mu \approx T^{+1.5}$ dependence, toward higher temperatures.

In the case of transfer doping, the density of ions in the Helmholtz layer (ionized impurity centers) is known, as it is exactly the number of holes in the accumulation layer. The negatively charged ions are distributed in the adsorbate film where they accumulate in the Helmholtz layer as shown schematically in Figure 5.15. To discuss the scattering problem in terms of absolute numbers, we take the density of holes detected at highest temperatures, where all holes are mobile, and assume that this is the density of "ionized impurities". Ionized impurity scattering dominates carrier propagation at low temperatures, so we take the mobilities detected at 125 and 150 K into account for this discussion. According to a well-established formula, ionized impurity scattering is inversely proportional to the density of ionized impurities (N_i) [64]:

$$\mu \approx \frac{1}{N_i}$$

Figure 5.16 shows the result, where the mobilities measured at $T = 125$ K and 150 K are plotted as a function of sheet-hole density (= ionized impurity density). This is in perfect agreement with the prediction (dashed line). We conclude,

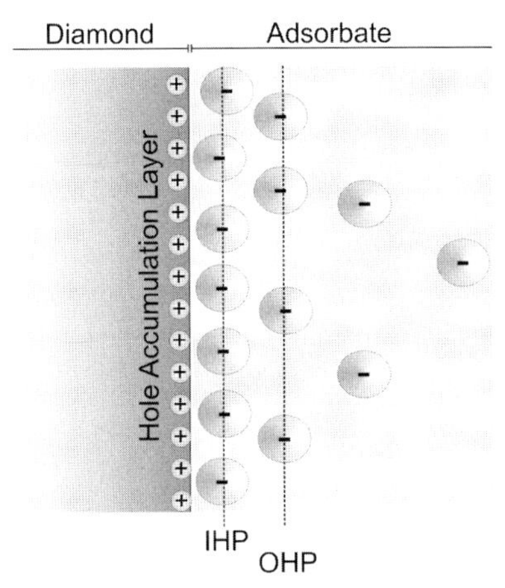

Figure 5.15 Schematic description of the interface of a H-terminated diamond with a hole accumulation surface layer and the Helmholtz layer of the adsorbate film which covers the diamond (IHP: inner Helmholtz plane, OHP: outer Helmholtz plane). The +-sign represents holes propagating at the surface and the −-sign indicates negatively charged ions in the liquid, which accumulated in the Helmholtz layer.

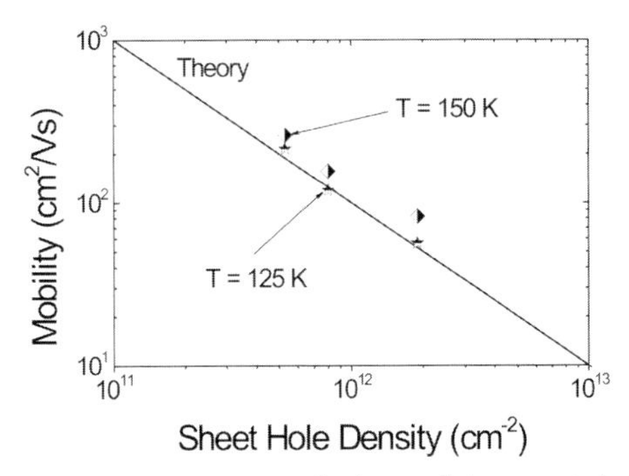

Figure 5.16 Experimentally detected decrease in hole mobilities with increasing hole sheet-density. The line represents the prediction following ionized impurity scattering.

therefore, that the mobility in the hole accumulation layer of H-terminated diamond is limited at low temperature by ionized impurity scattering. The impurities are in the Helmholtz layer of the adsorbate film, very close to the hole accumulation layer in diamond. The model has limits as the density of ions is not distributed homogenously in the bulk, but accumulated in a layer adjacent to the hole channel. As this layer, however, is very close to holes, the scattering may be comparable to a three-dimensional case. We applied other known scattering mechanisms, which are summarized in the article by Ando, Fowler and Stern [65];

however, only the scattering model discussed herein gives reasonable agreement with our experimental data. Scattering by adsorbate ions suggests limitations of electronic applications of such heterojunctions. The hole channel and the ionized impurities are too close to achieve ultrahigh mobilities as in the case of GaAs/AlGaAs high electron mobility transistors.

To summarize: We discussed the variation of mobilities using scattering laws derived for bulk material (3D case). This has been done by intent, as the mobilities are still so low that a 2D discussion makes no sense (for a review, see [65]). This may be a result of Coulomb disorder, which arises from the ions in the Helmholtz layer of the adsorbate film. These potential variations at the surface may be too strong for a truly 2D electronic system in the characterized temperature regime. To achieve further optimization, a spacer may be required, such as C_{60} or other forms of carbon [47].

5.3 Surface Electronic Properties of Diamond in Electrolyte Solutions

5.3.1 Redox Couple Interactions with Undoped H-Terminated CVD Diamond

Electrochemical experiments on H-terminated diamond are much more powerful for the investigation of transfer doping properties as the electrolyte solutions are well defined with respect to pH and redox couple chemical potentials. We used the same undoped hydrogen-terminated CVD diamond substrates as electrodes as discussed above in the Hall effect experiments. The contact arrangement, shown in Figure 5.17, has been optimized for the electrochemical experiments. We evaporated 200 nm thick Au films on the H-terminated surface with rotational

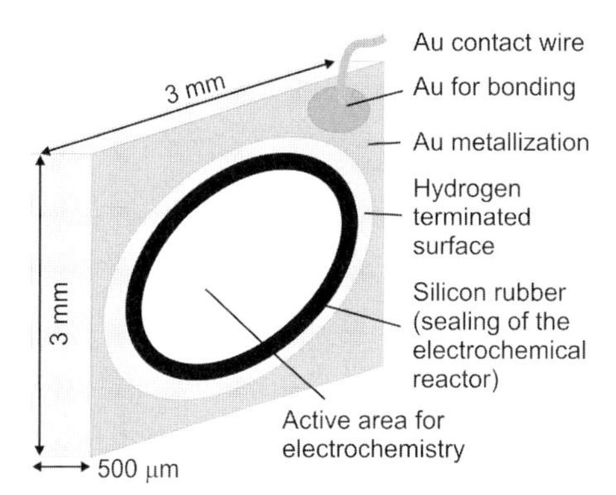

Figure 5.17 Contact configuration applied to characterize undoped H-terminated diamond by electrochemical experiments. The geometry is in-plane where Au is evaporated on H-terminated diamond to generate an ohmic contact on a 1 μm intrinsic CVD diamond grown on Ib diamond substrate.

symmetry. The electrochemically active area is about 0.8 mm^2. Pt is used as the counter electrode, a saturated calomel electrode (Hg/Hg_2Cl_2) as reference electrode and a voltammetric analyser is used for cyclic voltammetry experiments. The scan rate is 100 mV/s. To calculate the *RC*-time limitation of our experiments we take into account the active area of 0.8 mm^2, the experimentally detected series resistance of the surface conductive film of $10^5\,\Omega$, and the typical double layer capacitance of 5 $\mu F/cm^2$ as detected by M.C. Granger *et al.* [66]. Using $\tau = RC$, results in a time constant of about 4 ms. This is two to three orders of magnitude faster than our potential scan rate of 100 mV/s. We exclude therefore *RC*-limiting effects in our cyclic voltammetry experiments.

Redox molecules of 10 mM concentration has been mixed with the supporting 0.1 M Na_2SO_4 electrolyte solution. As redox analytes we used well characterized molecules like $Fe(CN)_6^{3-/4-}$ (formal potential (U^o_{FE}) is +0.46 V vs. SHE, chemical potential $\mu_{Fe} = -4.9$ eV below the vacuum level), $Ru(NH_3)_6^{2+/3+}$ ($U^o_{Ru} = +0.025$ V vs. SHE, $\mu_{Ru} = -4.46$ eV), methyl viologen $MV^{2+/1+}$ ($U^o_{MV2} = -0.48$ V vs. SHE, $\mu_{MV2} = -3.96$ eV), $MV^{+1/0}$ ($U^o_{MV1} = -0.81$ V vs. SHE, $\mu_{MV2} = -3.63$ eV) and $Co(sep)^{2+/3+}$ ($U_{Co} = -0.38$ V vs. SHE, $\mu_{Co} = -4.06$ eV) [67]. The redox energy levels with respect to vacuum and to H-terminated diamond are shown in Figure 5.18.

Cyclic voltammetry can be used to gather information about diamond film and surface qualities. In particular, this method can be applied to detect non-diamond carbon which is present at grain boundaries or graphitic deposits at the surface. It is known that sp^2 gives rise to catalytic effects on hydrogen and oxygen evolution which reduces the working potential window significantly. Figure 5.19 shows background cyclic voltammetry IV-curves as detected in 0.1 M H_2SO_4 (pH = 1) on our sample (DRC, full line in Figure 5.19). The data are compared with two high quality polycrystalline boron-doped diamond films from the literature [66].

On our undoped, H-terminated CVD diamond, the electron exchange reaction between the hole accumulation layer and the electrolyte reveals distinct features.

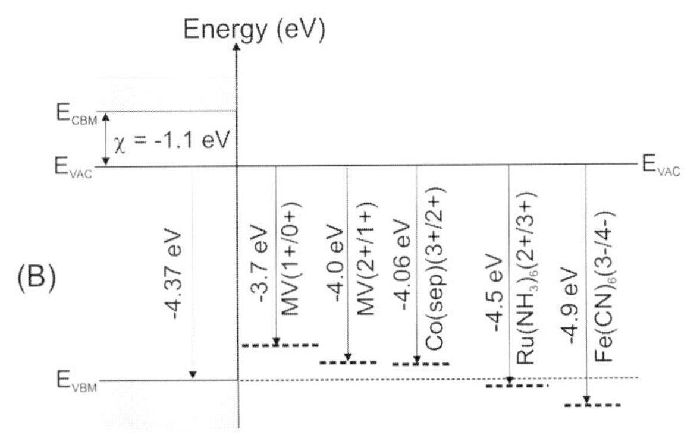

Figure 5.18 Redox chemical potentials of different redox couples with respect to the vacuum level. We assumed a negative electron affinity of −1.1 eV. Only $Fe(CN)_6^{3-/4-}$ and $Ru(NH_3)_6^{2+/3+}$ have levels below E_{CBM}.

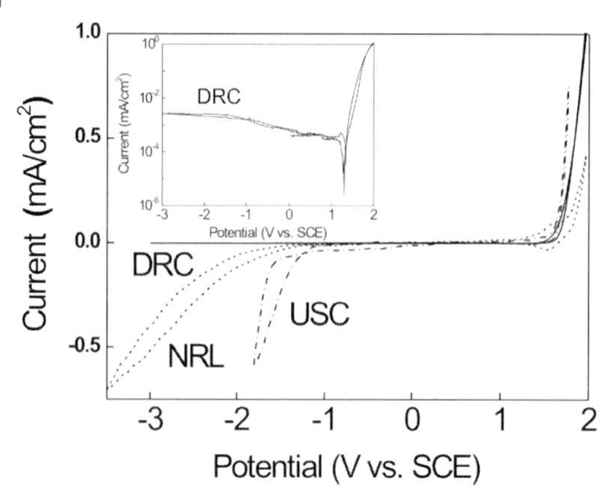

Figure 5.19 Background cyclic voltammetry IV-curves as detected in 0.1 M H_2SO_4 (pH = 1) on a hydrogen-terminated undoped single crystalline CVD diamond film (DRC, full line). The data are compared with two high quality polycrystalline boron-doped diamond films (NRL and USC) from the literature [66]. The inset shows the DRC-data on a log scale. The chemical window if DRC is larger than 4.6 V.

A large chemical potential window extends from less than −3 V (detection limit of our set-up) to +1.6 V (for pH 1). In this regime the background current density is in the $\mu A/cm^2$ range, as can be seen in the inset of Figure 5.19 where the background current is plotted on a log scale. We attribute the large window and the small current density to the absence of sp^2 carbon as no grain boundaries are at the surface [35, 67–69]. Hydrogen evolution cannot be detected down to −3 V, which is the limit of our set-up. The diamond surface can be oxidized at potentials larger than +1.6 V at pH 1, which removes hydrogen from the surface but does not produce any morphological damage to the surface. Such a large working potential window of diamond has not been reported in the literature for poly- or nanocrystalline diamond.

A summary and comparison of redox reactions with H-terminated diamond is shown in Figure 5.20. For $Fe(CN)_6^{3-/4-}$ and $Ru(NH_3)_6^{2+/3+}$ electron transfer from the redox couple into the diamond electrode (oxidation peak) and from the diamond to the redox couple (reduction peak) can be clearly detected. For methyl viologen $MV^{2+/1+}/MV^{+1/0}$ and $Co(sep)^{2+/3+}$ the result indicates no redox interaction.

To discuss these in a more general context, we compare our data with data published in the literature for H-terminated highly boron-doped (10^{19}–$10^{20}\,cm^{-3}$) diamond [66]. These films are three dimensional electrodes in contrast to our in-plane contact geometry. Cyclic voltammetry experiments on these polycrystalline films have been performed using 1 M KCl electrolyte solution and 0.05 V/s [66]. To overcome problems which arise by comparison in absolute terms, we discuss in the following normalized data as shown in Figure 5.21 ((a) $Fe(CN)_6^{3-/4-}$, (b) $Ru(NH_3)_6^{2+/3+}$, (c) $Co(sep)^{3+/2+}$), (d) $MV^{2+/1+}/MV^{+1/0}$).

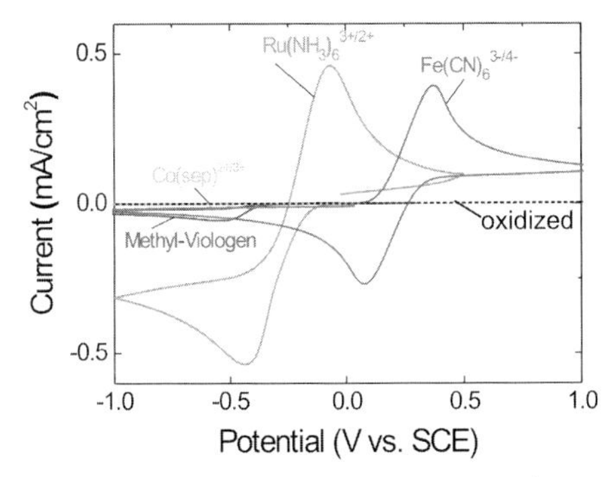

Figure 5.20 Cyclic voltammetric *I–V* curves for $Fe(CN)_6^{3-/4-}$, $Ru(NH_3)_6^{2+/3+}$, $Co(sep)^{3+/2+}$ and methyl viologen $MV^{2+/1+}/MV^{+1/0}$. After anodic oxidation the current decreased to zero (dashed line). Pt was used as counter electrode, a saturated calomel electrode (Hg/Hg_2Cl_2) as reference electrode and cyclic voltammetry experiments have been performed with scan rates of 0.1 V/s. Redox analyte of 10 mM concentration has been mixed with the supporting 0.1 M Na_2SO_4 electrolyte solution.

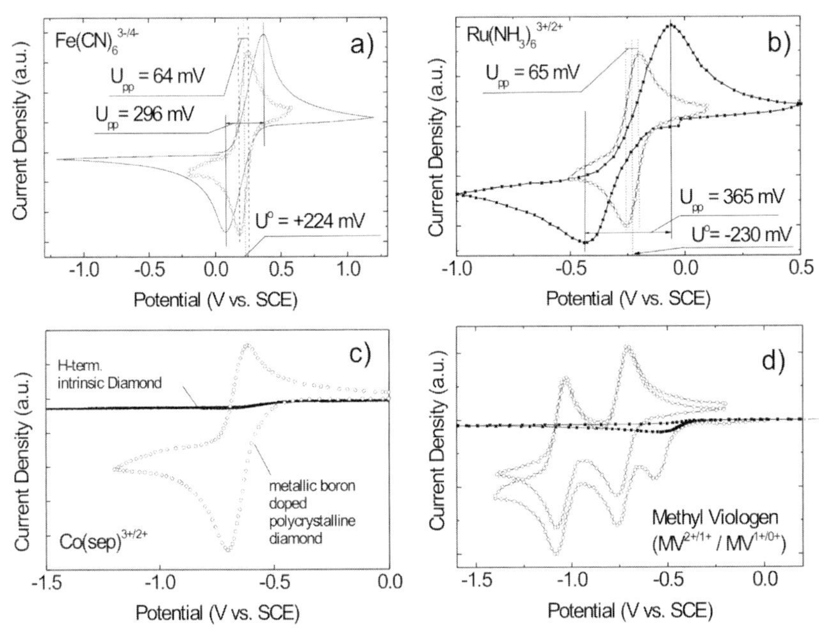

Figure 5.21 Normalized IV curves for $Fe(CN)_6^{3-/4-}$ (a), $Ru(NH_3)_6^{2+/3+}$ (b), $Co(sep)^{3+/2+}$ (c), and methyl viologen (d) as detected on intrinsic H-terminated diamond (full squares) and on boron-doped (metallic) H-terminated polycrystalline CVD diamond (open circles) [66, 67]. In (d), the open circles show cyclic voltammetric IV curves as measured on boron-doped diamond using methyl viologen. The cycles are slightly changing and are discussed in detail in Ref. [66].

For $Fe(CN)_6^{3-/4-}$ and for $Ru(NH_3)_6^{2+/3+}$, hydrogen-terminated intrinsic diamond acts like a metal electrode, with oxidation peak amplitudes of about 0.37 mA/cm^{-2} ($Fe(CN)_6^{3-/4-}$) and 0.46 mA/cm^2 ($Ru(NH_3)_6^{2+/3+}$). ΔU_{pp} for $Fe(CN)_6^{3-/4-}$ is 296 mV and for $Ru(NH_3)_6^{2+/3+}$ 365 mV compared to 64 mV and 65 mV as detected for metallic polycrystalline diamond [66]. The oxidation/reduction currents are reversible. The peak-shift and broadening indicate rate-limited electron transfer as we exclude RC-time constant effects. This is well described in the literature based on the Marcus–Gerischer model [70]. For $Fe(CN)_6^{3-/4-}$ and $Ru(NH_3)_6^{2+/3+}$ the reduction and oxidizing currents are symmetrically centered around the formal potential U° of +224 mV ($Fe(CN)_6^{3-/4-}$) and −230 mV ($Ru(NH_3)_6^{2+/3+}$). This is in good agreement with the data of Granger *et al.* [66] who detect +220 mV for $Fe(CN)_6^{3-/4-}$ and −217 mV for $Ru(NH_3)_6^{2+/3+}$. Taking into account the relation between electrode potential and electron energy with respect to the vacuum level we use [35]:

$$\varepsilon = (-e)U^{\circ} - 4.44\,\mathrm{eV}$$

where ε is the energy of electrons with respect to the vacuum level, e is the elementary charge, U° is the formal potential (vs. SHE), and 4.44 eV is the scaling parameter between electrode potential and electron energy. The chemical potential of redox couples $Fe(CN)_6^{3-/4-}$ and $Ru(NH_3)_6^{2+/3+}$ are calculated to be −4.9 eV ($Fe(CN)_6^{3-/4-}$) and −4.45 eV ($Ru(NH_3)_6^{2+/3+}$) which results in Fermi level positions of about 530 meV and 80 meV below the valence-band maximum at the surface, assuming a negative electron affinity of about −1.1 eV (see Figure 5.18) [34, 51]. For these two redox couples transfer doping conditions are fulfilled which gives rise to the insulator–metal transition of intrinsic diamond.

Cyclic voltammetry experiments on methyl viologen ($MV^{2+/1+}/MV^{+1/0}$) and $Co(sep)^{3+/2+}$ show distinct differences. The current densities are smaller than 5.5 $\times 10^{-5}$ A/cm^2, which is about one order of magnitude smaller than for $Fe(CN)_6^{3-/4-}$ and $Ru(NH_3)_6^{2+/3+}$. Methyl viologen is a two-level redox system with energetic levels at −3.7 eV (1+/0+ transition) and at −4 eV (2+/1+ transition) below the vacuum level [66]. The double reduction and oxidizing peaks as detected on boron-doped metallic polycrystalline diamond cannot be revealed on H-terminated intrinsic diamond. Instead, a broad current peak at about −550 mV is detected in reasonable agreement with the first scan data of Granger *et al.* [66]. In the case of $Co(sep)^{3+/2+}$ the currents are even smaller and show also very different variations compared to metallic boron-doped polycrystalline diamond electrode interactions. Generally, for these two redox couples the current–voltage variations do not resemble metallic electrode characteristics. We attribute this to the misaligned valence band maximum of diamond and the redox-potentials. The current variation may reflect some change in conductivity of the H-terminated surface, but further experiments are required to elucidate this problem.

Application of potentials larger than +1.6 V gives rise to oxidation of hydrogen-terminated diamond surfaces. To characterize the electronic properties in the transition region from perfectly hydrogen-terminated to fully oxidized, we have performed a sequence of voltammetric experiments using 10 mM $Fe(CN)_6^{3-/4-}$ as

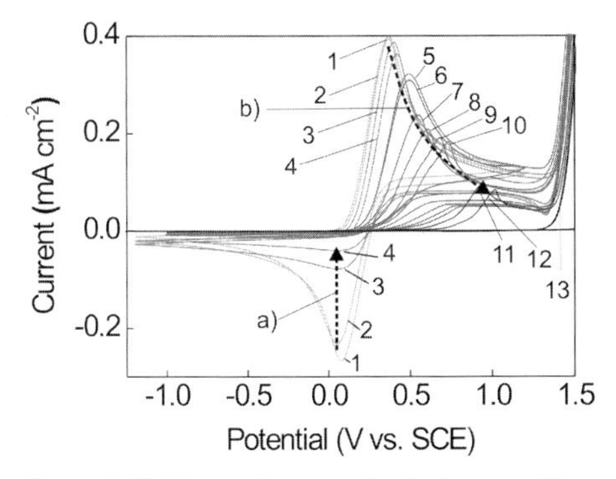

Figure 5.22 Electronic properties in the transition region from perfectly hydrogen-terminated to fully oxidized. A sequence of voltammetric experiments using 10 mM $Fe(CN)_6^{3-/4-}$ as redox couple in 0.1 M Na_2SO_4 electrolyte solution (pH = 6) has been applied (indicated by 1 to 13).

redox couple in 0.1 M Na_2SO_4 electrolyte solution (pH = 6). The scan rate of the voltammetric experiment was 0.1 V/s in the regime −1.2 V to +1.6 V. The results are shown in Figure 5.22. The application of a sequence of short time anodic oxidation can be summarized as follows:

1. The reduction peak marked by (a) in Figure 5.22 is decreasing rapidly, while the oxidizing peak remains nearly stable (steps 1 to 4).
2. After the disappearance of the reduction peak, the oxidizing peak, marked (b) in Figure 5.22, is decreasing and shifting to higher potentials (steps 5 to 12).

$$2H_2O \Rightarrow O_2 + 4H^+ + 4e^-$$

3. Finally, the redox-interaction completely disappears but oxygen evolution still can be detected. We attribute this to the chemical reaction: which has an energy level at −5.3 eV below the vacuum [35]. This is significantly deeper than for $Fe(CN)_6^{3-/4-}$ (energy level: −4.9 eV).
4. After prolonged anodic oxidation the current approaches zero due to the complete removal of hydrogen and of the surface conductivity.

5.3.2
Electrochemical Exchange Reactions of H-Terminated Diamond with Electrolytes and Redox Couples

The CVD diamond films under discussion are perfect insulators if measured in vacuum with a clean surface. In this case the Fermi level is in the band gap of diamond. The same films become metallic if immersed in redox-electrolyte solutions with chemical potentials below the valence band maximum. This can be ascribed to alignment of Fermi level and redox potential or chemical potential at the interface. It is a remarkable phenomenon, which requires "defect free" bulk properties and perfect hydrogen termination of the surface carbon dangling bonds to unpin the Fermi level. Both conditions seem to be fulfilled for these samples. Obviously, optimized growth and hydrogen termination of diamond generates high-quality subsurface and surface electronic properties. It is also certain that the insulator–metal transition is related to the presence of hydrogen at the surface as the electrochemical currents vanish after anodic oxidation.

A comparison with metallic boron-doped polycrystalline diamond shows that the current–voltage variations are broader and that the maxima are shifted to higher (oxidation) or lower (reduction) potentials with respect to the formal potential. Based on arguments given above we exclude RC-limitations as an origin for the shifts and broadening. We assume that these variations are generated by limited electron transfer rates at the interface. This is schematically illustrated in Figure 5.23. Metallic properties are established by occupied and empty electronic states in the valence band, separated by E_F. In the case of metals, the density-of-state distribution $\rho(E)$ is approximately constant around E_F whereas for diamond $\rho(E)$ is a function of energy. The application of external potentials shifts E_F, either up (negative potential) by filling of empty electronic levels or down (positive potential) by emptying of valence-band states. External potentials will, however, not change the confinement of holes (band-bending) significantly, as potentials are applied parallel to the conductive layer (perpendicular to the band bending).

The interaction of metal and semiconducting electrodes with redox couples has been described in the literature using the Marcus–Gerischer model [70], where the oxidizing density-of-states ($D_o(E)$) above the chemical potential μ and the reduction states ($D_R(E)$) below μ are Gaussian shaped. An upward shift of the Fermi energy, E_F, gives rise to electron transfer from the electrode into empty D_o

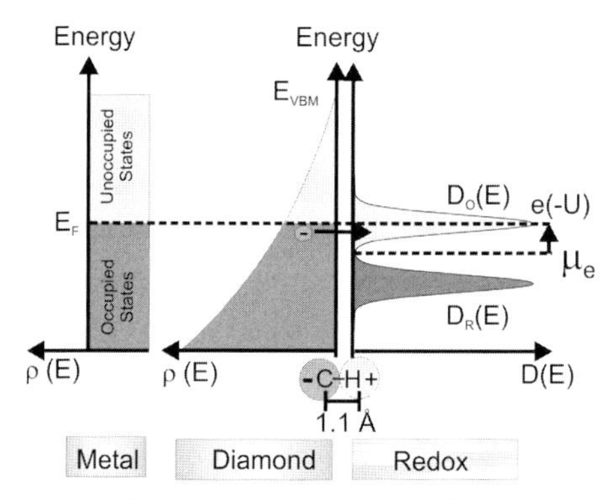

Figure 5.23 Schematic comaprison of a metal and a diamond electrode in contact with a redox system with chemical potential μ below E_{VBM}. We show the situation for a negative voltage applied to diamond which shifts the Fermi level E_F up compared to μ_e. This gives rise to a reduction electron flow into empty states of the redox couple.

states and vice versa. The local rate of reduction $r_R(E)$ or oxidation $r_O(E)$ are given by:

$$r_R(E) = A\{f(E)\rho(E)\varepsilon_{red}(E)D_O(E)dE\}/\Delta t$$

$$r_O(E) = A\{[1 - f(E)]\rho(E)\varepsilon_{ox}(E)D_R(E)dE\}/\Delta t$$

where A is the electrode area, Δt a given time interval, $f(E)$ the Fermi-distribution, $\rho(E)$ the density-of-state distribution (DOS) of the metal or semiconductor, $D_O(E)$ the DOS of the oxidizing redox states, $D_R(E)$ the DOS of the reduction states, $\varepsilon_{red}(E)$ and $\varepsilon_{ox}(E)$ the proportionality functions for reduction (red) and oxidation (ox), and dE the energy interval under discussion. The exchange rates are proportional to the DOS of the electrode $\rho(E)$.

The transition rate equations contain $\varepsilon_{red}(E)$ and $\varepsilon_{ox}(E)$. These parameters are governed by the wave-function overlap of electrons in the valence-band of diamond with empty D_O states and of electrons in the D_R states with empty states in the valence-band of diamond. Here, the hydrophobic properties of H-terminated diamond may play an important role as the gap between the redox–electrolyte molecules and the diamond surface will be a function of wetting characteristics. Holes in the surface conducting channel are confined by a dipole energy barrier of about 1.7 eV [57]. This conductive layer is very thin (10–20 Å), the continuous density-of-state distribution may therefore be transformed into a two-dimensional (2D) system with discrete energy levels [36–38]. It is, however, very likely that the 2D-states are broadened by disorder due to ionic interface scattering and surface roughness.

The experiments show that for two redox couples with chemical potential below the valence-band maximum oxidation and reduction currents can be detected with formal potentials like those for H-terminated metallic polycrystalline boron-doped diamond. This indicates that the electrochemical interaction of boron-doped diamond is also governed by energy alignment properties which are dominated by hydrogen termination and related negative electron affinities. The conductivity of a 3D electrode (metallic boron-doped diamond) is, however significantly higher than that of a conductive surface layer of 10 to 20 Å thickness. It affects the dynamic properties of cyclic voltammetry experiments as shown in Figures 5.20 and 5.21. In the case of a 3D metallic boron-doped film the peaks are narrow and separated by only about 65 mV.

We attribute the different results detected for methyl viologen and $Co(sep)^{3+/2+}$ to the non-formation of a conductive surface layer (insulator–metal transition does not occur), arising from the misalignment of redox potentials and the valence band maximum. A small surface conductivity may be generated for larger cathodic voltages, which induce electron transfer. The shape of the current–voltage variations as shown in Figures 5.20 and 5.21 are, however, very different and cannot be attributed to redox-couple interactions.

Finally we want to address the oxidation phenomena as detected by cyclic voltammetry. It is known that H-terminated diamond has a negative electron affinity in the range −1.1 to −1.3 eV and oxidized diamond shows a positive electron affinity [34, 51]. We therefore expect a vacuum level shift at the surface of diamond of more than $\Delta E \geq 1.1\,\text{eV}$. Our data indicate that partial oxidation generates a partial shift of the vacuum level, which gives rise to a redox-level shift as shown schematically in Figure 5.24. After initial partial oxidation the "reduction peak" (electron flow from the diamond to the redox-couple) disappears. This indicates that electrons in the surface conducting layer of diamond cannot be transferred into the redox couple as only occupied $D_R(E)$ states are aligned with the conductive layer. Further oxidation shifts the redox levels even higher, so that the "oxidizing peak" also decreases and finally vanishes. The redox-couple and the electronic levels of the valence band are approaching a situation which normally prevents the formation of a conductive layer as the redox energetic layers are higher than the valence band maximum of diamond. As the redox couple is dissolved in an electrolyte of 0.1 M Na_2SO_4 with pH 6, the chemical potential of the oxygen evolution at −5.3 eV, given by the reaction:

$$2H_2O \Leftrightarrow O_2 + 4H^+ + 4e^-$$

can accommodate electrons ("formation of surface conductivity") if O_2 is in the electrolyte. This is the case in our experiments as we applied anodic oxidation potentials (>+1.6 V). Figure 5.24 shows the energy diagram schematically. As long as this level is below the valence band maximum E_{VBM}, surface conductivity can be expected. Prolonged oxidation gives rise to insulating surface properties, as all

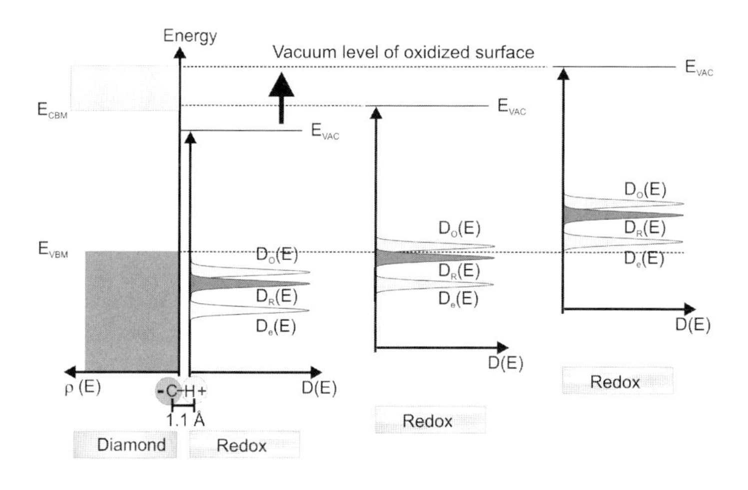

Figure 5.24 Diamond and redox-couple interaction ($D_O(E)$ and $D_R(E)$) as a funtion of (a) hydrogen termination, (b) partially oxidized and (c) fully oxidized. $D_e(E)$ is the chemical potential of the reaction $2H_2O \Leftrightarrow O_2 + 4H^+ + 4e^-$ the 0.1 M Na_2SO_4 electrolyte solution (pH = 6) which has an energy level of −5.3 eV below the vacuum.

H is chemically removed from the surface. The positive electron affinity misaligns energy levels significantly which prevents transfer doping.

Although the general feature of this model fits our experimental data, we are aware that it is a model which needs further discussions. It is interesting to note that a similar result has been reported by Cui and co-workers [51]. They applied photoelectron emission experiments on hydrogen-terminated diamond and removed the hydrogen termination partially by a sequence of annealing steps. They also found a continuous shift of the vacuum level from −1.3 eV to +0.38 eV for a hydrogen-free surface. This is in good agreement with our electrochemical results where we removed hydrogen by a sequence of oxidizing steps.

5.3.3 Ion Sensitive Field Effect Transistor (ISFET) from Undoped CVD Diamond

The redox couple and electrolyte interactions of diamond support the transfer doping model. In the following we want to address the pH sensitivity of the surface conducting layer as this is the last part of the transfer doping phenomena which needs to be addressed.

To measure the variation of surface conductivity in different pH liquids, we manufactured ion sensitive field effect (ISFET) structures. The geometric arrangement is shown in Figure 5.25 where a hydrogen-terminated surface area of $100 \times 500\,\mu m$ size connects two Au contacts of $130 \times 130\,\mu m$ which have been thermally evaporated on to diamond. Au wires are bonded to these pads and insulated by a resistive lacquer from contact to electrolyte solution.

The outer part has been oxidized by a soft oxygen plasma to generate insulating surface properties. The schematic arrangement of ISFET in electrolyte solution is shown in Figure 5.26. Platinum is used as the reference electrode between drain and source contacts. The pH was varied between 2 and 12 by use of Britton–Robinson buffer (0.04 M $H_3BO_3 + H_3PO_4 + CH_3COOH$, titrated by NaOH (0.2 M)).

Variations of the drain-source currents with pH, measured at constant drain-source potentials of 0.2 V are shown in Figure 5.27. The currents show transistor

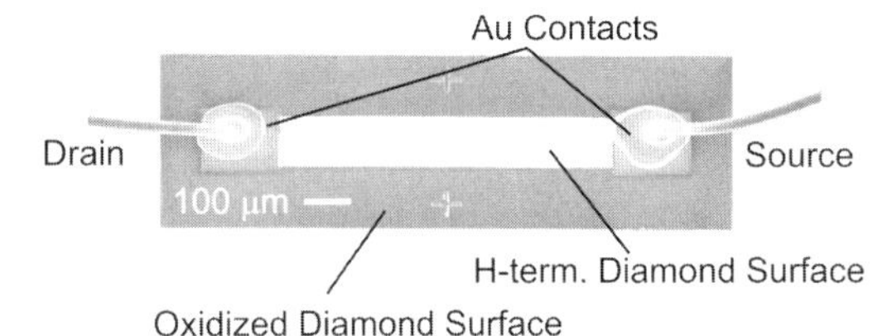

Figure 5.25 Top view of a realised ISFET structure as measured by electron microscopy. The length of the sensitive H-terminated area is 500 µm and the width 100 µm. On both ends, Au contacts have been thermally evaporated which act as drain and source electrodes, contacted by wire bonding. The H-terminated area is electrically insulated from the surrounding by an oxidized diamond surface. The Au contacts and wires are covered by a lacquer, which cannot be seen in this figure.

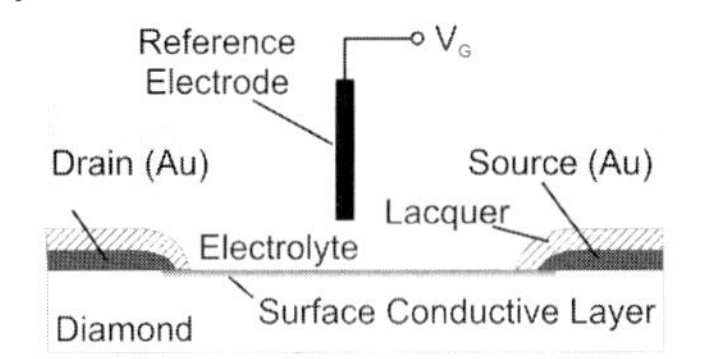

Figure 5.26 Schematic arrangement of an ion-sensitive field effect transistor favricated from diamond. Due to hydrogen termination of diamond a thin surface conductive layer is formed at the interface to the electrolyte. Au contacts serve as drain and source electrodes. Both are covered by a lacquer for insulation in the electrolyte solution.

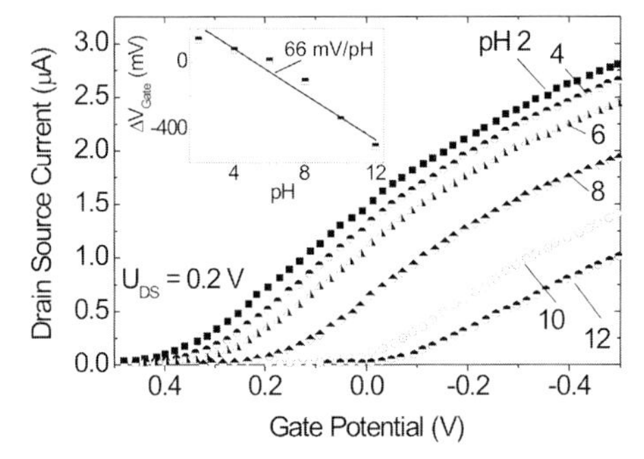

Figure 5.27 Drain source currents as a function of gate voltage of an ISFET structure on H-terminated diamond, measured in electrolyte solutions with pH in the regime 2 to 12. The pH sensitivity is shown in the inset for I_{DS} = 1 μA, which results in a pH sensitivity of 66 mV/pH.

characteristics with increasing amplitude for decreasing pH. For positive potentials the current is pinched off with thresholds, dependent on pH. To prevent degradation by partial oxidation, the maximum applied negative gate voltage was limited to −0.5 V. At this gate potential the drain-source resistance is in the range (7–13) × 10^4 Ω. Defining 1 μA drain source current as reference, a pH sensitivity of 66 mV/pH is calculated as shown in the inset of Figure 5.27. This is slightly higher than expected by the Nernst law (59 mV/pH). From the leakage current of the ISFET (current flow from surface channel of the diamond into the electrolyte) which is about 3 μA/cm^2 at −3 V, we calculate a gate resistance of 10^8 Ω. The drain source resistance is in the range of 10^5 Ω.

The gate properties can be best described by diode characteristics. Below a critical gate/channel voltage there is only a small leakage current flowing. Above the threshold voltage the diode opens and exponentially increasing currents are detected. The electronic circuit of a diamond ISFET is summarized in Figure 5.28. Gate leakage threshold voltages are shown in Figure 5.29 for different pH, measured in 0.1 M $HClO_4$ (A) (pH 1), 0.1 M Na_2SO_4 (B) (pH 6), and 0.1 M NaOH

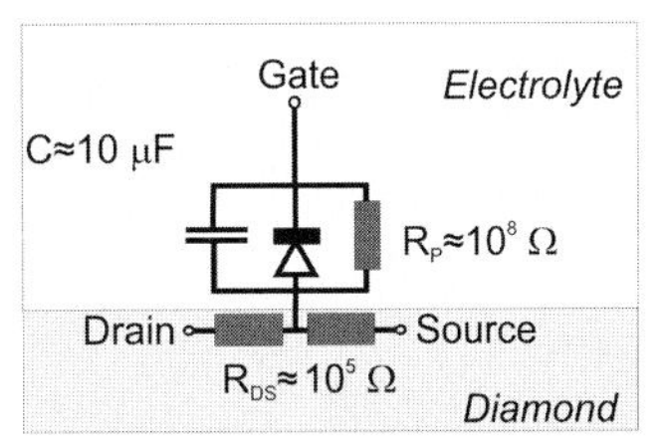

Figure 5.28 The diamond/electrolyte interface described in electronic terms. The interface is dominated by the Helmholtz capacitance, a leakage resistance in the range of $10^8\,\Omega$, and diode properties if a critical threshold voltage for anodic oxidation is applied. The drain source resistance is in the range $10^5\,\Omega$.

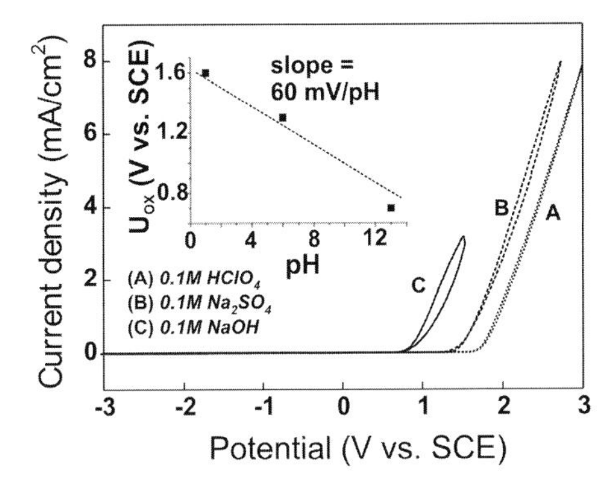

Figure 5.29 Cyclic voltammograms as detected on a H-terminated undoped diamond electrode in (A) 0.1 M $HClO_4$, (B) 0.1 M Na_2SO_4, and (C) 0.1 NaOH. The oxidizing threshold shifts with increasing pH to lower potentials with about 60 mV/pH.

(C) (pH 13). The threshold voltage of oxygen evolution at positive potentials (U_{OX}) shifts from +0.7 V (vs. SCE) at pH 13 to about +1.6 V at pH 1. The shift is about 60 mV/pH, which is in reasonable agreement with the Nernst prediction. Oxidation of H-termination occurs for potentials lager than U_{OX} which limits the application of pH sensors to gate potentials smaller than 0.7 V.

5.4 Discussion and Conclusions

The combination of results from ISFET characterization and from cyclic voltammetry experiments on the very same samples show that a surface conducting layer is generated in all investigatedelectrolytes, with pH ranging from 2 to 12. The drain-source currents of H-terminated single crystalline diamond ISFETs show high pH sensitivity, with about 66 mV/pH. The basic mechanism behind this phenomenon is the alignment of chemical potential and the Fermi level of diamond ("transfer doping model") as shown schematically in Figure 5.30a. In case of a low pH, the chemical potential with respect to the vacuum is large and therefore the Fermi level is deep in the valence band, giving rise to a high drain source current (Figure 5.30a), which is in agreement with our findings. With increasing pH the chemical potentials shifts up in energy as well as E_F and the drain source current is decreasing. For the site-binding model, where a partially oxidized surface is required [39, 40, 71], opposite characteristics are detected (see Figure 5.30b). We therefore conclude that our pH sensitivity favours the transfer doping model.

The leakage current at the interface diamond/electrolyte is in the range of several nano-amperes, for potentials which are below the oxidizing threshold U_{ox}. Above U_{ox} a strong rise in current is detected due to the oxidizing reaction:

$$4OH^- \rightarrow O_2 + 2H_2O + 4e^-.$$

The interface electronic properties are a complex issue. Here, we want to focus on two major parameters: (i) the interacting energy levels, which are empty and occupied states in the diamond electrode and in the electrolyte; and (ii) the interaction distance, as molecules must come close to the electrode to perform chemical reactions. The highly suppressed hydronium reduction reaction $2H^+ + 2e^- \rightarrow H_2$ indicates, that this mechanism is not the origin for surface conductivity of H-terminated diamond. The reaction is suppressed by a coulomb repulsion field, arising from the densely arranged (~10^{15} cm^{-2}) C–H-dipoles at the surface of diamond, indicated in Figure 5.31 by an enlarged distance Δx_{H+}. In addition, the interaction is also unlikely as the H$^+$-level ($2H^+ + 2e^- \leftrightarrow H_2$) shifts above the valence band maximum for increasing pH, as schematically shown in Figure 5.31. To

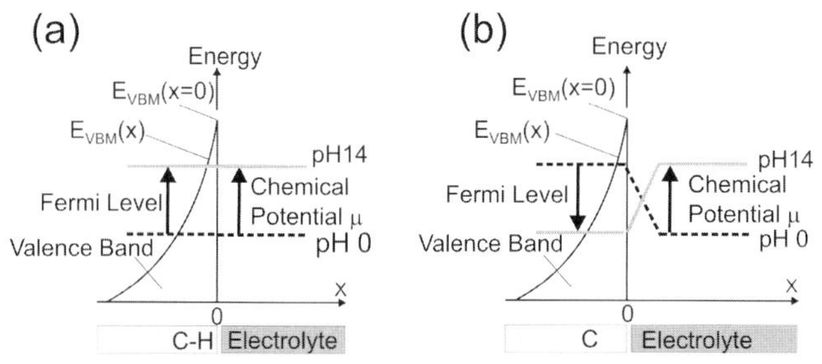

Figure 5.30 Schematic comparison of transfer doping (a) and site-binding (b) model. In case of (a) the shift of the chemical potential deeper in energy with decreasing pH gives rise to an increase in surface conductivity, because the Fermi level also moves deeper into the valence band. For the site-binding model (b), the opposite is expected ($E_{VBM\text{-}Surf}$. = valence-band maximum at the surface, $E_{VBM}(x)$ = valence-band maximum in diamond).

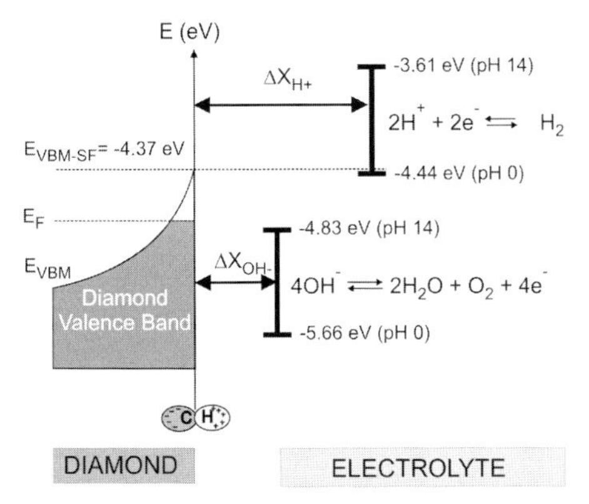

Figure 5.31 Schematic energy diagram of H-terminated diamond in contact with electrolytes with pH ranging from 0 to 14. The H^+ and OH^- levels are drawn in as function of pH. We also show an enlarged distance between hydronium (H^+) ions and the H-terminated diamond surface due to coulomb repulsion.

calculate the energy levels, we took into account a negative electron affinity of −1.1 eV [34], which results in a surface valence-band maximum ($E_{VBM\text{-}SF}$) of −4.37 eV below the vacuum level. The H^+ level (E_{H+}) varies from −4.44 eV (pH 0) to −3.61 eV (pH 14) [35]. Only for pH 0 and pH 1, E_{H+} is slightly below $E_{VBM\text{-}SF}$, which is the requirement for transfer doping.

Negatively charged OH^- molecules have levels well below the valence band maximum for all pH (see Figure 5.31) and can approach the electrode, but these are occupied states which can be oxidized only by application of potentials larger than U_{ox}.

The formation of surface conductivity in diamond requires, however, the opposite mechanism, namely electron transitions from occupied valence-band states into empty states of the electrolyte. This is possible if we follow the arguments of Angus *et al.* [35] and Ristein *et al.* [72] who assume that electrons will trigger the reduction of atmospheric oxygen or other atmospheric contaminants in the electrolyte, described by $O_2 + 2H_2O + 4e^- \rightarrow 4OH^-$. If we discuss this phenomenon like a redox couple interaction (O_2/OH^-), the O_2 is the empty state of the couple and energetically slightly above OH^-. As the energy level of OH^-, and therefore of O_2, for all pH is below the valence-band maximum the formation of a surface conducting film will take place in all electrolytes. The reaction will be limited either by a complete consumption of O_2 and/or by the electric field which builds up at the surface and which prevents unlimited electron flow. The consumption argument is of minor importance if we take into account absolute numbers. Considering data from Hall effect experiments we know that the typical sheet hole density is in the range of $10^{12}\,cm^{-2}$ for drain-source conductivities like in our experiments. The ISFET sensor area is $4 \times 10^{-4}\,cm^2$, which results in about 10^8 electrons which are exchanged at the interface to reduce O_2. This is a small number which may not generate saturation effects. Further experiments are, however, required to elucidate details of this phenomenon.

5.5 Summary

The formation of a conductive layer at the surface of undoped diamond is characterized using Hall effect, conductivity, contact potential difference (CPM), scanning electron microscopy (SEM), and cyclic voltammetry experiments. The experimental data show that due to electron transfer from the valence band into empty states of the electrolyte, a highly conductive surface layer is generated. Holes propagate in the layer with mobilities up to 350 cm^2/Vs. The sheet hole density in this layer is in the range 10^{11} to $5 \times 10^{12}\,cm^{-2}$, and depends on the pH of the electrolyte or adsorbate. This has been utilized to manufacture ion sensitive field effect transistors (ISFET) where the drain source conductivity is pH dependent, with about 66 mV/pH. The electronic properties of the diamond/electrolyte interface are governed by coulomb repulsion between positive ions in the electrolyte (hydronium ions) and the H^+—C^- dipole layer at the surface of diamond. This generates an enlarged tunnelling gap between diamond and positive ions in the electrolyte. Application of potentials larger than the oxidation threshold of +0.7 V (pH 13) to +1.6 V (pH 1) gives rise to exponentially increasing currents between diamond and electrolyte and to partial surface oxidation. The electronic interaction of diamond with redox couples and pH electrolyte solutions is well described by the transfer doping model which accounts for the detected properties of hydrogen-terminated undoped single crystalline CVD diamond.

We summarized theoretical calculations of the ideal surface electronic properties of H-terminated diamond in contact with an electrolyte film which show 2D properties, with discrete energy levels. The calculated energies are in reasonable agreement with contact potential difference experiments made. However, truly 2D properties can only be expected if the surface is atomically smooth, the hydrogen termination is perfect, and the ions in the Helmholtz layer of the adsorbate film do not interact too strongly by Coulomb scattering.

Transfer doping requires a spectacular Fermi level shift into the valence band, which seems to be possible only if hydrogen termination of the diamond surfaces is perfect and the sub-surface region is defect free. Such properties can be achieved by optimized CVD growth parameters.

Acknowledgments

The authors acknowledge the help of T. Yamamoto in fabricating ISFET structures and greatly appreciate discussions with Dr Okushi, Dr Yamasaki and with all members of the Diamond Research Center. Special thanks go to Dr Park who supported us with an electrochemical analyser. The authors also thank Elsevier,

New Diamond and Frontier Carbon Technology and American Physical Society for granting copyright permission for several figures.

References

1 Bergveld, P. (1990) IEEE Transduction Biomedical Engineering, BME-17, 70.

2 Bergveld, P. (2003) *Sensors and Actuators B*, **88**, 1.

3 Brajter-Toth, A. and Chamber, Q.I. (eds) (2002) *Electroanalytical Methods: Biological Materials*, Marcel Dekker, New York.

4 Xing, W.-L. and Cheng, J. (eds) (2003) *Biochips: Technology and Applications*, Springer Verlag.

5 Flynn, N.T., Tran, T.N.T., Cima, M.J. and Langer, R. (2003) *Langmuir*, **19**, 10909.

6 Kharatinov, A.B., Wasserman, J., Katz, E. and Willner, I. (2005) *Journal of Physics Chemistry B*, **105**, 4205.

7 Nazare, M.H. and Neves, A.J. (eds) (2001) *Properties, Growth and Applications of Diamond*, EMIS Datareview Series No. 26, INSPEC Publication.

8 Haymond, S., Babcock, G.T. and Swain, G.M. (2002) *Journal of American Chemical Society*, **124**, 10634.

9 Yang, W., Butler, J.E., Russel, J.N., Jr and Hamers, R.J. (2004) *Langmuir*, **20**, 6778.

10 Yang, W., Auciello, O., Butler, J.E., Cai, W., Carlisle, J.A., Gerbi, J.E., Gruen, D. M., Knickerbocker, T., Lasseter, T.L., Russell, J.N., Jr, Smith, L.M. and Hamers, R.J. (2002) *Nature Materials*, **1**, 253.

11 Tse, K.-Y., Nichols, B.M., Yang, W., Butler, J.E., Russell, J.N., Jr and Hamers, R.J. (2005) *Journal of Physics Chemistry B*, **109**, 8523.

12 Takahashi, K., Tanga, M., Takai, O. and Okamura, H. (2003) *Diamond and Related Materials*, **12**, 572.

13 Wang, J., Firestone, M.A., Auciello, O. and Carlisle, J.A. (2004) *Langmuir*, **20**, 11450.

14 Song, K.S., Degawa, M., Nakamura, Y., Kanazawa, H., Umezawa, H. and Kawarada, H. (2004) *Japanese Journal of Applied Physics*, **43**, L814.

15 Haertl, A., Schmich, E., Garrido, J.A., Hernando, J., Catharino, S.C.R., Walter, S., Feulner, P., Kromka, A., Steinmueller, D. and Stutzmann, M. (2004) *Nature Materials*, **3**, 736.

16 Tang, L., Tsai, C., Gerberich, W.W., Kruckeberg, L. and Kania, D.R. (1995) *Biomaterials*, **16**, 483.

17 Maier, F., Ristein, J. and Ley, L. (2001) *Physical Review B*, **64**, 165411.

18 Maier, F., Riedel, M., Mantel, B., Ristein, J. and Ley, L. (2000) *Physical Review Letters*, **85**, 3472.

19 Gi, R.I.S., Mizumasa, T., Akiba, Y., Hirose, H., Kurosu, T. and Iida, M. (1995) *Japanese Journal of Applied Physics*, **34**, 5550.

20 Maki, T., Shikama, S., Komori, M., Sakaguchi, Y., Sakuta, K. and Kobayashi, T. (1992) *Japanese Journal of Applied Physics*, **31**, 1446.

21 Hayashi, K., Yamanaka, S., Okushi, H. and Kajimura, K. (1996) *Applied Physics Letters*, **68**, 376.

22 Looi, H.J., Jackman, R.B. and Foord, J.S. (1998) *Applied Physics Letters*, **72**, 353.

23 Gi, R.I.S., Tashiro, K., Tanaka, S., Fujisawa, T., Kimura, H., Kurosu, T. and Iida, M. (1999) *Japanese Journal of Applied Physics*, **38**, 3492.

24 Williams, O.A. and Jackman, R.B. (2004) *Diamond and Related Materials*, **13**, 325.

25 Williams, O.A. and Jackman, R.B. (2003) *Semiconductor Science and Technology*, **18**, S34.

26 Williams, O.A., Whitfield, M.D., Jackman, R.B., Foord, J.S., Butler, J.E. and Nebel, C.E. (2001) *Diamond and Related Materials*, **10**, 423.

27 Williams, O.A. and Jackman, R.A. (2004) *Diamond and Related Materials*, **13**, 166.

28 Looi, H.J., Pang, L.Y.S., Molloy, A.B., Jones, F., Foord, J.S. and Jackman, R.B. (1998) *Diamond and Related Materials*, **7**, 550.

29 Jiang, N. and Ho, T. (1999) *Journal of Applied Physics*, **85**, 8267.

30 Hayashi, K., Yamanaka, S., Watanabe, H., Sekigachi, T., Okushi, H. and Kajimura, K. (1997) *Journal of Applied Physics*, **81**, 744.

31 Watanabe, H., Hayashi, K., Takeuchi, D., Yamanaka, S., Okushi, H. and Kajimura, K. (1998) *Applied Physics Letters*, **73**, 981.

32 Rezek, B., Sauerer, C., Nebel, C.E., Stutzmann, M., Ristein, J., Ley, L., Snidero, E. and Bergonzo, P. (2003) *Applied Physics Letters*, **82**, 2266.

33 Shirafuji, J. and Sugino, T. (1996) *Diamond and Related Materials*, **5**, 706.

34 Takeuchi, D., Kato, H., Ri, G.S., Yamada, T., Vinod, P.R., Hwang, D., Nebel, C.E., Okushi, H. and Yamasaki, S. (2005) *Applied Physics Letters*, **86**, 152103.

35 Angus, J.C., Pleskov, Y.V. and Eatonin, S.C. (2004) *"Thin Film Diamond II", Semiconductors and Semimetals*, Vol. 77 (eds C.E. Nebel and J. Ristein), Elsevier, p. 97.

36 Nebel, C.E., Rezek, B. and Zrenner, A. (2004) *physica status solidi (a)*, **11**, 2432.

37 Nebel, C.E., Rezek, B. and Zrenner, A. (2004) *Diamond and Related Materials*, **13**, 2031.

38 Nebel, C.E. (2005) *New Diamond and Frontier Carbon Technology*, **15**, 247.

39 Kawarada, H., Araki, Y., Sakai, T., Ogawa, T. and Umezawa, H. (2001) *physica status solidi (a)*, **185**, 79.

40 Garrido, J.A., Haertl, A., Kuch, S., Stutzmann, M., Williams, O.A. and Jackman, R.B. (2005) *Applied Physics Letters*, **86**, 73504.

41 Nebel, C.E. (2003) *Semiconductor Science and Technology*, **1** (S1), 18.

42 Nebel, C.E. (2003) *"Thin-Film Diamond I", Semiconductors and Semimetals*, Vol. 76 (eds C.E. Nebel and J. Ristein), Elsevier, p. 261.

43 Nesladek, M., Haenen, K. and Vanecek, M. (2003) *"Thin-Film Diamond I", Semiconductors and Semimetals*, Vol. 76 (eds C.E. Nebel and J. Ristein), Elsevier, p. 325.

44 Shin, D., Watanabe, H. and Nebel, C.E. (2006) *Diamond and Related Materials*, **15**, 121.

45 Shin, D., Watanabe, H. and Nebel, C.E. (2005) *physica status solidi (a)*, **202**, 2104.

46 Shin, D., Watanabe, H. and Nebel, C.E. (2005) *Journal of American Chemical Society*, **127**, 11236.

47 Nebel, C.E., Rezek, B., Shin, D., Watanabe, H. and Yamamoto, T. (2006) *Journal of Applied Physics* **99 (3)**, 033711.

48 Strobel, P., Riedel, M., Ristein, J. and Ley, L. (2004) *Nature*, **430**, 439.

49 Rezek, B. (2005) *New Diamond and Frontier Carbon Technology*, **15**, 275.

50 Sze, S.M.(1981) *Physics of Semiconductor Devices*, John Wiley & Sons, p. 251.

51 Cui, J.B., Ristein, J. and Ley, L. (1998) *Physical Review Letters*, **81**, 429.

52 Garrido, J.A., Nebel, C.E., Stutzmann, M., Snidero, E. and Bergonzo, P. (2002) *Applied Physics Letters*, **81**, 637.

53 Petrosyan, S.G. and Shik, A.Ya. (1989) *Soviet Physics–Semiconductors*, **23** (6), 696.

54 Gelmont, B. and Shur, M. (1992) *IEEE Transaction on Electron Devices*, **39** (5), 1216.

55 Willatzen, M., Cardona, M. and Christensen, N.E. (1994) *Physical Review B*, **50**, 18054.

56 Stern, F. and DasSarma, S. (1984) *Physical Review B*, **30**, 840.

57 Ristein, J., Maier, F., Riedel, M., Cui, J.B. and Ley, L. (2000) *physica status solidi (a)*, **181**, 65.

58 Nebel, C.E., Ertl, F., Sauerer, C., Stutzmann, M., Graeff, C.F.O., Bergonzo, P., Williams, O.A. and Jackman, R.B. (2002) *Diamond and Related Materials*, **11**, 351.

59 Sauerer, C., Ertl, F., Nebel, C.E., Stutzmann, M., Bergonzo, P., Williams, O. A. and Jackman, R.A. (2001) *physica status solidi (a)*, **186**, 241.

60 Watanabe, H., Takeuchi, D., Yamanaka, S., Okushi, H., Kajimura, K. and Sekiguchi, T. (1999) *Diamond and Related Materials*, **8**, 1272.

61 Okushi, H. (2001) *Diamond and Related Materials*, **10**, 281.

62 Okushi, H., Watanabe, H., Ri, S., Yamanaka, S. and Takeuchi, D. (2002) *Journal of Crystal Growth*, **237–239**, 1269.

63 Watanabe, H.R., Ri, S.-G., Yamanaka, S., Takeuchi, D. and Okushi, H. (2002) *New Diamond and Frontier Carbon Technology*, **12**, 369.

64 Seeger, K. (1999) *Semiconductor Physics*, Springer, Berlin.

65 Ando, T., Fowler, A.B. and Stern, F. (1982) *Review of Modern Physics*, **54**, 437.

66 Granger, M.C., Strojek, J.X.u, J.W. and Swain, G.M. (1999) *Analytica Chimica Acta*, **397**, 145.

67 Granger, M.C., Witek, M., Wang, J., Xu, J., Hupert, M., Hanks, A., Koppang, M.D., Butler, J.E., Lucazeau, G., Mermoux, M., Strojek, J.W. and Swain, G.M. (2000) *Analysis Chemistry*, **72**, 3739.

68 Swainin, G.M. (2004) *"Thin-Film Diamond II", Semiconductors and Semimetals*, Vol. 77 (eds C.E. Nebel and J. Ristein), Elsevier, p. 121.

69 Fujishima, A., Einaga, Y., Rao, T.N. and Tryk, D.A. (2005) *Diamond Electrochemistry*, Elsevier/BKC-Tokyo.

70 Gerischer, H. (1961) *Advanced Electrochemistry and Electrochemical Engineering*, **1**, 139.

71 Madou, M.J. and Morrison, S.R. (1989) *Chemical Sensing with Solid-State Devices*, Academic Press, Boston.

72 Ristein, J., Riedel, M. and Ley, L. (2004) *Journal of the Electrochemical Society*, **151**, E315.

6
Biosensors from Diamond

Christoph E. Nebel,[1] *Bohuslav Rezek,*[2] *Dongchan Shin,*[1] *Hiroshi Uetsuka*[1] *and Nianjun Yang*[1]
[1]Diamond Research Center, AIST, Central 2, Tsukuba 305-8568, Japan
[2]Institute of Physics, Academy of Sciences of the Czech Republic, Cukrovarnicka 10, CZ-162 53 Praha 6, Czech Republic

6.1
Introduction

Genomics research has elucidated many new biomarkers that have the potential to greatly improve disease diagnostics [1–3]. The availability of multiple biomarkers is important in diagnosis of complex diseases like cancer [4–6]. In addition, different markers will be required to identify different stages of disease pathogenesis to facilitate early detection. The use of multiple markers in healthcare will, however, ultimately depend upon the development of detection techniques that will allow rapid detection of many markers with high selectivity and sensitivity. Currently, extensive quests for proper transducer materials, for optimization of detection techniques and sensitivities, for realization of highly integrated sensor arrays, and for biointerfaces that show high chemical stability and that are required in high throughput systems, are ongoing. Most of the established substrate materials ("transducers") like latex beads, polystyrene, carbon electrodes, gold, and oxidized silicon or glass do not possess all of the desired properties like flatness, homogeneity, chemical stability, reproducibility and biochemical surface modifications [7–11]. In addition, future technologies will require integration of biofunctionalized surfaces with microelectronics or micromechanical tools, adding significant complexity to this topic [11–15], as most of microelectronic compatible materials, like silicon, SiO_x, and gold, show degradation of their biointerfaces in electrolyte solutions [15].

Diamond could become a promising candidate for bioelectronics as it shows good electronic [16–18] and chemical properties [19–21]. Figure 6.1 shows voltammograms for water electrolysis of various electrodes. The supporting electrolyte is 0.5 M H_2SO_4. Please note that each current/voltage scan has been shifted vertically for better comparison. Two polycrystalline films, B:PCD(NRL) with $5 \times 10^{19}\,B/cm^3$ and B:PCD(USU) with $5 \times 10^{20}\,B/cm^3$ (from [22, 23]) are compared with a single crystalline boron-doped diamond B:(H)SCD with $3 \times 10^{20}\,B/cm^3$ and with an

Physics and Applications of CVD Diamond. Satoshi Koizumi, Christoph Nebel, and Milos Nesladek

ISBN: 978-3-527-40801-6

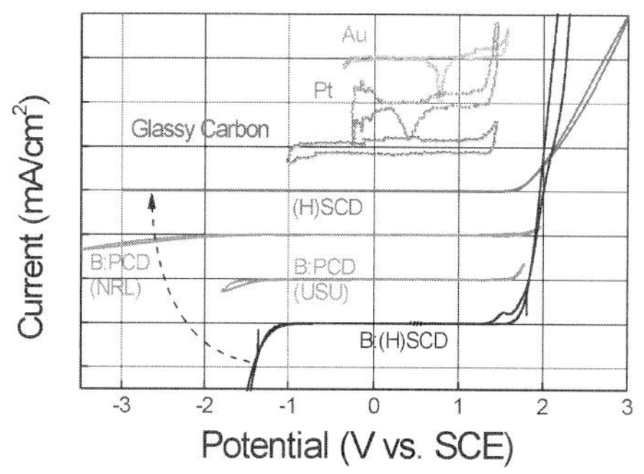

Figure 6.1 Voltammograms for water electrolysis on various electrodes. The supporting electrolyte is 0.5 M H_2SO_4. The graphs are shifted vertically for comparison. Two polycrystalline films, B:PCD(NRL) with 5×10^{19} B/cm^3 and B:PCD(USU) with 5×10^{20} B/cm^3 (from [22] and [23]), are compared with a single crystalline boron doped diamond B:(H)SCD with 3×10^{20} B/cm^3 and with an undoped diamond (H)SCD. Also shown are data for Pt, Au and glassy carbon from [19]. Oxidation reactions, for example, oxygen evolution, have positive currents and emerge around 1.8 V for all diamond samples. Reduction reactions, for example, hydrogen evolution, have negative currents and show very different properties. Note, the background current within the regime between hydrogen and oxygen evolution for diamond is very low, and the electrochemical potential window large, compared to glassy carbon, Pt and Au.

undoped diamond (H)SCD [24]. The electrochemical potential window of diamond is significantly larger and the background current within this regime considerably lower than conventional materials. In addition, by tuning the boron doping level, the onset of hydrogen evolution (rise of current at negative potentials) can be reduced or switched off completely by decreasing the boron doping level from extremely high with $>10^{20}$ cm^{-3} boron ("metallic") to "undoped" (= intrinsic diamond). There are some other parameters affecting the electrochemical potential window like crystal orientation [25], structural perfection of polycrystalline diamond [26], and surface termination [19]. Their discussion in this context is, however, beyond the scope of this chapter.

The surface induced conductivity of hydrogen-terminated undoped diamond in electrolyte solutions is another unique property which has attracted significant attention in recent years [27]. It is generated by transfer doping of hydrogen terminated diamond, immersed into electrolyte solution. The phrase "transfer doping" indicates that the surface conductivity in diamond arises from missing valence band electrons, as such electrons "transfer" into the electrolyte [28–30]. For such transitions, the chemical potential of an electrolyte must be below the energy level of the valence band maximum. For most semiconductors, this is not the case, as can be seen in Figure 6.2. Even for oxidized diamond, chemical potentials are mostly deep in the band gap of diamond. This changes drastically if the surface of diamond, which consists of about 2×10^{15} cm^{-2} carbon bonds, is terminated with hydrogen. Hydrogen–carbon bonds are polar covalent bonds (electronegativity of carbon: 2.5 and of hydrogen: 2.1), therefore, a dense surface dipole

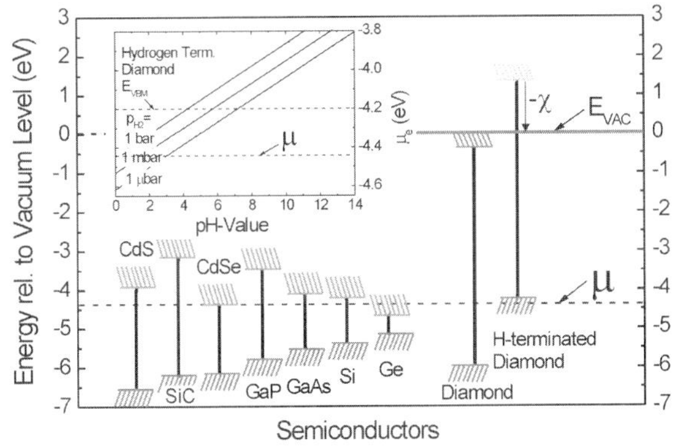

Figure 6.2 Energies of the conduction and valence band edges of a number of conventional semiconductors, and of hydrogen terminated and hydrogen free diamond relative to the vacuum level E_{VAC} are shown. Please note, H-terminated diamond shows a negative electron affinity ($-\chi$) as the conduction band edge is above the vacuum level of the electrolyte. The dashed horizontal line marks the chemical potential μ for electrons in an acidic electrolyte under conditions of the standard hydrogen electrode. The insert shows the chemical potential under general non-standard conditions as a function of pH and for different partial pressure of hydrogen in the atmosphere as given by Nernst's equation [30].

layer is generated with slightly negative charged carbon (C^-) and slightly positive charged hydrogen (H^+). From basic electrostatics, such a dipole layer causes an electrostatic potential step ΔV perpendicular to the surface over a distance of the order of the C—H bond length of 1.1 Å. Simple calculations show that the energy variation over this dipole is in the range of 1.6 eV (for a detailed discussion see [31]). This dipole energy shifts all energy levels of diamond about 1.6 eV up, with respect to the chemical potential of an electrolyte (see Figure 6.2). Conduction band states of the diamond are shifted above the vacuum level of the electrolyte. This scenario is called "negative electron affinity" (see Figure 6.2: clean diamond, where the vacuum level is about 0.3 eV above the conduction band minimum and H-terminated diamond, where the vacuum level is 1.3 eV below the conduction band minimum) [30, 32].

As all electronic states are shifted for the same dipole energy, occupied valence band states emerge above the chemical potential μ of electrolytes. Electrons from the diamond valence band (electronically occupied states) can, therefore, tunnel into empty electronic states of the electrolyte, until thermodynamic equilibrium between the Fermi level of the diamond and the electrochemical potential of the electrolyte is established. This is schematically shown in Figure 6.3a. Fermi level and chemical potential, μ, align and form a narrow valence band bending of 20 to 30 Å in width, which is, in effect, a confined hole accumulation layer [33, 34]. Such alignment requires defect free bulk and surface properties, as well as a perfect H-termination. During recent years, the growth of diamond has been optimized to such a level in combination with a perfect H-termination of the surface (for reviews see [35, 36]).

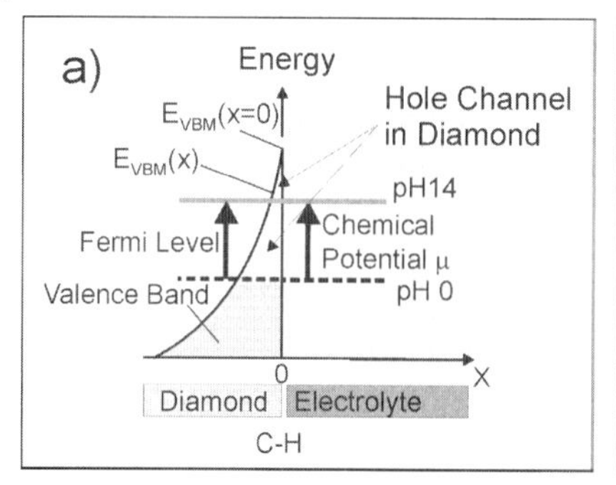

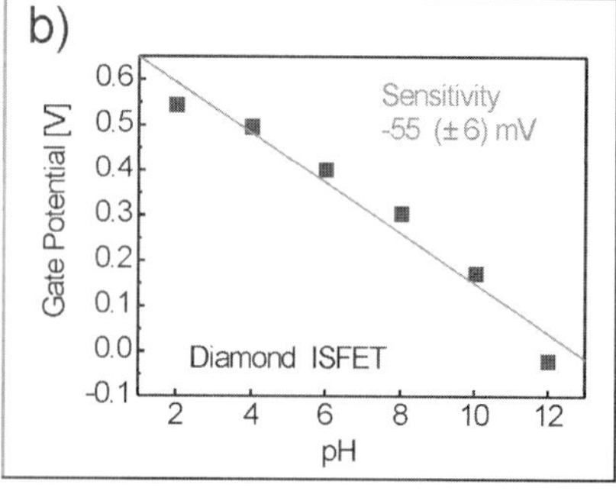

Figure 6.3 (a) Fermi level and chemical potential alignment at the interface diamond/electrolyte after equilibration. Due to transfer doping, electrons are missing in diamond, so that a thin, two dimensional (2D) hole accumulation layer is generated [33, 34]. The hole density in this layer depends on the chemical potential as indicated by arrows. (b) pH-sensitivity of a diamond ion sensitive field effect transistor (ISFET) [35, 36, 40–42]. The gate potential shift shows a pH dependence of 55 mV/pH, which is close to the Nernst prediction.

Evidence from theory and experimentation suggests that the electron affinity of diamond in contact with water is approximately 1 eV more positive than that observed in high vacuum [37–39]. A dominant interaction of diamond energy levels with the H_2/H^+ redox states seems to be, therefore, less likely. But, diamond valence band states will still scale with interactions to the O_2/H_2O couple, giving rise to the discovered phenomena. As the chemical potential of electrolytes is changing with pH-value, a variation of the surface conductivity can be detected experimentally, following closely the Nernst prediction with 55 mV/pH (see Figure 6.3b) [40–42].

Diamond is known to be biocompatible [43–45] and has, therefore, a potential for *in-vivo* electronic applications. When Takahashi *et al.* in 2000 [46, 47] first introduced a photochemical chlorination/amination/carboxylation process of the initially H-terminated diamond surface, a giant step towards the biofunctionalization of diamond was taken, as the obstacle of the chemical inertness of diamond had finally been removed. In 2002, Yang and coworkers introduced a new photochemical method to modify nanocrystalline diamond surfaces using alkenes [15], followed by the electrochemical reduction of diazonium salts which has been successfully applied to functionalize boron doped ultra-nanocrystalline diamond [48] and recently, a direct amination of diamond has been introduced [49]. Such functionalized surfaces have been further modified with DNA, enzymes and proteins, and characterized using fluorescence microscopy and impedance spectroscopy [15, 50, 51] voltammetry and gate potential shifts of ion sensitive field effect transistors [52, 53].

Perhaps the most influential argument for diamond applications in biotechnology has been given by Yang *et al.* in 2002 [15]. They characterized the bonding stability of DNA to nanocrystalline diamond and other substrates in hybridization/denaturation cycles using fluorescence microscopy investigations. The result

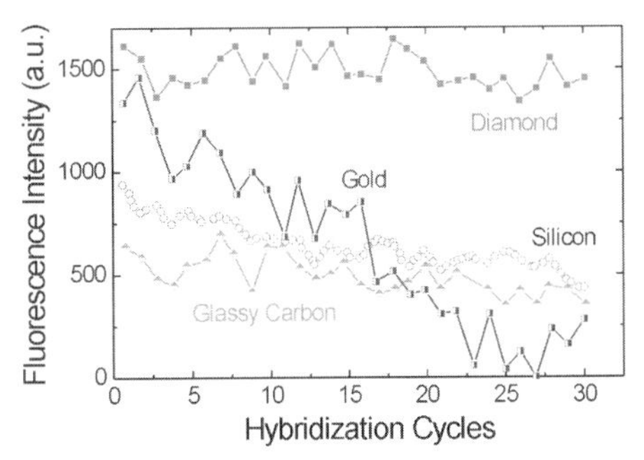

Figure 6.4 Stability of DNA bonding to ultrananocrystalline diamond, Au, Si and glassy carbon as detected during 30 successive cycles of hybridization and denaturation. In each case, the substrates were amine modified and then linked to thiol terminated DNA [15].

is shown in Figure 6.4 in comparison to Au, Si and glassy carbon. It demonstrates that DNA bonding to diamond is significantly better than to other substrates, as no degradation of fluorescence intensity could be detected. This long term bonding stability is especially important in multi-array sensor applications, which are costly to produce and which, therefore, need long term stability in high throughput systems.

Applications of diamond sensors will ultimately depend on the commercial availability of diamond films. This has improved significantly during recent years, since nano- and polycrystalline CVD diamond films can be grown by plasma enhanced chemical vapor deposition (CVD) heteroepitaxially on silicon and other substrates of large area. Growth parameters are currently optimized to deposit films at low temperature to allow integration into established silicon technology [54, 55] (M. Hasegawa, 2006, private communications). Single crystalline diamond produced by high temperature high pressure growth is commercially available, due to the increasing number of companies producing diamond. The size of these layers is relative small, typically 4 mm × 4 mm which is, however, large enough to be used as a substrate for homoepitaxial growth of high quality single crystalline CVD diamond ("electronic grade quality").

With respect to electronic applications, a careful selection of diamond material is required. Figure 6.5 summarizes the structural properties of nano-, poly- and single crystalline diamond. Ultranano-, nano- and polycrystalline diamond layers are dominated by grain boundaries which are decorated with sp^2 and amorphous carbon [56–58]. The volume fraction of sp^2 and grain boundaries depends on growth parameter and varies from layer to layer. In particular, ultra-nanocrystalline diamond contains a high volume fraction of up to 5% [55]. Amorphous carbon and sp^2 generate a continuous electronic density of states distribution in the gap of diamond. These states will affect sensor sensitivity and dynamic properties. Therefore, applications of polycrystalline diamond as a photo- or high energy particle detector show memory and priming effects which arise by metastable filling

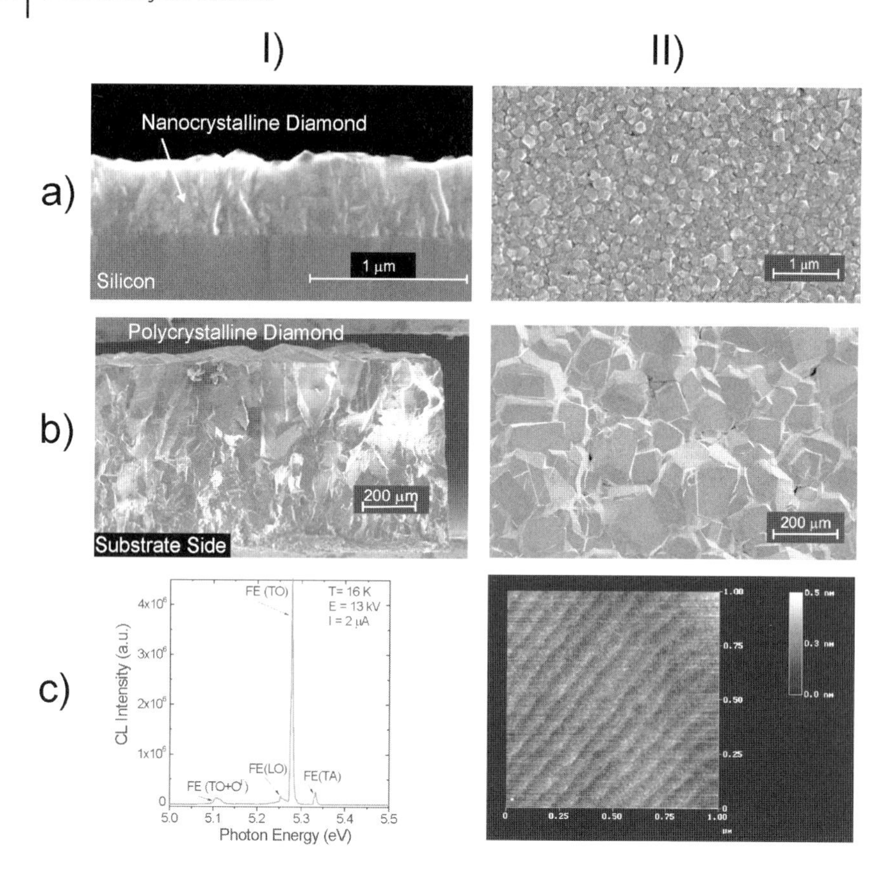

Figure 6.5 Comparison of different diamond films: (a) Ultranano-, nano- and (b) polycrystalline diamond layers are dominated by grain boundaries which are decorated with sp^2 and amorphous carbon [54, 56–58]. The volume-fraction of sp^2 and grain boundaries depends on growth parameter and varies from layer to layer. Amorphous carbon and sp^2 generate a continuous electronic density of states distribution in the gap of diamond. In addition, such diamond films show a significant surface roughness in the range of 30 to 50 nm for nanocrystalline diamond (see a, II) and micrometer to tens of micrometer for polycrystalline layers (b, II). On the other hand, over recent years, single crystalline CVD diamond has been optimized to electronic grade quality with atomically smooth surfaces (c, II) [63–65]. A typical cathodoluminescence spectrum measured at 16 K is shown in c, I.

of grain boundary states [59, 60]. In addition, such diamond films show a significant surface roughness in the range of 30 to 50 nm for nanocrystalline diamond, and micrometer to tens of micrometer for polycrystalline layers. Therefore, commercially available polycrystalline diamond is often mechanically polished, to achieve a smooth surface. However, this generates a thin, highly damaged diamond surface which cannot be tolerated in surface related electronic applications as surface defects, about 0.9 to 1.1 eV above the valence band maximum, will pin the Fermi level and will deteriorate heterojunction properties [61, 62].

On the other hand, single crystalline CVD diamond has been optimized over recent years to electronic grade quality with atomically smooth surfaces (see Figure 6.5c) [63–65]. Even at room temperature, these films show strong free exciton emissions at 5.27 eV and 5.12 eV, which are evidence of low defect densities, typically below 10^{15} cm^{-3}. The bulk resistivity of undoped films at 300 K is larger than 10^{15} Ωcm [56, 57]. Atomic force microscopy (AFM) characterization of such films shows surface morphologies which indicate atomically flat properties with step etch growth, where terraces run parallel to the (110) direction. After H-termination of such layers, heterojunction properties follow very well predicted properties of defect free diamond.

Diamond can be p-type doped by boron, which results in a doping level 360 meV above the valence band maximum [66]. Phosphorus doping has been introduced for n-type doping with the phosphorus doping level 0.6 eV below the conduction band minimum [67]. Both levels are basically too deep for room temperature electronic applications, which is the typical regime for bioelectronics. One way to overcome this problem is by the application of metallic doping where, in the case of boron, typically 10^{20} B/cm^3 or more atoms are incorporated into diamond [68]. This causes enough wave function overlap of holes in acceptor states to allow propagation in such states, without thermal activation to the valence band. Highly boron-doped diamond is, therefore, well established in electrochemistry. Applications of n-type diamond in electro- or biochemical sensors seem to be unfavorable, as the Fermi level (0.6 eV below the conduction band) and chemical potential of electrolytes (typically 4.5 eV below the vacuum level (see Figure 6.2) are too different, giving rise to energy barrier limited electronic interactions.

This brief introduction of the major properties of diamond show that it is, indeed, an interesting transducer material for biosensor applications. In the following, we will review our achievements with respect to interface properties of single crystalline CVD diamond to organic linker molecule layers and DNA films. We describe surface functionalization using amine and phenyl layers, which is currently attracting significant attention. There are other photochemical modifications of diamond available (for example see [46, 47] or [49]), which are based on direct or indirect surface amination. It is very likely that, sooner or later, the modification spectrum will become even broader. In the following, however, we want to focus on (i) amine and (ii) phenyl related modifications, as these techniques are established, are used by a growing number of scientists, and are characterized reasonably well.

For our experiments we used homoepitaxially grown, atomically smooth CVD diamonds, either undoped or metallically boron doped, which were free of grain boundaries, sp^2-carbon or other defects. We applied (i) photochemical attachment chemistry of alkene molecules to undoped diamond [69–71] and (ii) electrochemical reduction of diazonium salts [69, 72, 73], to form nitrophenyl linker molecules on boron-doped CVD diamond. The bonding mechanisms, kinetics, molecule arrangements and densities were introduced using a variety of experiments like X-ray photoelectron spectroscopy (XPS), scanning electron microscopy (SEM), atomic force microscopy (AFM), cyclic voltammetry, and several electronic characterization techniques. By use of a hetero-bifunctional crosslinker, thiol modified single stranded probe DNA (ssDNA) was bonded to diamond. Finally, such surfaces were exposed to fluorescence labeled target ssDNA, to investigate hybridization by use of fluorescence microscopy. We applied AFM in electrolyte solution, to gain information about geometrical properties of DNA, bonding strength, as well as the degree of surface coverage. Finally, in this chapter, we will introduce our first results with respect to label free electronic sensing of DNA hybridization using $Fe(CN_6)^{3-/4-}$ redox molecules as mediator in amperometric experiments, and variation of gate potential threshold shifts in DNA-FET structures. For a detailed summary of these results, see [69].

6.2
Materials and Methods

6.2.1
CVD Diamond Growth, Surface Modifications and Contact Deposition

High quality undoped, single crystalline diamond films of 200 nm thickness have been grown homoepitaxially on 3 mm × 3 mm (100) oriented synthetic Ib substrates, using microwave plasma chemical vapor deposition (CVD). Growth parameters were: substrate temperature 800 °C, microwave power 750 W, total gas pressure 25 Torr, total gas flow 400 sccm with 0.025% CH_4 in H_2. Note, the used substrates have been grown commercially by high pressure high temperature (HPHT) techniques and contain typically up to $10^{19}\,cm^{-3}$ dispersed nitrogen (type Ib diamond). To achieve H-termination after growth, CH_4 is switch off and diamond is exposed to a pure hydrogen plasma for 5 minutes, with otherwise identical parameters. After switching off the hydrogen plasma, the diamond layer is cooled down to room temperature in H_2 atmosphere. A detailed discussion of sample growth properties can be found in [63–65]. Layers are highly insulating with resistivities larger than $10^{15}\,\Omega cm$. Surfaces are smooth, as characterized by atomic force microscopy (AFM). A typical result is shown in Figure 6.6. The root mean square (RMS) surface roughness is below 1 Å.

Boron doped single crystalline diamond films have been grown homoepitaxially on synthetic (100) Ib diamond substrates with 4 mm × 4 mm × 0.4 mm size, using microwave plasma assisted chemical vapor deposition (CVD). Growth parameters are: microwave power 1200 W which generate a substrate temperature around 900 °C, gas pressure 50 Torr, gas flow 400 sccm with 0.6% CH_4 in H_2. B_2H_6, as boron source, is mixed in CH_4, where the boron/carbon atomic ratio (B/C) was 16 000 ppm. Typically 1 µm thick films have been grown within 7 hours. H-termination has been achieved in the same way as described above. To measure bulk properties, boron-doped diamond is wet-chemically oxidized by boiling in a mixture of H_2SO_4 and HNO_3 (3 : 1) at 230 °C for 60 minutes. Figure 6.7 shows a typical result of conductivity, σ, which is in the range of 200 $(\Omega cm)^{-1}$ at 300 K, showing a negligible activation energy of 2 meV ("metallic properties"). It is

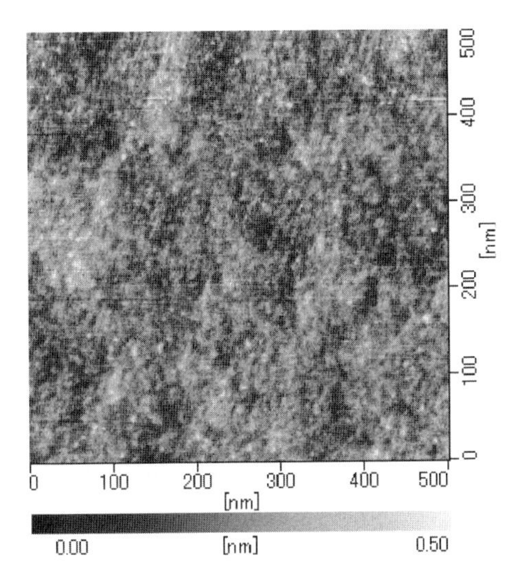

Figure 6.6 Typical AFM surface morphology of a single crystalline CVD diamond surface, as detected by AFM and used in these studies. The root mean square (RMS) roughness below 1 Å.

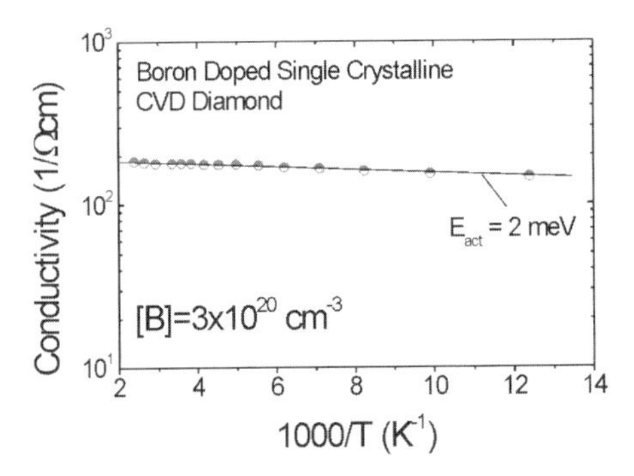

Figure 6.7 Typical temperature dependent conductivity of a metallically boron-doped CVD diamond. The doping level is in the range of $5 \times 10^{20}\,B/cm^3$. σ is activated with 2 meV which indicates hopping propagation of holes in the acceptor band.

achieved by ultra-high doping of diamond with $3 \times 10^{20}\,cm^{-3}$ boron acceptors as detected by secondary ion mass spectroscopy (SIMS). The crystal quality is not deteriorated by this high boron incorporation. A series of X-ray diffraction (XRD) and Raman experiments have been applied to investigate details of crystal quality which will be discussed elsewhere.

To obtain patterns of H- and O-termination on diamond surfaces, we apply photolithography, using photoresist as a mask to protect H-terminated areas while uncovered surface parts are exposed to a 13.56 MHz RF oxygen plasma. Plasma

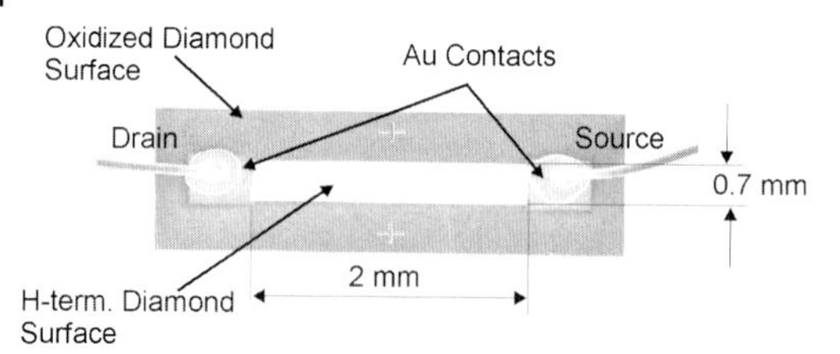

Figure 6.8 Typical geometry and arrangement of realized CVD diamond DNA ions sensitive field effect transistor (FET) structures (DNA-ISFET). The image has been generated by scanning electron microscopy (SEM), where the sensor area ("gate") is H-terminated and surrounded by oxidized diamond. Drain and source contacts are from Au. The H-terminated area is photochemically modified to bond ssDNA marker molecules covalently to diamond.

parameters are: oxygen (O_2) gas pressure 20 Torr, plasma power 300 W, and duration 2.5 minutes. Wetting angle experiments of H-terminated surfaces show angles >94° indicating strong hydrophobic properties. After plasma oxidation the wetting angle approaches 0°, as the surface becomes hydrophilic.

For electronic characterization or realization of DNA-field effect transistors (DNA-FET) we deposited Ohmic contacts on H-terminated diamond by thermal evaporation of 200 nm thick Au onto photoresist patterned diamond, followed by a lift off process. In the case of highly boron-doped diamond Ti (100 Å)/ Pt (100 Å)/ Au (2000 Å), contacts have been realized using e-beam evaporation. Figure 6.8 shows a DNA-FET from diamond as measured by scanning electron microscopy (SEM). The sensor area of size 2 mm × 0.7 mm is originally H-terminated diamond which connects drain and source Au contacts. This area is surrounded by insulating diamond which has been oxidized. The H-terminated surface is chemically modified, as described below, to covalently bond DNA to it. For experiments in electrolyte solution, drain and source contacts are insulated by silicon rubber. We use Pt as gate electrode (not shown) in buffer solutions.

Electrochemical experiments on boron-doped diamond are performed on typical areas of 3 mm^2 size. Ohmic contacts to boron-doped diamond are evaporated outside of this area and sealed with silicon rubber.

6.2.2
Photochemical Surface Modification of Undoped Diamond

Undoped single crystalline diamond surfaces are modified by photochemical reactions with 10-amino-dec-1-ene molecules protected with trifluoroacetic acid group ("TFAAD") [70, 71]. Restricted Hartree–Fock calculation of theoretical geometric properties of TFAAD molecules were performed with the Gaussian 98 package with density functional theory (B3LYP/6-31G(d)) [74]. A typical result is shown in Figure 6.9. The length of the molecule is 11.23 Å, and the diameter 5.01 Å. It is interesting to note that the protecting cap molecule shows a tilted arrangement. In the case of an upright arrangement of molecules on diamond, one therefore expects a monolayer thickness of around 11 to 15 Å.

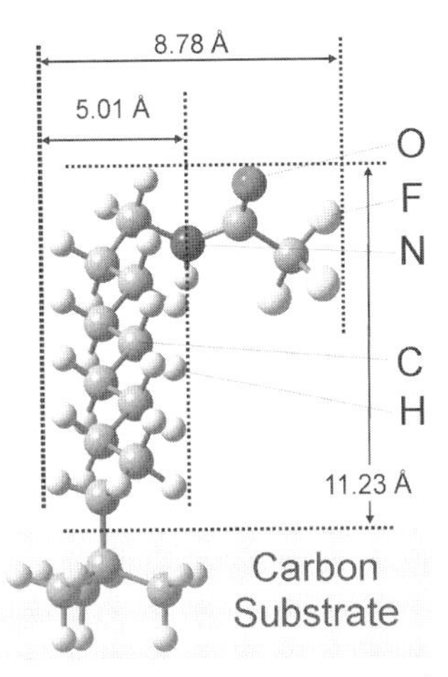

Figure 6.9 10-amino-dec-1-ene molecule protected with trifluoroacetic acid group ("TFAAD"), as determined by Molecular Orbital Calculations [74].

The chemically reactive end of TFAAD is terminated with an olefin (C=C), the other is protected from reactions using a trifluoroacetic cap. Chemical attachment is accomplished by placing four microliters of TFAAD on the diamond substrate. Then, the TFAAD is homogeneously distributed by spin-coating with 4000 rounds/min in air for 20 seconds which forms a 5 micrometer thick liquid TFAAD layer. After accomplishment of spin coating, samples are sealed into a chamber with a quartz window in a nitrogen atmosphere. Then, UV illumination is switched on for a given period of time. The ultraviolet light is generated in a high pressure mercury lamp with emission at 250 nm of 10 mW/cm^2 intensity.

6.2.3 Electrochemical Surface Functionalization

Electrochemically induced covalent attachment of nitrophenyl molecules has been performed using an Electrochemical Analyzer 900 (CHI instruments), and a three electrode configuration, with a platinum counter electrode and an Ag/Ag$^+$ (0.01 M) reference electrode (BAS, Japan) [72, 73]. The active area of the boron-doped diamond working electrode is about 0.03 cm^2. Electrolyte solution for the reduction of 4-nitrobenzene diazonium tetrafluoroborate is 0.1 M tetrabutylammonium tetrafluoroborate (NBu_4BF_4) in dehydrated acetonitrile (Wako chemicals, H_2O: <50 ppm). The diazonium salts reduction is performed in a N_2-purged glove box. Nitrophenyl modified diamond surfaces are then sonicated with acetone and acetonitrile. XPS, AFM and voltammetric experiments have been applied to characterize the surface bonding properties, and to reduce the nitrophenyl groups to aminophenyl groups. The nitrophenyl groups grafted on single crystalline diamond substrate can be considered as covalently bonded free nitrobenzene to diamond as shown in Figure 6.10 (Molecular Orbital Calculations).

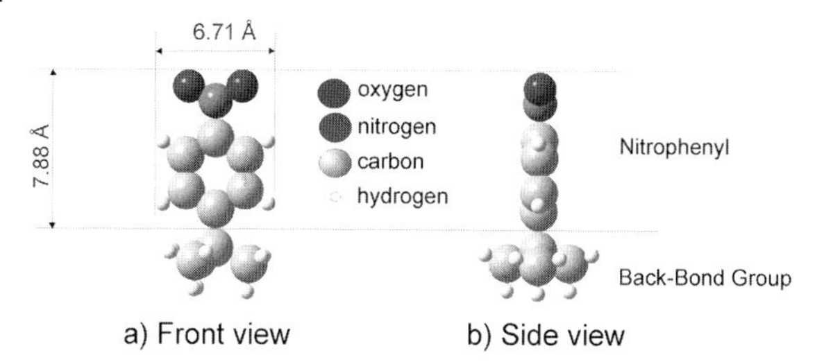

Figure 6.10 Front (a) and side view (b) of a nitrophenyl molecule, as calculated by Molecular Orbital Calculations [74].

6.2.4 HeteroBifunctional CrossLinking and DNA Attachment

To provide chemically reactive amine groups to the photochemically treated diamond samples, the trifluoroacetamide protecting group was removed by refluxing the TFAAD modified surface in 2:5 $MeOH/H_2O$ with 7% (w/w) K_2CO_3.

The electrochemically modified surface of boron-doped diamond with nitrophenyl groups ($-C_6H_5NO_2$) are electrochemically reduced to aminophenyl ($-C_6H_5NH_2$) in 0.1 M KCl solution of $EtOH-H_2O$ to provide reactive aminophenyl groups.

To attach DNA, the amine or the phenyl layer is then reacted with 14 nM solution of the heterobifunctional crosslinker sulphosuccinimidyl-4-(N-maleimidomethyl) cyclohexane-1-carboxylate in 0.1 M pH 7 triethanolamine (TEA) buffer for 20 minutes at room temperature in a humid chamber. The NHS-ester group in this molecule reacts specifically with the -NH_2 groups of the linker molecules to form amide bonds. The maleimide moiety was then reacted with (2–4) μl thiol modified DNA (300 μM thiol DNA in 0.1 M pH 7 TEA buffer) by placing the DNA directly onto the surface in a humid chamber and allowing it to react for given times between 10 minutes to 12 hours at room temperature. As probe ssDNA we used the sequence S1 (=5′-HS-C_6H_{12}-T_{15}-GCTTATCGAGCTTTCG-3′) and as target ssDNA the sequence F1 (=5′FAM-CGAAAGCTCGATAAGC-3′), where FAM indicates the presence of a fluorescence tag of fluorescein phosphoramidite. To investigate mismatched interactions, a four bases mismatched target ssDNA (5′-FAM-CGATTGCTCCTTAAGC-3′) has been used. For some fluorescence experiments the green label (FAM) has been replaced by red fluorescence markers (Cy5). All DNA molecules have been purchased from TOS Tsukuba OligoService (http://www.tos-bio.com, Japan). A schematic summary of chemical modification schemes is shown in Figure 6.11.

Denaturation of samples has been performed in 8.3 M urea solution for 30 minutes at 37 °C, followed by rinsing in deionized water. Samples are then hybridized again for another DNA cycle.

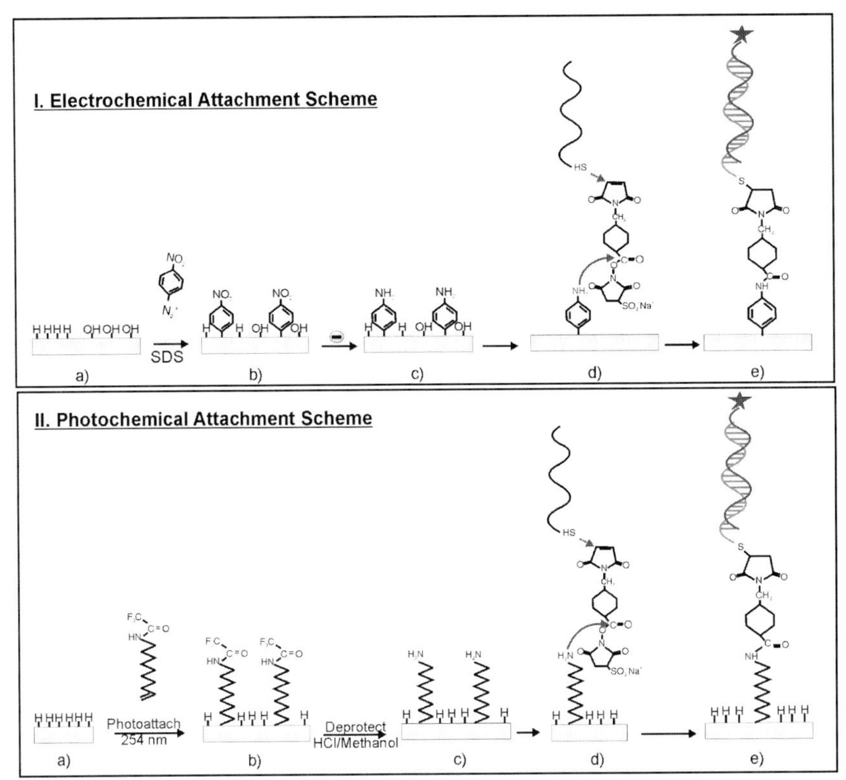

Figure 6.11 Electrochemical (I) and photochemical (II) boning schemes: Scheme I: (a) Nitrophenyl linker molecules are electrochemically bonded to H- or O-terminated diamond. Nitrophenyl is reduced to aminophenyl and reacted with a heterobifunctional crosslinker. Finally, thiol modified ssDNA is attached. Scheme II: Amine molecules are photochemically covalently attached on H-terminated diamond. The linker molecules are then deprotected and reacted with heterobifunctional crosslinker molecules and thiol modified ssDNA.

6.2.5
X-Ray Photoelectron Spectroscopy (XPS), Atomic Force Microscopy (AFM) and Fluorescence Microscopy (FM)

The chemical attachment is characterized using an X-ray photoelectron spectroscopy (XPS) system (Tetra Probe, Thermo VG Scientific) with a monochromatized AlKα source (1486.6 eV) at a base pressure of 10^{-10} Torr. Unless otherwise noted, electrons are collected ejected between 25° and 80° degree with respect to the surface normal. (atomic sensitivity factors: C, 0.296; F, 1; N, 0.477; O, 0.711). The mean free path of electrons is 36 Å for perpendicular excitation.

Microscopic morphology and structural properties of amine, nitrophenyl, and DNA layers have been characterized by atomic force microscopy (AFM) (Molecular

Imaging PicoPlus) [75–77]. For DNA characterization, layers were immersed into a SSPE buffer (300 mM NaCl, 20 mM NaH_2PO_4, 2 mM ethylenediaminetetraacetic acid (EDTA), 6.9 mM sodium dodecyl sulphate (SDS), titrated to pH 7.4 by 2 M NaOH). The buffer solution enables DNA to assume natural conformation and avoids effects of water meniscus around the AFM tip.

Surface morphologies are investigated in oscillating mode AFM (O-AFM), where the tip-surface interaction is controlled by adjusting the tip oscillating amplitude to a defined value (AFM set point ratio measurements) [75–77]. The set point ratio is defined as $r_{SP} = A_O/A_{SP}$, where A_O is the amplitude of free cantilever oscillations and A_{SP} the amplitude of the tip, approaching the surface. Measurements are made typically with A_O of 6 and 10 nm. In addition, we used cantilever phase shift detection (phase lag of cantilever oscillation with respect to oscillation of the excitation piezo element) to enhance the material contrast between diamond and DNA.

Molecule bonding properties (mechanical properties) of linker and DNA layers have been characterized by contact mode AFM where we applied different loading forces to the AFM tip (C-AFM) in the range 6 to 200 nN. The scan rate was (10–20) μm/s. For forces above a critical threshold, linker and DNA molecules are removed. The difference in height is then measured in O-AFM. Doped silicon AFM cantilevers are used in these experiments with a spring constant of 3.5 N/m. The cantilever resonance frequency is 75 kHz in air and 30 kHz in buffer solution.

Fluorescence microscopy has been applied using a Leica Fluorescence Imaging System DM6000B/FW4000TZ where the fluorescence intensity is evaluated by grayscale analysis (Leica QWin software). Please note, that we have characterized all diamond layers before surface modifications to detect fluorescence emission arising form the bulk of diamond, like from nitrogen/carbon vacancy complexes, for example. Those samples have been excluded from our experiments. The shown fluorescence is therefore truly from fluorescence labeled DNA.

6.2.6 DNA-Field Effect Transistors

To realize in-plane gate DNA-field effect transistors (DNA-FET), undoped CVD diamonds with atomically smooth surfaces have been grown by microwave plasma assisted chemical vapor deposition (see above) on Ib substrates. The layer thickness is typically 200 nm. After hydrogen termination of the diamond surface, a H-terminated sensor area of 2 mm × 0.7 mm size has been processed by photolithography and plasma oxidation. Two Au contacts (0.7 mm × 0.5 mm) evaporated to each end of the H-terminated surface serve as drain and source contacts (see Figure 6.8).

Alkene crosslinker molecules are then attached by photochemical means for (12–20) hours, followed by the attachment of probe ssDNA. The density of ssDNA has been varied between 10^{12} and $10^{13}\,cm^{-2}$ for these experiments. Samples are then transferred into polyetheretherketon (PEEK) sample holders as shown in Figure 6.12a. Please note that during the transfer, the ssDNA layer is covered by sodium chloride buffer solution (1 M with 0.1 M phosphate pH 7.2). The drain/

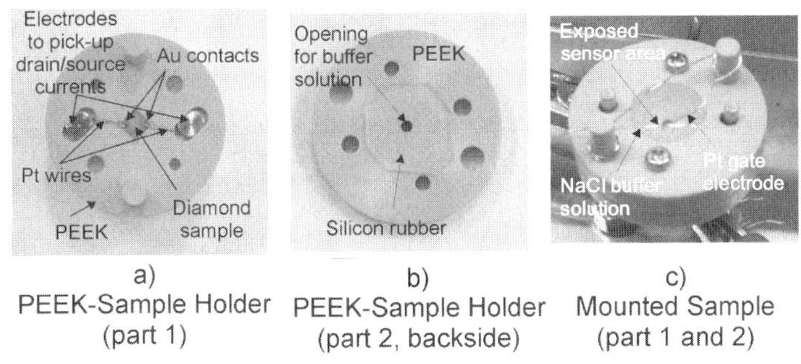

Figure 6.12 DNA-FET sample holder arrangement. The diamond sensor is mounted in a PEEK plate as shown in (a). The drain and source pads are contacted using Pt wires. A second PEEK part (b) with silicon rubber is mounted on to of part (a) and closed to seal the drain source pads from electrolyte buffer (c). A Pt wire is used as gate electrode and the exposed sensor area is 0.7 mm × 1 mm.

source contacts are sealed against contact with buffer solution by a silicon rubber (see Figure 6.12b) which is also used to press platinum wires to the drain and source Au contacts (Figure 6.12a). A top view of the setup is shown in Figure 6.12c. The PEEK top part has been designed in a way that 1 mm of the sensor area is exposed to buffer solution. To apply well defined gate potentials we use a thin Pt wire of 0.2 mm diameter which is immersed into the sodium-chloride buffer. Drain source currents are measured as function of gate potential that has been varied between 0 and –0.6 V. The gate threshold potential of DNA-FETs has been characterized applying several cycles where firstly properties of the FET with probe ssDNA have been determined, followed by determination of FET properties after hybridization with complementary target ssDNA. Then the sample has been denatured and characterized again.

6.3 Results

6.3.1 Photochemical Surface Modifications of Undoped Single Crystalline Diamond

The H-terminated samples are photochemically reacted with long chain ω-unsaturated amine, 10-aminodec-1-ene, that has been protected with the trifluoroacetamide functional group [69–71]. We refer to this protected amine as TFAAD. A variety of attachment experiments on H-terminated and oxidize diamond surfaces show that this process only works on H-terminated diamond.

To characterize the molecule arrangement and layer formation of TFAAD on hydrogen terminated CVD diamond, we first applied AFM scratching experiments with the aim of (i) identifying the threshold force for mechanical removal of

TFAAD molecules, and (ii) of using the step edge of scratched areas to measure the thickness of the TFAAD layers. Please note that diamond is ultra-hard (100 GPa hardness). Therefore, the AFM cantilever will not penetrate diamond even at the highest applied loading forces. Such contact mode AFM experiments carve out rectangular trenches in the amine layer. AFM tapping mode line scans are then used to measure the height profile across scratched areas.

Forces below 100 nN remove only a fraction of TFAAD molecules, while forces equal or larger than 100 nN give rise to complete removal (for details see [77]). A typical result of AFM characterization applied to an amine layer that has been photochemically attached for 1.5 hours is shown in Figure 6.13. The TFAAD molecules were removed by 200 nN applied loading force from an area of $2 \times 2\,\mu m^2$ to truly remove the amine layer. The upper image shows topography, and the lower image a typical line profile. Within the cleaned area, the surface roughness of the diamond is in the range 1.2 to 2.3 Å. On the amine film, we detect a roughness between 3.8 to 9.3 Å. The TFAAD layer thickness is determined to be in the range 5 to 13 Å, which indicates dispersed layer properties as for a closely packed, dense layer. The height is expected to be around 11 Å. The variation of film properties as a function of UV-photochemical attachment times is shown in Figure 6.14. For times shorter than 4 hours, the average layer thickness increased from zero (0 hour) to 9.25 (1.5 hours) and 13.3 Å (4 hours), which is about the length of upright standing TFAAD molecules on diamond. Illumination times between 4 and 10 hours do not result in a remarkable enlargement of the layer thickness, which saturates around 14 Å, but the average roughness decreases. These films are

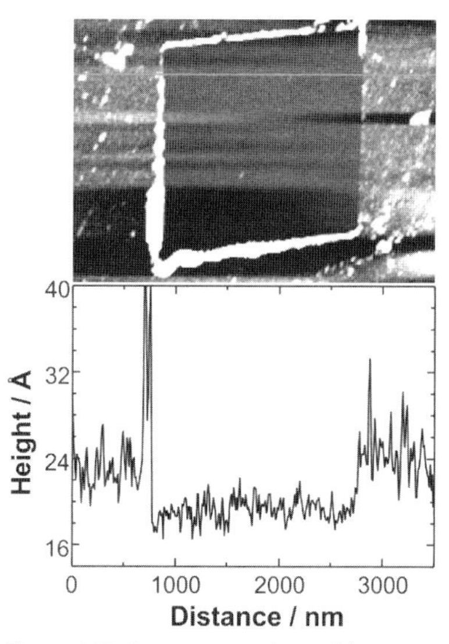

Figure 6.13 AFM topography and line scan of a TFAAD film attached by 1.5 hours of illumination to diamond. The scratched area is $2\,\mu m \times 2\,\mu m$ and the loading force was 200 nN.

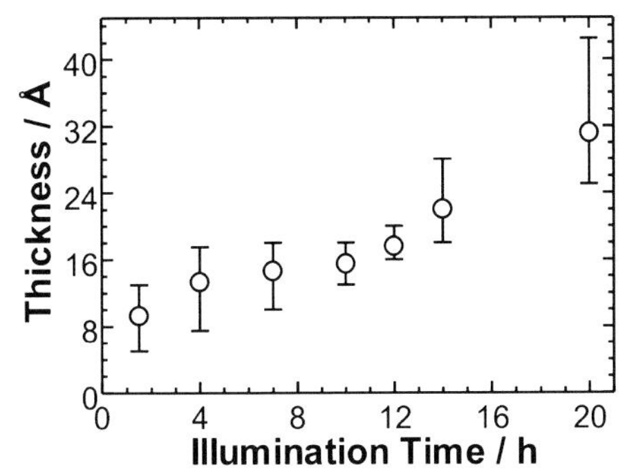

Figure 6.14 Thickness of TFAAD films, attached photochemically to H-terminated single crystalline diamond, as a function of illumination time. Each averaged thickness was measured at four different scratched areas. Please note that the indicated bars represent the width of height variations.

closed, with no structural defects like pinholes, for example. Illumination longer than 10 hours then gives rise to further enlargement, the film becomes thicker, for example 31.2 Å after 20 hours, which is about three times the length of TFAAD molecules. Notice, the bars in Figure 6.14 indicate an average roughness of films. After short term attachment (<10 hours), the roughness is large (about ±4.0 Å), while layers attached for 10 to 12 hours are relative smooth. Longer attachment times, again, give rise to enlarged roughness (±8.0 Å after 20 hours). We attribute this to: (i) 2D formation of a dispersed submolecular layer for times <10 hours, (ii) 2D formation of a dense monomolecular layer for times between 10 and 12 hours and (iii) cross polymerization and 3D growth for times longer then 12 hours.

To investigate the chemical bonding of TFAAD molecules to diamond X-ray photoelectron spectroscopic measurements (XPS) have been applied. Figure 6.15a shows XPS survey spectra of a clean hydrogen-terminated single crystalline diamond surface before and after exposed to TFAAD and 10 mW/cm^2 UV illumination intensity (254 nm) for 2 hours. Before XPS measurements, samples were rinsed in chloroform and methanol (each 5 minutes in ultrasonic). The overall spectrum shows a strong fluorine peak with a binding energy of 689 eV, an O(1s) peak at 531 eV, a N(1s) peak at 400 eV and a large C(1s) bulk peak at 284.5 eV. Please note that on clean H-terminated diamond, no oxygen peak can be detected. The C(1s) spectrum reveals two additional small peaks at 292.9 eV and 288.5 eV (see Figure 6.15b) which are attributed to carbon atoms in the CF_3 cap group and in the C=O group, respectively. From these experiments, we conclude that UV light of about 250 nm (5 eV) initiates the attachment of TFAAD to H-terminated single crystalline diamond. The ratio of the F(1s) XPS signal (peak area) to that of the total C(1s) signal (R_{FC}) as a function illumination time is shown in Figure 6.15c. The time dependence of R_{FC} follows approximately an exponential law: R_{FC}

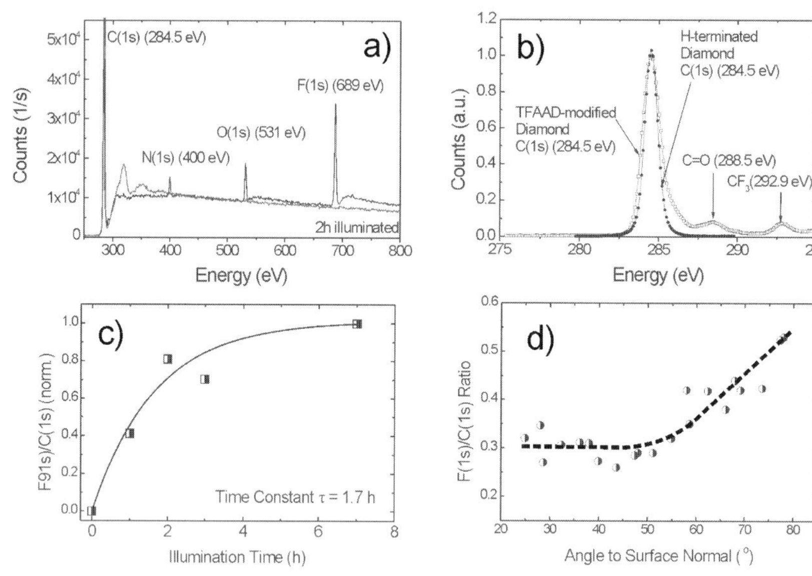

Figure 6.15 (a) XPS survey spectrum of a hydrogen-terminated single crystalline diamond surface that was exposed to TFAAD and 20 mW/cm^2 UV illumination (250 nm) for 2 hours. (b) The C(1s) spectrum reveals two additional small peaks at 292.9 eV and 288.5 eV which are attributed to carbon atoms in the CF_3 cap group and in the C=O group, respectively. (c) The ratio of the F(1s) signal (peak area) to that of the total C(1s) signal as a function illumination time is time dependent, follows an exponential increase (dashed line), with a characteristic time constant τ of 1.7 hours. (d) Angle resolved (with respect to the surface normal) XPS experiments show an increase of the F(1s)/C(1s) peak intensities, rising from 48° to 78°.

$= A\{1 - \exp(-t/\tau)\}$, with a characteristic time constant τ of 1.7 hours. Saturation of the area ratio F(1s)/C(1s) is achieved after about 7 hours.

Angle resolved XPS experiments shown in Figure 6.15d are used to calculate the density of bonded TFFAD molecules to diamond in absolute units. As parameters we used the following data: density of carbon atoms = 1.77×10^{23} atoms/cm^3, atomic sensitivity factors for C (0.296), and for F (1), and a mean free path for perpendicular illumination = 36.7 Å. Taking into account the area ration F(1s)/C(1s) results in about 2×10^{15} cm^{-2} TFAAD molecules bonded after 7 hours. This corresponds to the formation of a monolayer TFAAD as the surface density of carbon bonds on diamond is 1.5×10^{15} cm^{-2}. However, the TFFAD layer itself consists of 12 carbon atoms which contribute to the signal, so that the real coverage is smaller than this number (see discussion below).

Figure 6.16 shows a combination of AFM thickness data (Figure 6.14) and XPS results (Figure 6.15c). The combined data support our three step formation model of TFAAD film growth on diamond which is governed by: (i) formation of a submonomolecular layer ($t < 8$ hours); (ii) formation of single molecular layer (8 hours $< t < 12$ hours); (iii) slow, but continuous multilayer formation by cross-polymerization ($t > 12$ hours).

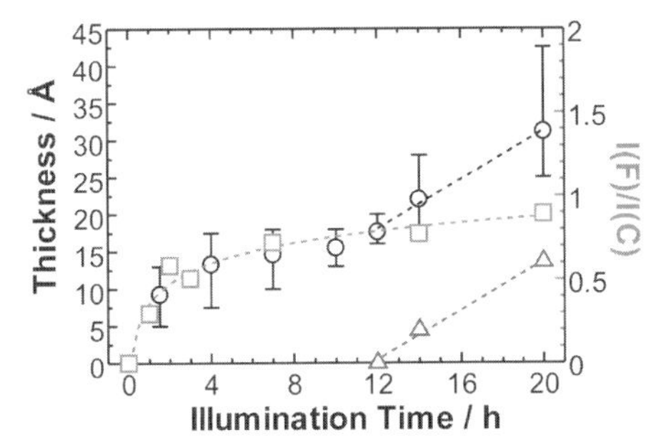

Figure 6.16 Variation of TFAAD film thickness as detected by AFM (circles) and the variation of the integrated peak intensity ratio F(1s)/C(1s) (squares) from XPS experiments as a function of illumination time. Triangles show the onset of crosspolymerization of TFAAD molecules.

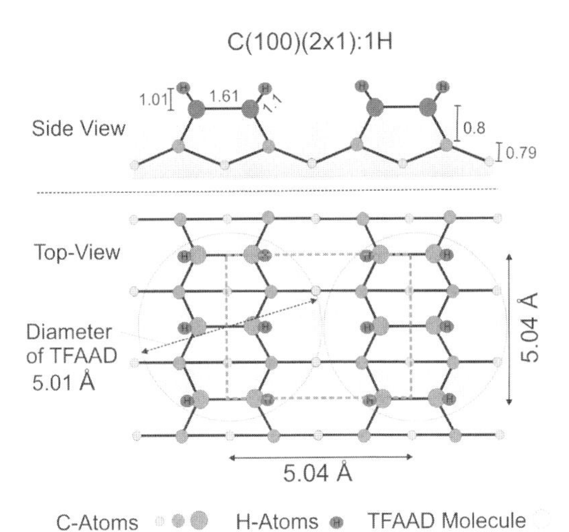

Figure 6.17 Side and top view of a reconstructed diamond (100) (2 × 1):1H surface (data from [78]). Also shown are typical areas of TFAAD molecules, assuming a diameter of 5.04 Å as shown in Figure 6.9.

A perfectly smooth surface of (100) oriented diamond contains about $1.5 \times 10^{15}\,cm^{-2}$ carbon bonds that are terminated by hydrogen, as shown schematically in Figure 6.17 [78]. The diameter of the 10-amino-dec-1-ene molecules (not taking into account the trifluoroacetic acid top) is 5.01 Å so that in the case of a closely packed TFAAD film the upright standing molecule will require an area of about $2 \times 10^{-15}\,cm^{-2}$ (if we assume rotational symmetry). This gives a closed layer density of $5 \times 10^{14}\,cm^{-2}$. Each TFAAD molecule covers the area of six hydrogen atoms. Only one of those hydrogen bonds needs to be broken to bond an amine molecule.

Therefore, most of the diamond surface may be still H-terminated after generation of a densely packed TFAAD layer.

As hydrogen cannot be detected by XPS we applied an additional experiment to investigate if the surface is still terminated with hydrogen. It is known that H-terminated undoped CVD diamond shows a surface conductivity in buffer solution which arises from transfer doping [30, 35, 36]. We have, therefore, measured the drain source current variation before and after TFAAD attachment, using a typical ISFET geometrical arrangement in SSPE buffer solution. Figure 6.18 shows the conductivity of a perfect H-terminated single crystal diamond before (open circles) and after (open squares) photochemical attachment of TFAAD for 20 hours in SSPE buffer solution. A perfectly (100%) H-terminated diamond gives rise to a drain source current of approximately 7.5 μA at $U_G = -0.6$ V. After photoattachment of TFAAD molecules for 20 hours, the drain source current decreased to 3 μA, which is 40% of the initial drain source conductivity. In the case of 100% H-removal by photoattachment, the drain source conductivity would completely disappear. As the wetting properties of TFAAD covered diamond surfaces are also changing, which may affect transfer doping properties of the surface, we leave a quantitative interpretation of this result to further research activities. The highest theoretical packing density of TFAAD of $5 \times 10^{14}\,cm^{-2}$ would require the removal of only 17% of all hydrogen atoms terminating the surface. Following recently published data in the literature, the packing density is most likely less than this number, and in the range of $2 \times 10^{14}\,cm^{-2}$ [71, 77, 79]. Our experimental data, as well as these theoretical considerations, indicate that the diamond TFAAD interface is still reasonably well terminated with hydrogen. This is promising with respect to sensor applications as high sensitivity will require a defect free interface.

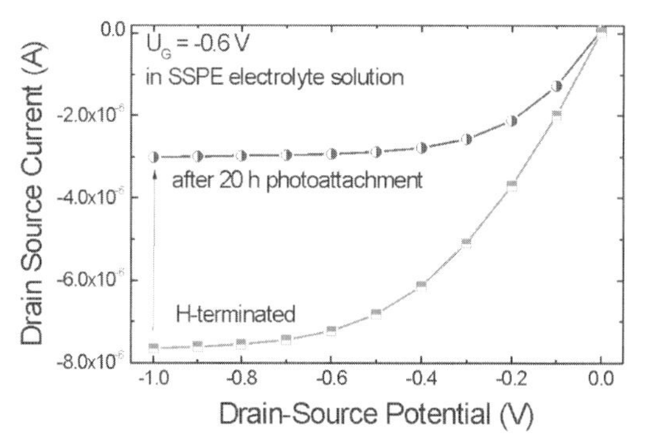

Figure 6.18 Comparison of diamond ion sensitive field effect transistor properties (ISFET) measured in SSPE buffer with perfectly H-termination of the surface (circles) and after photoattachment of amine molecules for 20 hours (squares). The drain source current is decreasing for about 60%, but the surface of diamond remains conductive after photoattachment.

Detailed in situ characterization of the attachment process indicates that electron emission by sub-bandgap light triggers the covalent bonding of TFAAD molecules to diamond [70, 71]. In this process, valence band electrons are optically excited into empty hydrogen induced states slightly above the vacuum level as shown schematically in Figure 6.19 [31]. From there, they can reach unoccupied π^* states of TFAAD molecules, generating a nucleophilic situation in the C=C bonding structure.

These radical anions may abstract hydrogen from the surface, as shown schematically in Figure 6.20, creating a carbon dangling bond at the surface which itself is very reactive towards olefins. However, to obtain covalently bonded ordered monomolecular layers on diamond requires some additional features. In the case of random self-assembly of molecules on diamond, a chaotically organized layer that would be generated as the basic requirement of surface mobility to allow intermolecular forces to play their ordering role is missing. Such a process would resemble a dart game, since the grafted moieties would be irreversibly immobilized on the surface, due to the formation of strong covalent C—C bonds. In our case, ordered formation of monolayers were detected, which is a strong argument in favor of the model of Cicero *et al.* [80]. In their investigation of olefin addition to H-terminated silicon, they deduce a model which requires the formation of alkene anions that abstract H atoms from the surface, thereby creating carbon surface dangling bonds that covalently bond to other alkene molecules in the

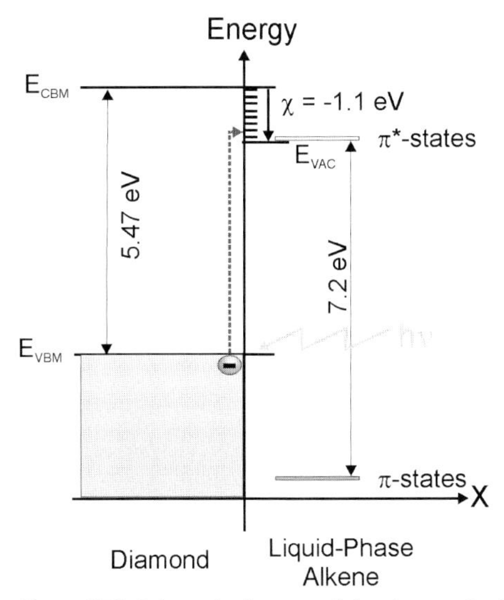

Figure 6.19 Schematic diagram of the photoexcitation mechanism at the surface of diamond in contact with TFAAD (from [70, 71]). Valence band electrons are photoexcited into empty surface states of diamond and then into empty electronic states of TFAAD molecules, which generates nucelophilic properties.

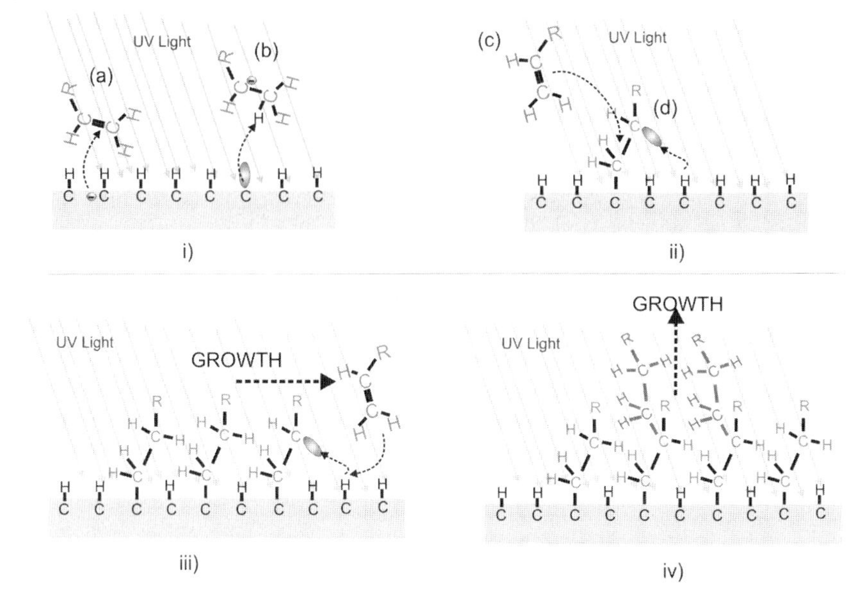

Figure 6.20 Schematic grafting mechanism of the diamond surface by amines (from [77, 80]). (i) shows the generation of radical anions by electron transfer from diamond to the olefin (a). The nucleophilic properties of the radical causes a hydrogen abstraction (b) which results in a surface carbon dangling bond; (ii) The dangling bond reacts with an olefin molecule (c) to form a diamond-carbon/olefin-carbon bond. This olefin abstracts a hydrogen atom from the diamond surface (d) which is a new site for olefin addition; (iii) The hydrogen abstraction reaction results in a chain reaction; (iv) In the case of extended illumination (>10 hours), a 3D growth sets in due to cross-polymerization of olefin molecules.

liquid. Surface dangling bonds are reactive towards alkenes, as demonstrated by Cicero *et al.* [80]. They showed that the olefin addition on H-terminated Si (111) surfaces follows a chain reaction, initiated at isolated Si dangling bonds. These surface dangling bonds further react with olefins to form carbon centered radical bonds to diamond. These radicals abstract hydrogen from neighboring H–C bonds and thereby, regenerate surface dangling bonds which then propagate the reaction, as depicted schematically in Figure 6.20. Such an ordered and self-organized layer formation seems to be a very reasonable model for our findings. The surfaces of grafted diamonds after 10 hours of attachment appear closed with no pinholes. However, some modulation is evident, which may reflect the fact that the ordered growth will break down and compete with other spots where the same process has been triggered. Therefore, one can expect some disorder due to competing domain growth.

The number of electrons photoexcited from diamond into the olefin film is huge, as only one out of about 1500 triggers chemical bonding to diamond [71]. These electrons create radical anions which give rise to cross polymerization and particle formation [79]. The cross polymerization on the monomolecular amine layer bonded to diamond is, however, slow, and may arise from the fact that the amine layer on diamond prevents the generation of radical anions, as TFAAD molecules cannot approach the surface for a electron transition. The radical anions are generated predominantly on diamond surface areas which are not grafted. These radicals then need to diffuse, driven by statistical properties, to find reaction partners and to cross polymerize. It seems less likely that they will bond to immobilized TFAAD molecules, than that they will react with other molecules in the near vicinity in liquid phase.

6.3.2 DNA Bonding and Geometrical Properties

To attach DNA, we applied the recipe introduced by Yang *et al.* 2002 [15], and Hamers *et al.* 2004 [81], where the protected amine is firstly deprotected, leaving behind a primary amine. The primary amine is then reacted with the heterobifunctional crosslinker sulphosuccinimidy l-4-(N-maleimidomethyl) cyclohexane-1-carboxylate, and finally, reacted with thiol modified DNA to produce the DNA modified diamond surface. In our experiments, we use a 4 μl droplet which is placed on the diamond layer and which covers a circular area of about 2 mm in diameter, thereby covering oxidized and H-terminated surface parts. To assess whether DNA modified diamond surfaces have been generated, such surfaces have been exposed to complementary oligonucleotides that were labeled with fluorescence tags FAM. Figure 6.21 shows the result of intense green fluorescence (=100%) from originally H-terminated regions, and less intense fluorescence (≅70%) from oxidized surface areas. There are two reasons for the weak fluores-

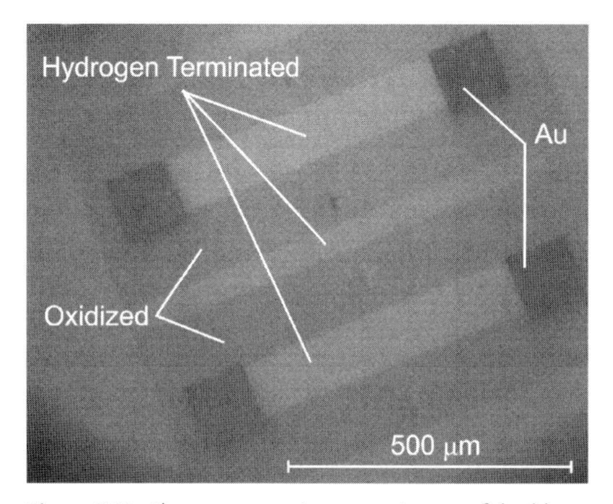

Figure 6.21 Fluorescence microscopy image of double stranded (ds) DNA helices on diamond using green fluorescence tags (Cy5) attached to complementary DNA. The bright areas arise from initially H-terminated diamond, and the less intense regions were originally oxidized. Black areas are Au contacts.

cence contrast. The first is that noncovalently bonded DNA is attached to oxidized diamond. This will be discussed in the following section, using atomic force microscopy experiments (AFM). The second reason is that transparent diamond gives rise to light trapping, so that the transparent diamond appears green in case of fluorescence emission.

6.3.3 Atomic Force Characterization (AFM) of DNA in Single Crystalline Diamond

DNA functionalized and hybridized surfaces are characterized by AFM measurements in 2× SSPE/0.2% SDS (sodium dodecyl sulphate) buffer solution [75, 82]. By performing contact mode AFM scratching experiments, DNA can be removed and at the interface between clean and DNA covered diamond, and the height of DNA can be measured. In addition, the force required to penetrate and remove DNA has be determined, giving insight into the mechanical stability of the bonding. Scratching experiments were performed with different tip loading forces between 10 and 200 nN. A typical result of such experiments on a diamond surface modified with double strand (ds) DNA is shown in Figure 6.22a. For each force, an area of 2 μm × 10 μm has been scratched (scan rate 20 μm/s). Forces around 45 nN (±12 nN) generate a surface which appears to be clean of DNA, as also detected by fluorescence microscopy shown in Figure 6.22b.

C-AFM experiments on the boundary between initially oxidized and H-terminated diamond are shown in Figure 6.23. With C-AFM, we detect non-covalently bonded DNA on oxidized diamond. These molecules can be removed with forces around 5 nN. This is five times lower than DNA bonded to the H-terminated surface. The layer is also significantly thinner as on H-terminated diamond.

By measuring across the boundary between the DNA functionalized and the cleaned surface, using O-AFM, the DNA layer thickness can be obtained as shown in Figure 6.24. O-AFM is preferable to C-AFM on soft layers as the tip–surface

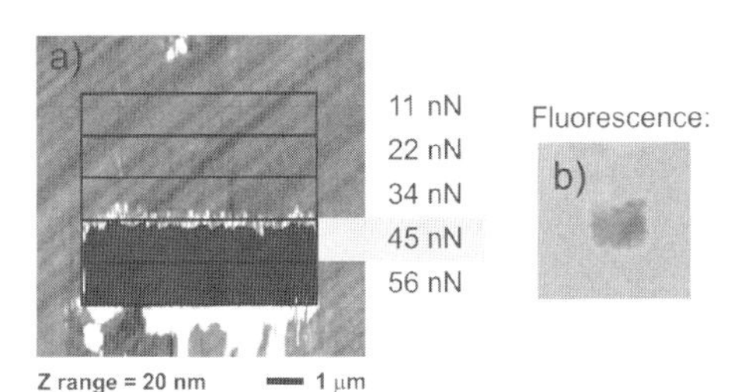

Figure 6.22 Typical surface morphology of a DNA film on diamond after application of contact mode AFM (a), where increasing AFM tip loading forces have been applied. Forces larger than 45 nN gives rise to DNA removal on photochemically treated and initially H-terminated diamond. After removal of DNA from the surface it appears dark in fluorescence microscopy (b).

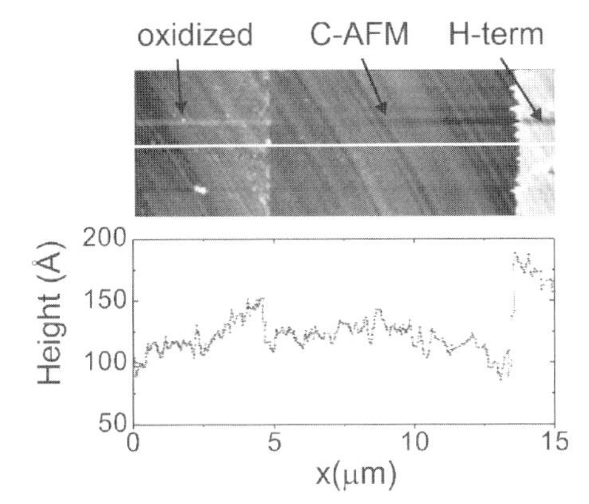

Figure 6.23 Oscillatory AFM measurement applied at the boundary of cleaned diamond surface to initially oxidized (left) and initially H-terminated (right) diamond show that on both areas DNA molecules are present. The height on O-terminated diamond is however lower than on H-terminated diamond. Molecules on oxidized diamond can be removed with forces of about 5 nN.

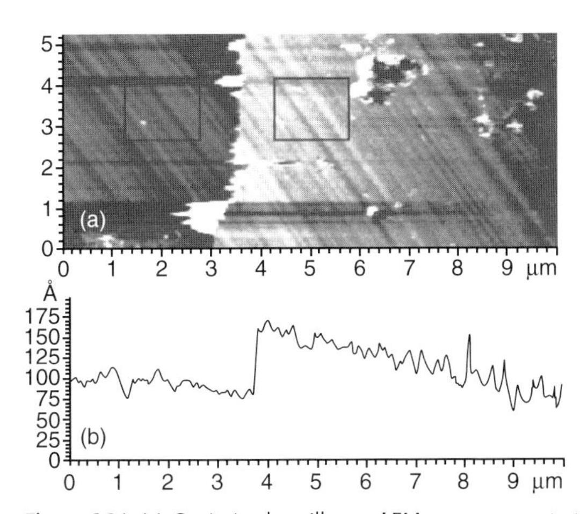

Figure 6.24 (a) Optimized oscillatory AFM measurement at the boundary of a cleaned diamond surface to diamond with attached double stranded DNA molecules. The squares denote the regions where AFM phase shifts were evaluated. (b) AFM height profile across the boundary reveals a DNA layer thickness of 76 Å (see also [75]).

interaction can be minimized by monitoring the phase shift of the cantilever oscillations [83]. The phase shift was measured as a function of the set-point ratio, $r_{sp} = A_o/A_{sp}$, where A_{sp} is the set-point amplitude and A_o is the amplitude of free cantilever oscillation, on DNA functionalized and cleaned diamond surface regions (see squares in Figure 6.24a). Figure 6.25 summarizes results of phase contrast

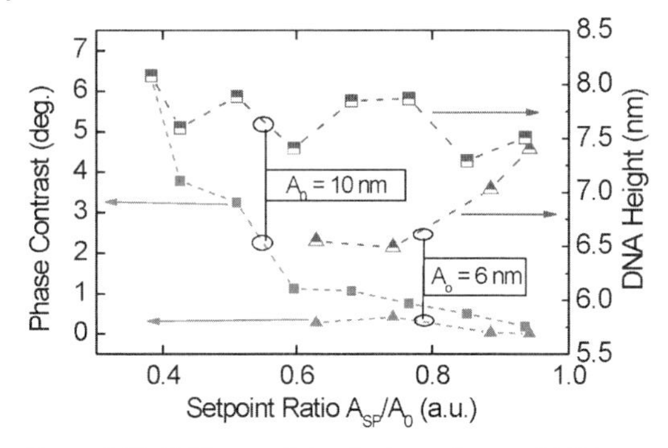

Figure 6.25 AFM set point ratio dependence of AFM measurements of DNA height and phase contrasts across a DNA functionalized and cleaned diamond surface for free oscillation amplitudes (A_o) of 6 and 10 nm. Extrapolation to set-point ratio 1 results in a DNA height of about 76 Å.

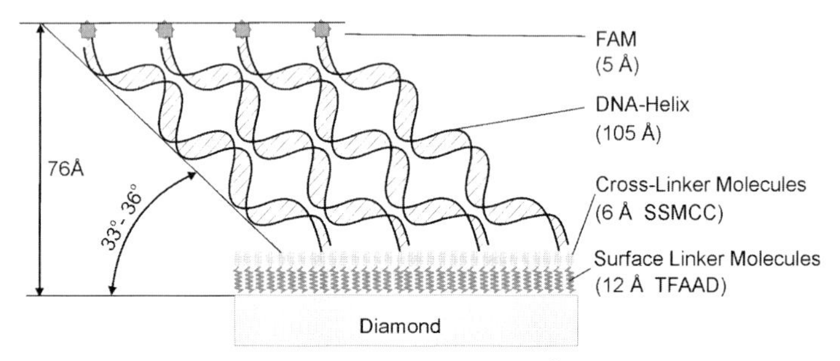

Figure 6.26 From AFM, a compact DNA layer of 76 Å height is resolved by optimizing phase and height contrast measurements. The axis of the double helices DNA is therefore tilted by about 30° to 36° with respect to the diamond surface as shown schematically in this figure.

and set-point ratio measurements. The phase contrast between diamond and DNA is positive and approaches zero for set-point ratios approaching 1, that is, for increasing tip–surface distance. For a phase contrast near zero, which corresponds to minimized tip–surface interaction, the DNA layer thickness reaches about 75 to 78 Å. Simple O-AFM measurements results in about 70 Å (see Figure 6.24) which is slightly smaller than the real DNA thickness. The height of 75 to 78 Å is, however, still significantly lower than the expected height of about 130 Å for upright standing DNA (105 Å) on linker (12 Å) and cross linker molecules (6 Å) and fluorescence marker FAM6 (5 Å). We attribute this to a tilted arrangement of DNA molecules as shown in Figure 6.26 (schematic view of molecules). Using triangular geometry, the tilt angle is around 33° to 36°, which is similar to results of DNA bonded to gold [82, 83]. There, the detected film thickness of DNA layers has been attributed to orientational properties of DNA duplexes in the monolayer.

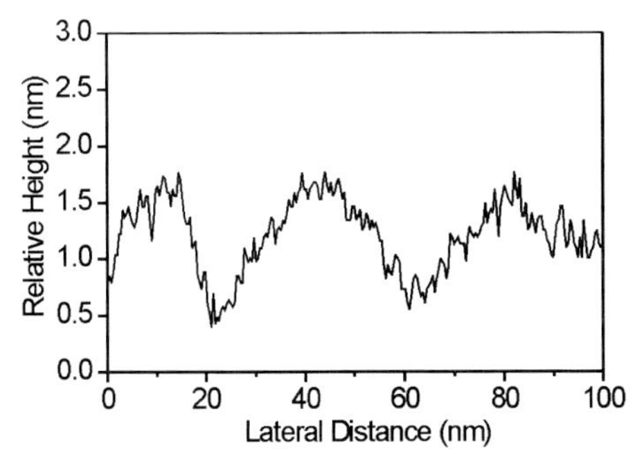

Figure 6.27 AFM height profile shows a dense DNA layer with a RMS height modulations of ±5 Å.

Due to the large charge density of the DNA backbone (2-/base pair without condensed counter-ions), the orientation of the individual helices is very sensible towards neighboring molecules and to surface charges. As a consequence, small changes in applied electrochemical potential can cause drastic changes in helical orientation (for a summary, see [82]). In our case, the tilted average helical packing orientation arrangement of 35° certainly reflects the minimization of Coulomb repulsion forces between individual duplexes. Effects from diamond surface charges (C—H dipole) or from externally applied electric fields on DNA arrangements are currently investigated by variations of buffer solution ionicities and by application of externally applied electric fields to the diamond transducer.

A topographic surface profile of DNA double helix molecules, bonded on diamond is shown in Figure 6.27. It reveals broad undulations due to collective interaction of several DNA oligomers with the tip. The height is modulated with a periodicity of about 30 to 50 nm. The DNA surface roughness is around ±5 Å. No pinholes can be detected in the layer. Obviously, a closed DNA film has been synthesized on diamond using photochemical attachment.

For sensor applications, dilute DNA films in the range 10^{12} to $10^{13}\,cm^{-2}$ are required [84]. To decrease the bonding density we reduce the time of marker DNA attachment. A saturated and very dense film is achieved after 12 hours exposure. Following the arguments of Takahashi *et al.* 2003 [46, 47] this will result in a DNA density of about $10^{13}\,cm^{-2}$. By decreasing the time of marker ssDNA attachment, the density can be decreased to $10^{12}\,cm^{-2}$ as shown in Figure 6.28. Here, we have evaluated the change in fluorescence intensity after hybridization, where the attachment of ssDNA maker molecules has been varied between 10 minutes to 12 hours. The attachment kinetics is well described empirically by an exponential function with a time constant, τ, of about 2 hours.

6.3.4 Electrochemical Surface Modifications of Boron -oped Single Crystalline Diamond

In the following we summarize the electrochemical modification of highly conductive p-type single crystalline CVD diamond [72, 73, 76]. The general chemical

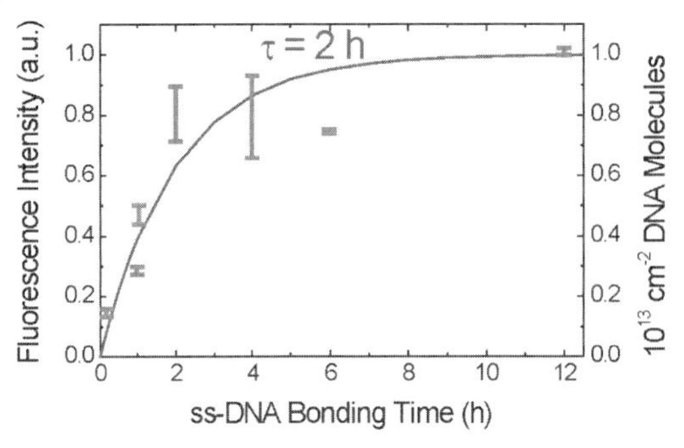

Figure 6.28 Variation of the ssDNA marker molecule attachment time, between 10 minutes and 12 hours gives rise to an exponentional increasing fluorescence intensity. The fluorescence intensity of hybridized DNA follows an activated property (full line) with a time constant of 2 hours. The DNA density on diamond varies between 10^{12} and $10^{13}\,cm^{-2}$.

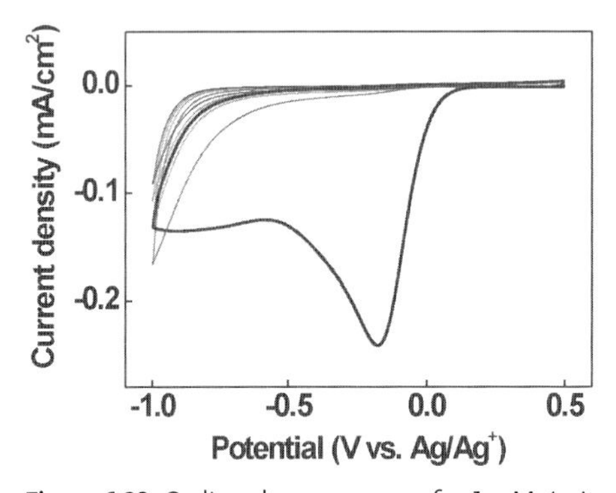

Figure 6.29 Cyclic voltammograms for 1 mM 4-nitrobenzene diazonium tetrafluoroborate on highly B doped single crystalline diamond (DRC). Electrolyte solution: 0.1 M NBu_4BF_4 in CH_3CN. scan rate: 0.2 V/s (for details see [72]).

scheme is shown in Figure 6.11 (I) where hydrogen-terminated or oxidized diamond is modified with phenyl molecules (step a to b). Figure 6.29 shows cyclic voltammograms of 4-nitrobenzene diazonium salts (1 mM) reactions on highly B doped single crystalline CVD diamond film in 0.1 M NBu_4BF_4 acetonitrile solution (san rate: 0.2 V/s). An irreversible cathodic peak of the first sweep at −0.17 V (vs. Ag/AgCl) indicates nitrophenyl group attachment by diazonium salt reduction [72, 73, 76]. The reduction peak on single crystalline diamond films decreases rapidly with increasing number of scans within +0.5 to −1.0 V (vs. Ag/AgCl), due to increasing surface passivation with nitrophenyl molecules.

The electrochemical attachment works on H-terminated diamond but also on oxidized diamond surfaces as can be seen in Figure 6.30. The voltammetric peak

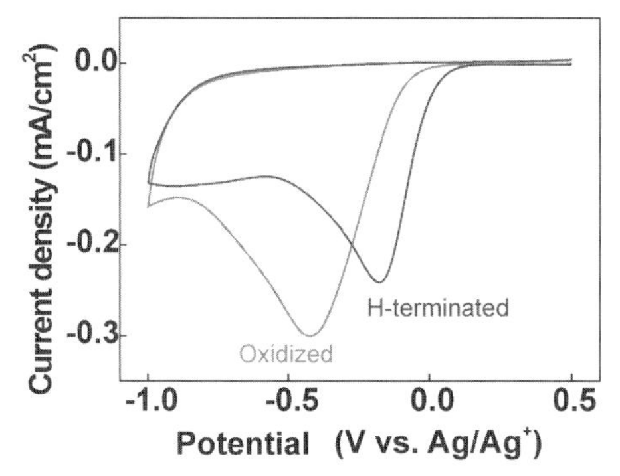

Figure 6.30 Comparison of cyclic voltammorgams for 1 mM 4-nitrobenzene diazonium tetrafluoroborate measured on H-terminated (red) and oxidized (green) highly B doped single crystalline diamond. Please note that the attachment of phenyl molecules to oxidized diamond occurs at slightly more negative potentials of about −410 mV compared to H-terminated diamond at −170 mV (vs. Ag/AgCl).

is shifted from −0.17 V for H-termination to −0.41 V (vs. Ag/AgCl) for oxidized diamond. This potential shift of $\Delta U = 240\,\text{mV}$ arises by a change of heterojunction properties at the solid/liquid interface. One significant change arises by the variation of the electron affinity as H-terminated diamond has a negative electron affinity of around −1.1 eV, while oxidized diamond shows positive affinities [32, 61]. In addition, it is known that oxidized diamond surfaces show surface Fermi level pinning due to surface defects [85]. The termination of carbon bonds with oxygen instead of hydrogen adds additional complexity as the bonding energies are changing from C—H of 413 kJ/mol to 360 kJ/mol for C—O and to 805 kJ/mol for C=O, respectively.

After electrochemical derivatization, the diamond substrates are sonificated in acetonitrile, acetone, and isopropanol, in order to investigate properties of attached nitrophenyl layers in 0.1 M NBu_4BF_4 solution. Nitrophenyl groups grafted on single crystalline diamond films show two reversible electron transfer steps, which are reproducibly detected in all potential sweeps (see Figure 6.31). The generalized reversible redox reactions of this system are summarized in Figure 6.32. To estimate the surface coverage (Γ) of nitrophenyl groups on diamond, we use the transferred charge of the electron transfer reaction at −1.17 V (see shaded area in Figure 6.31). This results in $3.8 \times 10^{-7}\,\text{C/cm}^2$ or 8×10^{13} molecules/cm^2, indicating the formation of a sub-monolayer on diamond (5% coverage) [72, 76].

Detailed investigations of phenyl layer formation on other electrodes show that this simple interpretation is misleading [86, 87]. In most cases, multilayers are deposited. We have, therefore, applied additional experiments to characterize the growth and thickness of the phenyl layer on diamond [76]. Generally, all performed contact- and oscillatory-mode AFM experiments on deposited nitrophenyl layers reveal layer thicknesses in the range 28 to 68 Å. Taking into account the length of

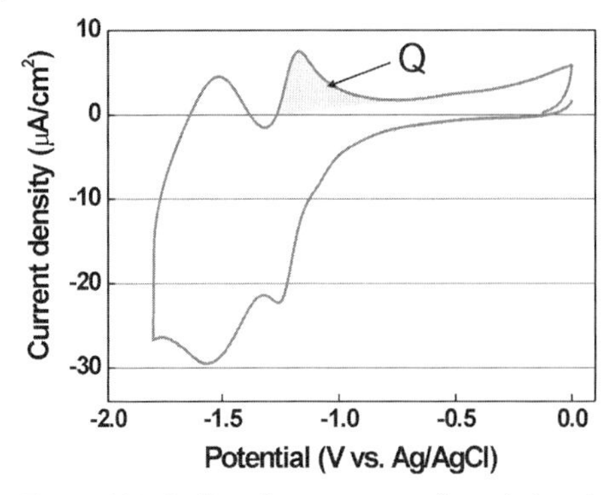

Figure 6.31 Cyclic voltammograms from 4-nitrophenyl modified single crystalline diamond in blank electrolyte solution. Electrolyte solution: 0.1 M NBu_4BF_4 in CH_3CN. For details see [72].

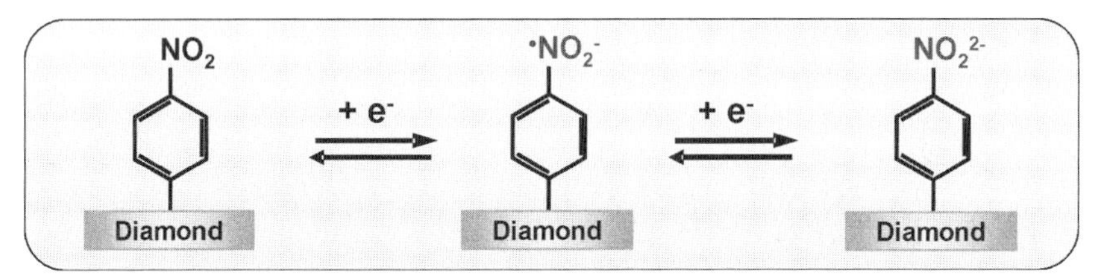

Figure 6.32 Schematic reduction/oxidation reactions of nitrophenyl bonded to diamond, giving rise to a two electron transfer reaction mechanism [72].

nitrophenyl molecules of about 8 Å (Figure 6.10), it indicates clearly multilayer formation by cyclic voltammetry deposition.

A more effectively controlled growth of phenyl films on diamond can be achieved by electrochemical means, applying fixed potentials for a given period of time instead of potential cycles [87]. In our case we applied −0.2 V (vs. Ag/AgCl) and measured the transient current during the deposition. The result is shown in Figure 6.33. In case of unlimited electron transfer and diffusion limited attachment the dynamics follow the Cottrell law ($i(t) \approx t^{-0.5}$) as introduced by Allongue *et al.* in 2003 [87]. At the very beginning of the transient current, such a characteristic can be detected. However, for longer times, the current decays faster then predicted by this law. We assume that the growing phenyl layer limits an effective electron transfer, slowing down the bonding process. The density of electrons involved in this reaction saturates around $4 \times 10^{15}\,cm^{-2}$.

To verify the layer formation, contact mode AFM has been applied. With increasing force to the tip, the phenyl-layer can be removed from diamond. A typical result is shown in Figure 6.34, where the nitrophenyl layer has been attached by one cyclic voltammetry scan (from +0.5 V to −1.0 V vs. Ag/AgCl at a scan rate of

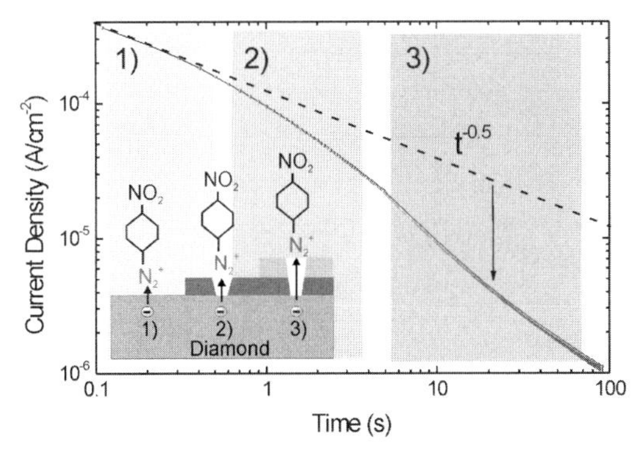

Figure 6.33 Transient current as detected during nitrophenyl attachment at a constant potential of −0.2 V (vs. Ag/AgCl). Also shown is the theoretical decay following a $t^{-0.5}$ time dependence. The inset shows a schematic growth model where the phenyl layer starts to grow (1), to become thicker (2) and finally terminates the growth (3), as the electron tunneling transition through the phenyl layer decreases to zero [76].

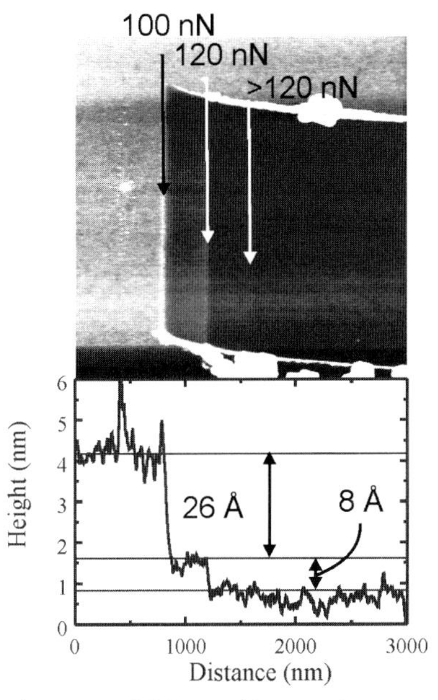

Figure 6.34 AFM scratching experiments on nitrophenyl modified diamond. With forces >100 nN most of the phenyl-layer can be removed while forces >120 nN are required to remove the linker layer to diamond completely [76].

200 mV/s). Forces below 100 nN do not damage to the phenyl film. Above 100 nN, a layer of 26 Å thickness is removed and forces above 120 nN give rise to the removal of a thin layer, 8 Å in height. We assume that firstly a random oriented phenyl layer is removed, while forces above 120 nN are required to remove phenyl linker molecules bonded to diamond.

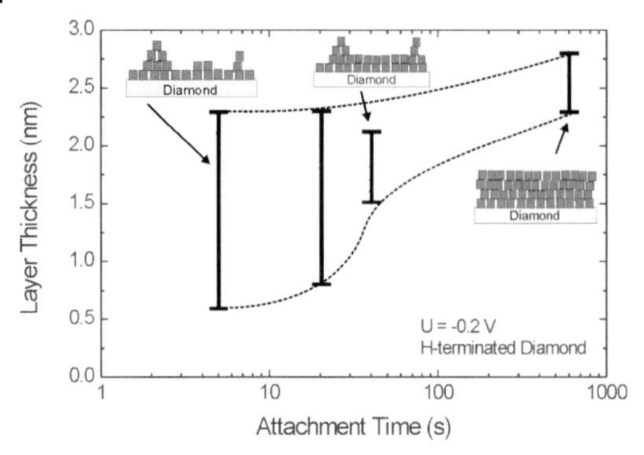

Figure 6.35 Nitrophenyl layer growth during constant potential attachment experiments with −0.2 V (vs. Ag/AgCl) is governed by three dimensional (3D) growth properties. After short time attachment the layer thickness varies strongly, whereas after 90 s the variations become much smaller, indicating a dense layer formation of about 25 Å thickness (for details see [76]).

AFM characterization on phenyl layers, which have been grown at a constant potential of −0.2 V for different times, indicate 3D growth as shown in Figure 6.35 (applied tip force: 200 nN, scan rate: 4 μm/s). After short time attachment (5 s) the thickness of the layer is already between 8 and 23 Å. The thickness variation decreases with increasing attachment time, saturating around 25 Å. Taking into account the saturated electron density of $4 \times 10^{15}\,cm^{-2}$ and the final thickness of the phenyl layer of 25 Å, the phenyl molecule density in the layer is about $2 \times 10^{21}\,cm^{-3}$.

The orientation of phenyl molecules has been characterized by angle resolved XPS experiments. The integrated peak intensities of O(1s) to C(1s) shows a strong angle dependence for attachments at −0.2 V for times up to 40 s (see Figure 6.36). We attribute this to an oriented growth of nitrophenyl, with NO_2 molecules preferentially located on the growing top of the layer. This is different in the case of much thicker layers (30 to 65 Å), attached by 5 cycles in the range +0.5 V to −1.0 V vs. Ag/AgCl at a scan rate of 200 mV/s. Here the XPS angle variation is weak; molecules are arranged in a more disordered structure.

From these experiments, we conclude that the formation of phenyl layers on diamond is governed by 3D growth, with preferential alignment of NO_2 cap molecules on the top of growing films, if films are not grown too thick. Growth saturates at a layer thickness of about 25 Å, using constant potential attachment (−0.2 V), while significantly thicker layers in the range 35 to 65 Å are detected after cyclic attachment (+0.5 V to −1 V). Properties of such thick layers are governed by a more random molecule orientation. This is schematically summarized in Figure 6.37.

Subsequently, nitrophenyl groups are electrochemically reduced to aminophenyl ($-C_6H_5NH_2$) in 0.1 M KCl solution of EtOH—H_2O solvent (see Figure 6.38).

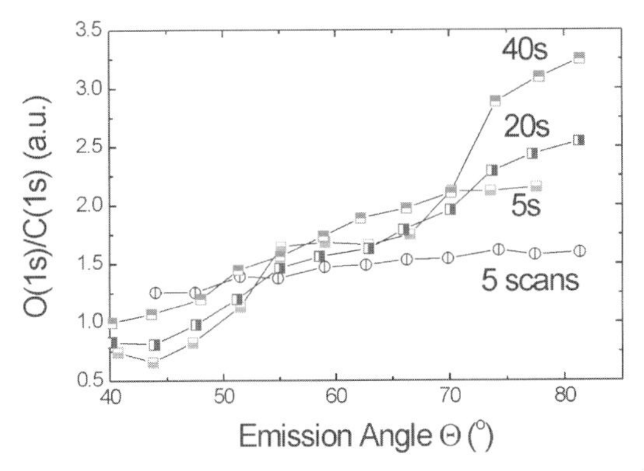

Figure 6.36 Angle resolved XPS experiments show oriented growth of nitrophenyl layers, grown with constant potential. Cyclic potential attachment method gives rise to significantly thicker layers of typically 30 to 70Å with less pronounced molecule arrangement.

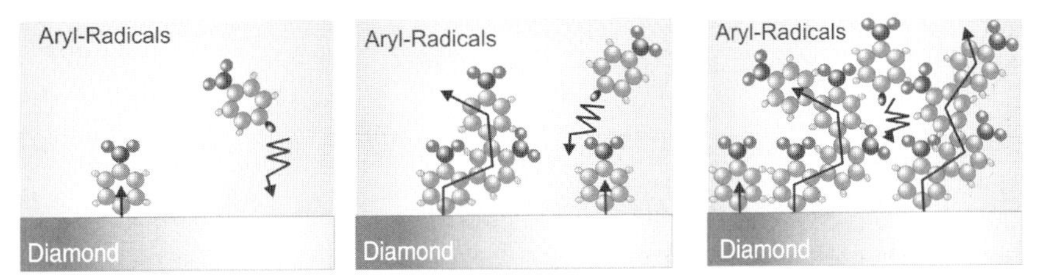

Figure 6.37 Nitrophenyl groups at an initial stage of attachment grow three dimensional (3D) as shown here schematically, forming layers of varying heights and densities. Layer thicknesses of up to 80Å are detected for cyclic voltammetry attachment after five cycles, whereas the layer becomes denser and only about 25Å thick in case of constant potential attachment.

The first voltammetry sweep gives rise to an irreversible reduction peak at –0.94 (vs. SCE), which is not detected in the second and higher cyclic voltammetry cycles. Such modified surfaces are then used for chemical bonding to heterobifunctional crosslinker molecules, SSMCC and thiol modified probe DNA oligonucleotides.

Figure 6.39 shows a fluorescence image of S1 ssDNA marker molecules bonded to diamond after hybridization with its complementary ssDNA target molecules F1 labeled with Cy5. The image shows DNA bonding to H-terminated diamond, and to oxidized diamond. The laid "T" shape pattern in Figure 6.39 arises from surface oxidation. The fluorescence from this area is about 10% darker than from hydrogen terminated diamond. As the light intensity is proportional to the density of fluorescence centers, the density of DNA bonded to oxidized diamond is about 10% smaller than on H-terminated diamond. No fluorescence can be detected using a 4-base mismatched ssDNA target molecule for hybridization.

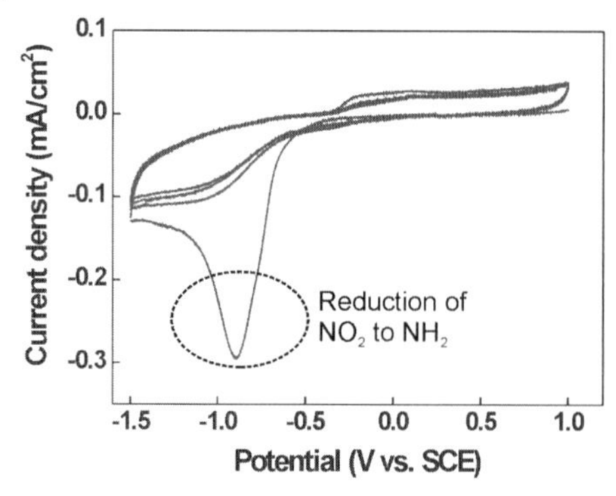

Figure 6.38 The nitro groups (NO_2) are electrochemically changed to amino groups (NH_2) by cyclic voltammograms in 0.1 M KCl solution with 10 : 90 (v/v) EtOH—H_2O. Scan rate: 0.1 V/s. During the first sweep, a pronounced peak is detected which reflects the reduction of NO_2 to NH_2. The variation in the second and third sweep is minor compared to the first sweep reduction.

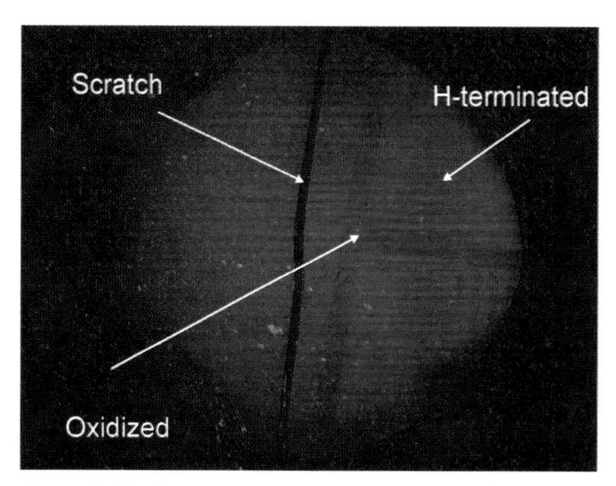

Figure 6.39 Fluorescence microscopy image of a dsDNA functionalized single crystalline diamond electrode after DNA hybridization with complementary oligonucleotides terminated with Cy5 dye molecules. The layer was originally H-terminated but a T-shape has been oxidized. Here, the fluorescence intensity appears weaker. To generate a contrast with respect to clean diamond, a scratched area has been realized.

6.3.5
Atomic Force Characterization of DNA Bonded Electrochemically to Boron-Doped Single Crystalline Diamond

Geometrical properties, as well as the density and bonding strength of DNA bonded by this electrochemical technique to boron doped diamond, have been characterized by AFM experiments as described in paragraph 2.5 and 3.3.

The height of DNA layers is detected to be around 90 Å. This is slightly higher then in the case of photoattachment, and arises from the thicker linker molecule layer which is about 25 Å in the case of phenyl, and 12 Å for amine linkers. Again, a tilted arrangement is deduced, comparable to results from photoattachment (≈35°). The layer is dense with no pinholes. The removal forces are between 60 and 122 nN, the statistical average around 76 nN. It is interesting to note that on initially oxidized diamond, the forces are lower, namely in the range of 34 nN. A comparison of forces is shown in Figure 6.40. These results indicate strong bonding of DNA to diamond for both, photo- and electrochemical surface modifications. Removal forces are about two times higher then detected on Au and Mica as summarized in Figure 6.41 [88–91]. This is promising with respect to diamond biosensor applications where exceptional chemical stability is required.

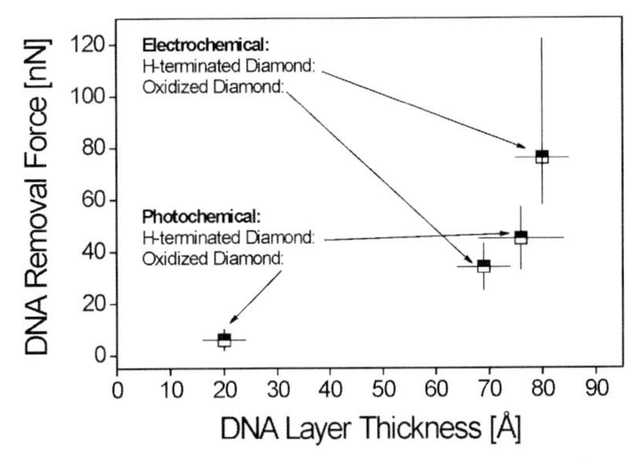

Figure 6.40 Comparison of critical removal forces of electrochemically attached dsDNA on H-terminated and oxidized diamond and of photochemical attached dsDNA on H-terminated and oxidized diamond. Attachment to initially H-terminated diamond of both linker molecules systems gives rise to stronger bonding than to initially oxidized diamond surfaces. DNA bonded with phenyl linker molecules show the strongest bonds.

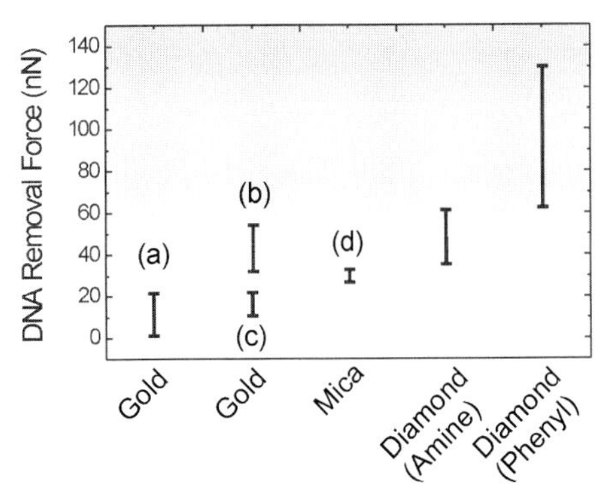

Figure 6.41 Comparison of DNA removal forces as detected in our experiments on diamond and compared with DNA bonding to Au (a: [83], b: [84], c: [85]) and Mica (d: [86]).

6.3.6
Electronic Detection of DNA Hybridization

Most DNA detection techniques are based on DNA hybridization events. In DNA hybridization, the target ssDNA is identified by a probe ssDNA which gives rise to hybridization. This reaction is known to be highly efficient and extremely specific. Commonly used DNA detection techniques (radiochemical, enzymatic, fluorescent) are based on the detection of various labels or reagents and have been proven to be time consuming, expensive, and complex to implement. For fast, simple, and inexpensive detection, direct methods are required. In the following section, we want to introduce recently achieved results with respect to field effect and voltammetric sensing, using single crystalline CVD diamond as transducer.

6.3.6.1 DNA-FET

Diamond ion-sensitive field effect transistors show sensitively variation of pH with about 55 mV/pH [40–42]. This is close to the Nernst limit of 60 mV/pH. The effect arises from transfer doping so that no gate insulator layer is required. Therefore, the separation between surface channel and electrolyte is very small. The application of such a FET system for DNA hybridization detection is new, and we will show some typical results, achieved on structures with 10^{12} to $10^{13}\,cm^{-2}$ ssDNA markers bonded to diamond. A schematic figure of the device with hybridized dsDNA is shown in Figure 6.42. We used a 1 M NaCl solution (containing 0.1 M phosphate with pH 7.2) with a Debye length of 3 Å as calculated by [84].

$$\lambda_D = \left(\frac{\varepsilon_{el}\varepsilon_O kT}{2z^2q^2l} \right)^{1/2}$$

where k is the Boltzmann constant, T the absolute temperature, ε_o the permittivity of vacuum ε_{el} the dielectric constant of the electrolyte, z is the valency of ions in the electrolyte, q is the elementary charge and I represents the ionic strength, for a 1-1 salt, it can be replaced by the electrolyte concentration n_o. As the linker molecule is 10 to 15 Å (amine) and the crosslinker molecule 5 Å long, the DNA is not in touch with the Helmholz layer in our experiments.

Figure 6.43a shows a comparison of drain source currents (I_{DS}) measured at a fixed drain-source potential of −0.5 V as function of gate potential for ssDNA marker molecules attached on the gate, for hybridized marker and target DNA on the gate and after removal of DNA from diamond. The initial ssDNA density bonded to diamond is $4 \times 10^{12}\,cm^{-2}$. The drain source current increases by hybridization, as also detected for a sensor where the initial ssDNA density is slightly smaller ($10^{12}\,cm^{-2}$) or larger ($10^{13}\,cm^{-2}$). The gate potential variations from ssDNA to complementary dsDNA bonding vary between 30 to 100 mV as shown in Figure 6.43b . There is a clear trend that with decreasing DNA density, the potential shift becomes larger (as predicted by Poghossian *et al.* 2005 [84]). Taking into account the ion sensitivity of diamond ISFETs of 55 mV/pH, this reflects a decrease of pH of the buffer solution of about 1 to 1.4 by hybridization. Poghossian *et al.* 2005

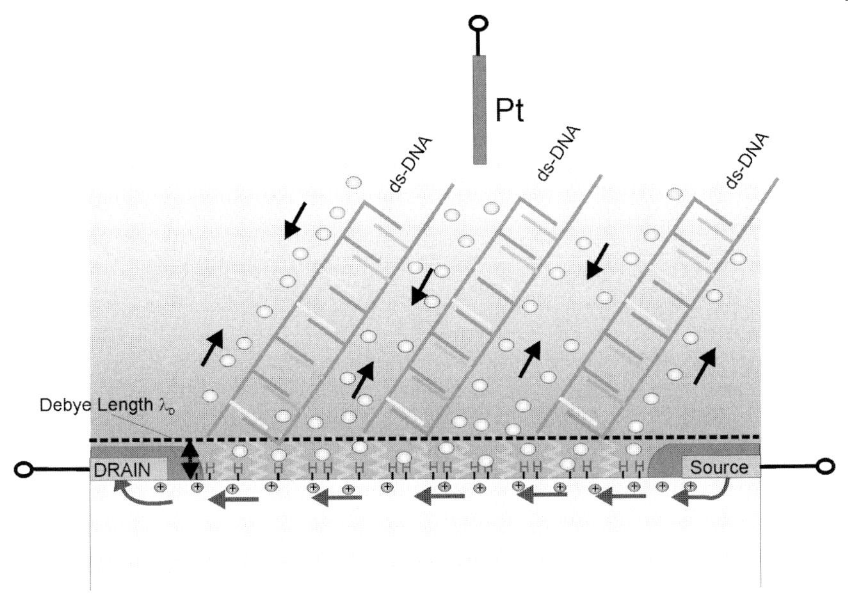

Figure 6.42 Schematic description of DNA hybridized on a diamond DNA ion sensitive field effect transistor (DNA-ISFET) sensor. The surface conductivity of diamond will be change by accumulation of compensating cations in the DNA layer which is caused by the negatively charged backbone structure of DNA (for details see [84]). The 1 M NaCl buffer shrinks the Debye length in our experiments to about 3 Å.

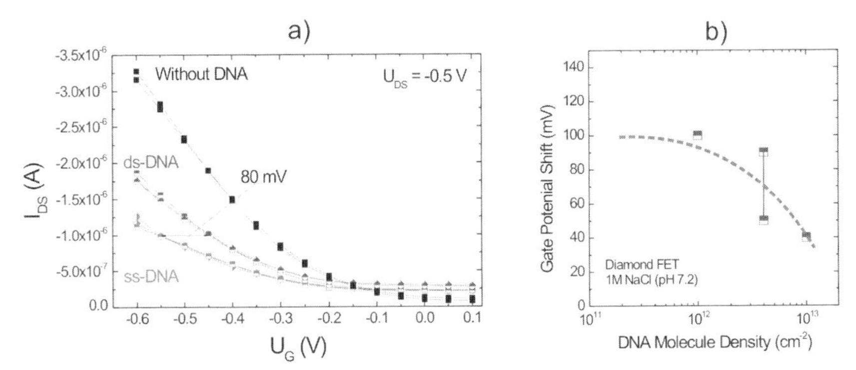

Figure 6.43 (a) Drain source current variations measured as a function of gate potential at a fixed drain source potential of −0.5 V for ssDNA (marker DNA), after hybridization with complementary target ssDNA to form dsDNA and after removal of DNA by washing. A gate potential shift of about 80 mV is detected on this DNA-ISFET with about $4 \times 10^{12}\,cm^{-2}$ molecules bonded to the gate. (b) Gate potential shifts as detected on diamond transistor structures with $10^{12}\,cm^{-2}$, $4 \times 10^{12}\,cm^{-2}$ and $10^{13}\,cm^{-2}$ ssDNA marker molecules bonded to the gate area. The threshold potential is increased towards less dense grafted diamond gates areas.

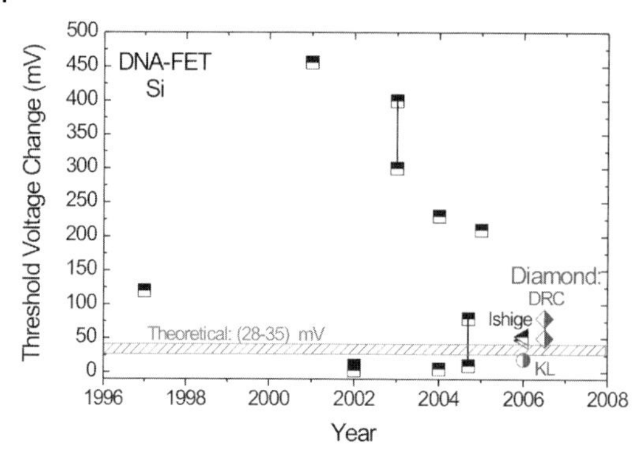

Figure 6.44 Comparison of silicon based DNA-ISFET sensitivities as a function of data publication (black half filled squares from [79], and half filled triangles from [87]) with diamond sensitivities as shown in this paper (half filled diamonds, DRC) and with data deduced on polycrystalline CVD diamond films from [88] (half filled circle). The shaded area indicates the theoretically predicted sensitivity, following the model of Poghossian *et al.* 2005 [84].

[84] have calculated that the average ion concentration within the intermolecular spaces after hybridization can be more than three to four times higher for cations than before hybridization. In case of a Nernstian slope of the sensor, they predict a gate potential shift of 28 to 35 mV. Our results indicate a stronger change. The increase in drain source current with hybridization can be well described by the transfer doping model as the increase in cation density will cause a decrease of pH. Therefore, the chemical potential will increase giving rise to enhanced surface conductivity.

A summary of published sensitivities of DNA FET's from silicon is shown in Figure 6.44 using data from [84, 92, 93]. The initial sensitivities reported before 2004 of more than 100 meV threshold potential shifts by hybridization have not been reproduced. On the contrary, sensitivities are decreasing towards experimentally reproducible, as well as theoretical predictable, values in the range 30 to 80 meV.

6.3.6.2 Cyclic Voltammetry and Impedance Spectroscopy

For amperometric detection of DNA hybridization we use $Fe(CN_6)^{3-/4-}$ as mediator redox molecule [94]. The detection principle is shown in Figure 6.45. In case of ssDNA bonded to diamond (Figure 6.45a), the relatively small negatively charged redox molecules can diffuse through the layer of DNA and interact with the diamond electrode to cause a redox current. By hybridization of DNA, the intermolecular space shrinks which leads to Coulomb repulsion between negatively charged $Fe(CN_6)^{3-/4-}$ and negatively charged sugar-phosphate backbones of hybridized DNA. The redox reaction with diamond will, therefore, decrease (Figure 6.45b).

The result is shown in Figure 6.46 where we used cyclic voltammetry on H-terminated metallically doped (p-type) single crystalline diamond in 0.5 mM

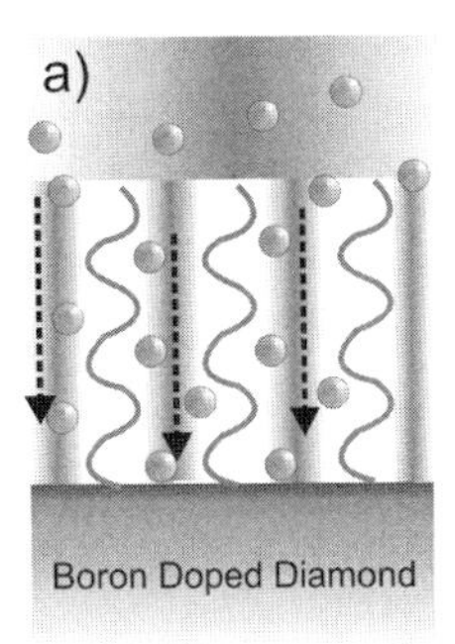

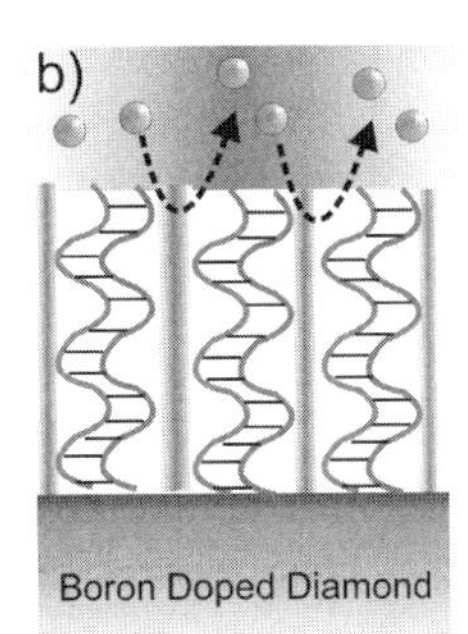

ss-DNA ds-DNA

Figure 6.45 Schematic DNA hybridization detection mechanism using $Fe(CN_6)^{3-/4-}$ as mediator redox molecules. In case of ssDNA (a), the negatively charged redox molecules $Fe(CN_6)^{3-/4-}$(blue balls) can diffuse through the DNA layer. After hybridization (b), the space between individual dsDNA molecules becomes to small for negatively charge molecules to overcome the repulsive Coulomb forces from the negatively charge backbone of dsDNA.

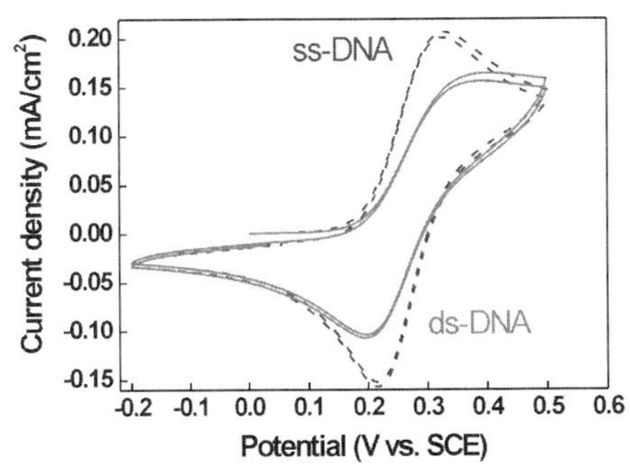

Figure 6.46 Cyclic voltammograms on ss and dsDNA grafted metallically boron doped (p-type) single crystalline diamond in 0.5 mM $Fe(CN_6)^{3-/4-}$, 100 mM KCL, 100 mM KNO_3 measured with respect to Ag/AgCl with a scan rate of 100 mV/s. The oxidation and reduction peaks are decreasing for about 50 μA/cm^2 by hybridization.

$Fe(CN_6)^{3-/4-}$, 100 mM KCL, 100 mM KNO_3 measured with respect to Ag/AgCl with a scan rate of 100 mV/s. The H-terminated diamond shows a well pronounced oxidation peak at +280 mV and a corresponding reduction wave with a peak at +126 mV (not shown here). These characteristics are well known, and have been published in the literature [22–24]. After electrochemical attachment of phenyl linker molecules and ssDNA marker molecules, the redox amplitude is decreased to about 30% of the clean diamond surface. The peaks are not significantly shifted in potential or broadened by chemical modifications of the electrode (see Figure 6.46). By hybridization, peaks are slightly shifted towards higher oxidation and lower reduction potentials. The change in amplitude is about $50\,\mu A/cm^2$. This change in voltammetric signal is reproducible and can be detected for several hybridization/denaturation cycles. In the future, this technique needs to be characterized in depth to evaluate the sensitivity, durability, and reproducibility of this sensor array.

Yang *et al.* reported in 2004 [50] about cyclic voltammetry experiments using $Fe(CN_6)^{3-/4-}$ on boron doped nanocrystalline diamond films coated with amines as linker and DNA. After ssDNA marker attachment, their redox currents decreased drastically and they concluded that the application of cyclic voltammetry is inhibited by the highly insulating nature of the molecular amine layer linking DNA molecules to diamond. Our voltammetric experiments show that a detailed control of phenyl molecule deposition is required. Insulating properties, with respect to $Fe(CN_6)^{3-/4-}$, are detected if the phenyl layer is grown slowly and thick enough to form a dense scaffold on diamond. By short time attachment, using constant potential attachment technique, dispersed layers are generated, so that diffusion of $Fe(CN_6)^{3-/4-}$ is not suppressed.

Alternatively, Yang *et al.* 2004 [50], Hamers *et al.* 2004 [81] and Gu *et al.* [51] applied impedance spectroscopy, where they detected hybridization induced variations at low frequency. We have also applied such a detection scheme on our grafted diamond layers. A typical result is shown in Figure 6.47. A clear difference between single strand DNA bonded to diamond and double strand DNA is detected. This technique allows discriminating hybridization of matched and mismatched DNA. However, the interpretation of these data requires us to know details about dielectric variations of ssDNA and dsDNA layers, about thickness variations by applied external electric fields [82], as well as the effect of redox molecules like $Fe(CN_6)^{3-/4-}$ on the dielectric and conductivity properties of DNA films on diamond.

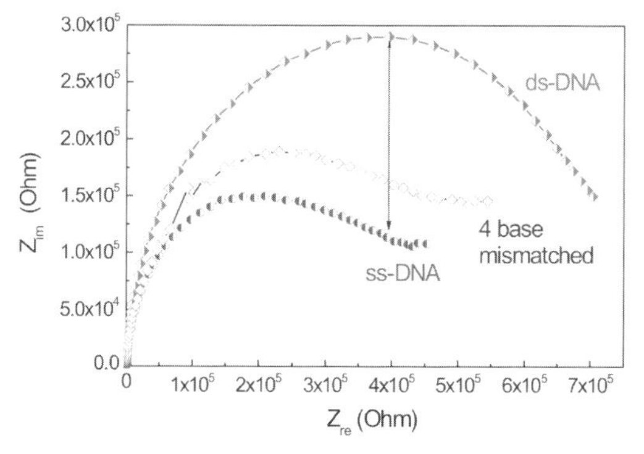

Figure 6.47 Impedance spectroscopic properties of DNA-modified nanodiamond films. The impedance is shown in the complex plane as detected for ssDNA, for exposure to 4-base mismatched DNA and after exposure to complementary DNA in pH 7.4 phosphate buffer containing 1 mM $Fe(CN_6)^{3-/4-}$.

6.4 Bonding and Detection of Enzymes and Proteins on Diamond

After the establishment of surface functionalization techniques for diamond using photo- and electrochemical means, bonding of molecules like enzymes [48] and proteins [49] has been demonstrated, relatively quickly. A detailed characterization of the sensing properties of these devices is, however, still missing. As the use of carbon based transducers like glassy carbon, highly oriented pyrolytic graphite (HOPG), diamond like carbon, and boron-doped diamond is well established in electrochemistry, increasing biosensing applications using such materials can be expected. While the past has been dominated by the search to find chemical treatments to achieve surface modifications of diamond, the future will be dominated by optimization of multilayer structures for biosensing. Whether diamond will be the material of choice for biosensors will depend on the outcome of these research activities. The window of opportunity for diamond is now wide open, however, to compete with other transducer materials which are cheaper, less sophisticated in growth, and compatible with silicon technology will require optimized use of the core advantages of diamond which are: (i) chemical stability, (ii) strong bonding of organic molecules, (iii) low electrochemical background current, (iv) wide electrochemical potential window, (v) no gate insulator for DNA-FETs from diamond, and (vi) perfect surface electronic properties, which includes control of interfacial energy alignments by H- or O-termination to optimize electron transfer reactions.

6.5 Conclusions

In this chapter, we have summarized recently achieved results with respect to photo- and electrochemical surface modifications of diamond. Both techniques have been optimized to a level that has allowed the realization of first generation electrochemical and field effect DNA sensors. By AFM experiments, we have demonstrated the formidable bonding stability of biomolecular arrangements on diamond. This confirms earlier reports by Yang *et al.* 2002 [15] about exceptional DNA bonding stability in hybridization cycles. Therefore, applications of diamond biosensors in high throughput systems, where especially high bonding stability is required, will be of significant interest in the future. Our experiments have also shown that a basic understanding of growth mechanisms, of the electronic and chemical properties of each layer of the composite biorecognition film (for example: amine/crosslinker/ssDNA) is required to achieve progress with respect to optimization of sensor performance.

Finally, the proper selection of diamond transducer materials for biosensor applications is of comparable importance. Electronic detection of bonding events in electrolyte solutions requires high quality diamond with minimized defect densities, no grain boundaries and no sp^2, and with a defect free surface termination. Such diamond transducers are, and will be, more expensive than established semiconducting materials like Si. We assume therefore, that future applications of diamond will be in clinical, high throughput systems which require high bonding stability. Multi-array sensors from single crystalline diamond can be realized using established technologies and chemistry. As prices for established DNA multi-array optical sensors are rather expensive (typically >5000 Euro, see http://www.affymetrix.com), the costs for diamond substrates of typically 150 to 250 Euro will not be a strong argument against the use of diamond for such applications. In any case, diamond needs to show that its device properties and performances are superior, compared to other transducer layers, to be commercially successful.

There are several fields where nano- and polycrystalline diamond seems to become a promising leading application. Nanosize diamond particles with a typical diameter of 10 nm are currently being investigated as core material for rapid, low volume solid phase extraction of analytes, including proteins and DNA, from a variety of biological samples [95–100]. These particles are unique as they are optically transparent, carry specific functional groups, and can mix rapidly with sample solutions upon agitation. After extraction of the target analyte by these particles, the diamond nanoparticles can be recollected and analyzed. Second, nanodiamond is capable to be used as color center marker, as it has several interesting defect related light emitting centers (color centers), like the nitrogen–vacancy (N—V) complex and others, which makes nanodiamond unique with respect to fluorescence microscopy applications. These color centers are resistant against photobleaching and can be taken up by mammalian cells with minimal cytotoxicity [95, 100].

The increasing demand for secure, mobile, wireless communication has stimulated interest in technologies capable of reducing the size and power consumption of wireless modules, and enhancing the bandwidth efficiency of communication networks. Nanodiamond microelectrical mechanical systems (MEMS) are at the leading edge of this field, as the frequency (f) quality (Q) product of such nanodiamond oscillators reached $f = 1.51$ GHz and the quality factor $Q = 11\,555$ [101–103]. In the field of biosensing, such high quality mechanical oscillation systems are also of significance for improved detection and sensibility [104, 105].

Direct manipulation of living cells, or transfer of molecules into cells (cell surgery), is an emerging and increasingly important technology in biology. It requires new tools with dimensions below the micrometer regime. A typical gene surgery tip should be about 40 μm long with a diameter of around 400 nm (for a review see [106]). In addition, the material should not poison the cell during manipulation, the surface should be optimized with respect to friction of cell membranes and it should allow the tip to electrostatically bond with, or release DNA fragments. The best candidates for these applications are nano- and polycrystalline diamond as they show (i) the required mechanical stability (hardness 50 100 GPa), (ii) the surface can be adjusted to optimize friction, and (iii) if properly designed, the core of the tip can be made conductive by boron doping.

Based on these arguments, we are convinced that bioapplications of diamond, either monocrystalline for electronic sensing or poly- and nanocrystalline for mechanical techniques, represent very realistic opportunities for this promising material. Diamond will surely find its place in the rapidly growing field of biology and biotechnology.

Acknowledgments

The authors wish to thank T. Nakamura for the synthesis of TFFAD molecules, which contributed significantly our progress, as well as to T. Yamamoto, who helped to fabricate DNA-FET detectors. We also want to thank Dr Park, who supported our activities with an electrochemical analyzer system, Dr Watanabe, Dr Ri and Dr Tokuda for growth of excellent diamond films. The authors thank *The Journal of the Royal Society Interface* for partial copyright transfer of our review paper with the title: Diamond and biology (2009) Journal of the Royal Society Interface 4, p. 439.

References

1 Etzioni, R., Urban, N., Ramsey, S., McIntosh, M., Schwartz, S., Reid, B., Radich, J., Anderson, G. and Hartwell, L. (2003) The case of early detection. *Nature Reviews Cancer*, **3**, 243–52.

2 Sander, C. (2000) Genomic medicine and the future of health care. *Science*, **287**, 1977–78.

3 Srinivas, P.R., Kramer, B.S. and Srivastava, S. (2001) Trends in biomarker research for cancer detection. *The Lancet Oncology*, **2**, 698–704.

4 Brawer, M.K. (2001) *Prostate Specific Antigen*, Marcel Dekker, New York.

5 Buriak, J.M. (2002) *Chemical Reviews*, **102**, 1271.

6 Wulfkuhle, J.D., Liotta, L.A. and Petricoin, E.F. (2003) Proteomic applications for the early detection of cancer. *Nature Reviews Cancer*, **3**, 267–75.

7 Kremsky, J.N., Wooters, J.L., Dougherty, J.P., Meyers, R.E., Collins, M. and Brown, E.L. (1987) Immobilization of DNA via oligonucleotides containing an aldehyde or carboxylic-acid group at the 5′ terminus. *Nucleic Acids Research*, **15** (7), 2891–909.

8 Rasmussen, S.R., Larsen, M.R. and Rasmussen, S.E. (1991) Covalent immobilization of DNA onto polystyrene microwells – the molecules are only bound at the 5′ end. *Analytical Biochemistry*, **198** (1), 138–42.

9 Millan, K.M., Spurmanis, A.J. and Mikkelsen, S.R. (1992) Covalent immobilization of DNA into glassy-carbon electrodes. *Electroanalysis*, **4** (10), 929–32.

10 Hashimoto, K., Ito, K. and Ishimori, Y. (1994) Sequence-specific gene detection with a gold electrode modified with DNA probes and an electrochemical active dye. *Analytical Chemistry*, **66** (21), 3830–3.

11 Strother, T., Cai, W., Zhao, X.S., Hamers, R.J. and Smith, L.M. (2000) Synthesis and characterization of DNA-modified silicon (111) surfaces. *Journal of the American Chemical Society*, **122** (6), 1205–9.

12 Buriak, J.M. (2002) Organometallic chemistry on silicon and germanium surfaces. *Chemical Reviews*, **102** (5), 1271–308.

13 Bousse, L., de Rooji, N.F. and Bergeveld, P. (1983) Operation of chemically sensitive field-effect sensors as a function

of the insulator electrolyte interface. *IEEE Transaction on Electron Devices*, **30** (10), 1263–70.

14 Linford, M.R., Fenter, P., Eisenberg, P.M. and Chidsay, C.E.D. (1995) Alkyl monolayers on silicon prepared from 1-alkenes and hydrogen-terminated silicon. *Journal of the American Chemical Society*, **117** (11), 3145–55.

15 Yang, W., Auciello, O., Butler, J.E., Cai, W., Carlisle, J.A., Gerbi, J.E., Gruen, D. M., Knickerbocker, T., Lasseter, T.L., Russell, J.N., Jr, Smith, L.M. and Hamers, R.J. (2002) DNA-modified nanocrystalline diamond thin-films as stable, biologically active substrates. *Nature Materials*, **1**, 253–7.

16 Nebel, C.E. and Ristein, J. (eds) (2003) *"Thin-Film Diamond I", Semiconductors and Semimetals*, Vol. 76, Elsevier Academic Press.

17 Nebel, C.E. and Ristein, J. (eds) (2004) *"Thin-Film Diamond II", Semiconductors and Semimetals*, Vol. 77, Elsevier AcademicPress.

18 Jackman, R. (ed.) (2003) *Semiconductor Science and Technology, Diamond Electronics*, Vol. **18** (3).

19 Angus, J.C., Pleskov, Y.V. and Eaton, S.C. (2004) Electrochemistry of diamond, in *"Thin-Film Diamond II", Semiconductors and Semimetals*, Vol. 77 (eds C.E. Nebel and J. Ristein), Elsevier Academic Press, p. 97.

20 Swain, G.M. (2004) Electroanalytical applications of diamond electrodes, in *"Thin-Film Diamond II", Semiconductors and Semimetals*, Vol. 77 (eds C.E. Nebel and J. Ristein), Elsevier Academic Press, p. 121.

21 Fujishima, A., Einaga, Y., Rao, T.N. and Tryk, D.A. (2005) *Diamond Electrochemistry*, Elsevier/BKC-Tokyo.

22 Granger, M.C., Xu, J., Strojek, J.W. and Swain, G.M. (1999) Polycrystalline diamond electrodes: basic properties and applications as amperometric detectors in flow injection analysis and liquid chromatography. *Analytica Chimica Acta*, **397** (1–3), 145–61.

23 Granger, M.C., Witek, M., Xu, J.S., Wang, J., Hupert, M., Hanks, A., Koppang, M.D., Butler, J.E., Lucazeau, G., Mermoux, M., Strojek, J.W. and Swain, G.M. (2000) Standard electrochemical behavior of high-quality, boron-doped polycrystalline diamond thin-film electrodes. *Analytical Chemistry*, **72** (16), 3793–804.

24 Shin, D., Watanabe, H. and Nebel, C.E. (2005) Insulator-metal transition of intrinsic diamond. *Journal of the American Chemical Society*, **127**, 11236–7.

25 Kondo, T., Einaga, Y., Sarada, B.V., Rao, T.N., Tryk, D.A. and Fujishima, A. (2002) Homoepitaxial single-crystal boron-doped diamond electrodes for electroanalysis. *Journal of the Electrochemical Society*, **149** (6), E179–84.

26 Pleskov, Y.V., Evstefeeva, Y.E., Krotova, M.D., Elkin, V.V., Mazin, V.M., Mishuk, V.Y., Varnin, V.P. and Teremetskaya, I.G. (1998) Synthetic semiconductor diamond electrodes: the comparative study of the electrochemical behaviour of polycrystalline and single crystal boron-doped films. *Journal of the Electrochemical Society*, **455** (1–2), 139–46.

27 Landstrass, M.I. and Ravi, K.V. (1998) Resistivity of chemical vapor-deposited diamond films. *Applied Physics Letters*, **55** (10), 975–7.

28 Gi, R.S., Mizumasa, T., Akiba, Y., Hirose, Y., Kurosu, T. and Iida, M. (1995) Formation mechanism of p-type surface conductive layer on deposited diamond films. *Japanese Journal of Applied Physics*, **34**, 5550–5.

29 Shirafuji, J. and Sugino, T. (1996) Electrical properties of diamond surfaces. *Diamond and Related Materials*, **5** (6–7), 706–13.

30 Maier, F., Riedel, M., Mantel, B., Ristein, J. and Ley, L. (2000) Origin of surface conductivity in diamond. *Physical Review Letters*, **85** (16), 3472–5.

31 Sque, S.J., Jones, R. and Briddon, P.R. (2006) Structure, electronics, and interaction of hydrogen and oxygen on diamond surfaces. *Physical Review B*, **73** (8), 085313.

32 Takeuchi, D., Kato, H., Ri, G.S., Yamada, T., Vinod, P.R., Hwang, D., Nebel, C.E., Okushi, H. and Yamasaki, S. (2005) Direct observation of negative electron affinity in hydrogen-terminated diamond surfaces. *Applied Physics Letters*, **86** (15), 152103.

33 Nebel, C.E., Rezek, B. and Zrenner, A. (2004) 2D-hole accumulation layer in hydrogen terminated diamond. *physica status solidi (a)*, **11**, 2432–8.
34 Nebel, C.E., Rezek, B. and Zrenner, A. (2004) Electronic properties of the 2D-hole accumulation layer on hydrogen terminated diamond. *Diamond and Related Materials*, **13**, 2031–6.
35 Nebel, C.E., Rezek, B., Shin, D. and Watanabe, H. (2006) Surface electronic properties of H-terminated diamond in contact with adsorbates and electrolytes. *physica status solidi (a)*, **203** (13), 3273–98.
36 Nebel, C.E. (2005) Surface transfer-doping of H-terminated diamond with adsorbates. *New Diamond and Frontier Carbon Technology*, **15**, 247–64.
37 Chakrapani, V., Eaton, S.C., Anderson, A.B., Tabib-Azar, M. and Angus, J.C. (2005) *Electrochemical and Solid-State Letters*, **8**, E4–8.
38 Piantanida,. G., Breskin, A., Chechik, R., Katz, O., Laikhtman, A., Hoffman, A. and Coluzza, C. (2001) *Journal of Applied physiology*, **89**, 8259.
39 Rao, T.N., Tryk, D.A., Hashimoto, K. and Fujishima, A. (1999) *Journal of the Electrochemical Society*, **146**, 680.
40 Nebel, C.E., Rezek, B., Shin, D., Watanabe, H. and Yamamoto, T. (2006) Electronic properties of H-terminated diamond in electrolyte solutions. *Journal of Applied Physiology*, **99**, 33711.
41 Nebel, C.E., Kato, H., Rezek, B., Shin, D., Takeuchi, D., Watanabe, H. and Yamamoto, T. (2006) Electrochemical properties of undoped hydrogen terminated CVD diamond. *Diamond and Related Materials*, **15**, 264–8.
42 Rezek, B., Shin, D., Watanabe, H. and Nebel, C.E. (2007) Intrinsic hydrogen-terminated diamond as ion-sensitive field effect transistor. *Sensors and Actuators B*, **122**, 596–9.
43 Tang, L., Tsai, C., Gerberich, W.W., Kruckeberg, L. and Kania, D.R. (1995) Biocompatibility of chemical-vapor-deposited diamond. *Biomaterials*, **16** (6), 483–8.
44 Mathieu, H.J. (2001) Bioengineered material surfaces for medical applications. *Surface and Interface Analysis*, **32**, 3–9.
45 Cui, F.Z. and Li, D.J. (2000) A review of investigations on biocompatibility of diamond-like carbon and carbon nitride films. *Surface and Coatings Technology*, **131**, 481–7.
46 Takahashi, K., Tanga, M., Takai, O. and Okamura, H. (2000) DNA bonding to diamond. *BioIndustry*, **17** (6), 44–51.
47 Takahashi, K., Tanga, M., Takai, O. and Okamura, H. (2003) DNA preservation using diamond chips. *Diamond and Related Materials*, **12** (3–7), 572–6.
48 Wang, J., Firestone, M.A., Auciello, O. and Carlisle, J.A. (2004) functionalization of ultrananocrystalline diamond films by electrochemical reduction of aryldiazonium salts. *Langmuir*, **20** (26), 11450–6.
49 Zhang, G.-J., Song, K.S., Nakamura, Y., Ueno, T., Funatsu, T., Ohdomari, I. and Kawarada, H. (2006) DNA Micropatterning on polycrystalline diamond via one-step direct amination. *Langmuir*, **22** (8), 3728–2734.
50 Yang, W., Butler, J.E., Russel, J.N., Jr and Hamers, R.J. (2004) Interfacial electrical properties of DNA-modified diamond thin films: intrinsic response and hybridization-induced field effects. *Langmuir*, **20**, 6778–87.
51 Gu, H., Su, X. and Loh, K.P. (2005) Electrochemical impedance sensing of DNA hybridization on conducting polymer film-modified diamond. *The Journal of Physical Chemistry B*, **109**, 13611–18.
52 Song, K.S., Degawa, M., Nakamura, Y., Kanazawa, H., Umezawa, H. and Kawarada, H. (2004) Surface-modified diamond field-effect transistors for enzyme-immobilized biosensors. *Japanese Journal of Applied Physics, Part 2-Letters and Express Letters*, **43**, (6B), L814–17.
53 Härtl, A., Schmich, E., Garrido, J.A., Hernando, J., Catharino, S.C.R., Walter, S., Feulner, P., Kromka, A., Steinmueller, D. and Stutzmann, M. (2004) Protein-modified nanocrystalline diamond thin films for biosensor applications. *Nature Materials*, **3** (10), 736–42.
54 Williams, O.A. and Nesladek, M. (2006) Growth and properties of nanocrystalline

diamond films. *physica status solidi (a)*, **203** (13), 3375–86.

55 Carlisle, J.A. and Auciello, O. (2003) *Ultrananocrystalline Diamond*, The Electrochemical Society, Interface, Springer, pp. 28–31.

56 Nebel, C.E. (2003) Transport and defect properties of intrinsic and boron-doped diamond, in *"Thin-Film Diamond I"*, *Semiconductors and Semimetals*, Vol. 76 (eds C.E. Nebel and J. Ristein, Elsevier, p. 261.

57 Nebel, C.E. (2003) Electronic properties of CVD diamond. *Semiconductor Science and Technology 1*, **18** (3), S1–11.

58 Nesladek, M., Haenen, K. and Vanecek, M. (2003) *"Thin-Film Diamond I"*, *Semiconductors and Semimetals*, Vol. 76 (eds C.E. Nebel and J. Ristein), Elsevier, p. 325.

59 Bergonzo, P., Tromson, D. and Mer, C. (2003) Radiation detection devices made from CVD diamond. *Semiconductor Science and Technology*, **18**, S105–12.

60 Bergonzo, P. and Jackman, R. (2004) Diamond-based radiation and photon detectors, in *"Thin-Film Diamond I"*, *Semiconductors and Semimetals*, Vol. 76 (eds C.E. Nebel and J. Ristein), Elsevier, pp. 197–309.

61 Cui, J.B., Ristein, J. and Ley, L. (1998) Electron affinity of the bare and hydrogen covered single crystal diamond (111) surface. *Physical Review Letters*, **81** (2), 429–32.

62 Kono, S., Shiraishi, M., Goto, T., Abukawa, T., Tachiki, M. and Kawarada, H. (2005) An electron-spectroscopic view of CVD diamond surface conductivity. *Diamond and Related Materials*, **14**, 459–65.

63 Watanabe, H., Takeuchi, D., Yamanaka, S., Okushi, H., Kajimura, K. and Sekiguchi, T. (1999) Homoepitaxial diamond film with an atomically flat surface over a large area. *Diamond and Related Materials*, **8** (7), 1272–6.

64 Takeuchi, D., Yamanaka, S., Watanabe, H., Sawada, S., Ichinose, H., Okushi, H. and Kajimura, K. (1999) High quality homoepitaxial diamond thin film synthesis with high growth rate by a two-step growth method. *Diamond and Related Materials*, **8** (6), 1046–9.

65 Okushi, H. (2001) High quality homoepitaxial CVD diamond for electronic devices. *Diamond and Related Materials*, **10** (3–7), 281–8.

66 Collins, A.T., Dean, P.J., Lightowler, E.C. and Sherman, W.F. (1965) Acceptor-impurity infrared absorption in semiconducting synthetic diamond. *Physical Review*, **140**, A1272–4.

67 Koizumi, S., Kamo, M., Sato, Y., Ozaki, H. and Inuzuka, T. (1997) Growth and characterization of phosphorous doped {111} homoepitaxial diamond thin films. *Applied Physics Letters*, **71** (8), 1065–7.

68 Borst, T.H. and Weis, O. (1996) Boron-doped homoepitaxial diamond layers: fabrication, characterization, and electronic applications. *physica status solidi (a)*, **154** (1), 423–44.

69 Nebel, C.E., Shin, D., Rezek, B., Tokuda, N. and Uetsuka, H. (2007) Diamond and biology. *Journal of the Royal Society Interface*, Issue: FirstCite Early Online Publishing. Interface 4, p. 439.

70 Nebel, C.E., Shin, D., Takeuchi, D., Yamamoto, T., Watanabe, H. and Nakamura, T. (2006) Photochemical attachment of amine linker molecules on hydrogen terminated diamond. *Diamond and Related Materials*, **15** (4–8), 1107–12.

71 Nebel, C.E., Shin, D., Takeuchi, D., Yamamoto, T., Watanabe, H. and Nakamura, T. (2006) Alkene/diamond liquid/solid interface characterization using internal photoemission spectroscopy. *Langmuir*, **22** (13), 5645–53.

72 Shin, D.C., Tokuda, N., Rezek, B. and Nebel, C.E. (2006) Periodically arranged benzene-linker molecules on boron-doped single-crystalline diamond films for DNA sensing. *Electrochemistry Communications*, **8** (5), 844–50.

73 Shin, D., Rezek, B., Tokuda, N., Takeuchi, D., Watanabe, H., Nakamura, T., Yamamoto, T. and Nebel, C.E. (2006) Photo- and electrochemical bonding of DNA to single crystalline CVD diamond. *physica status solidi (a)*, **203** (13), 3245–72.

74 Frisch, M.J., Trucks, G.W., Schlegel, H.B., Scuseria, G.E., Robb, M.A., Cheeseman, J.R., Zakrzewski, V.G., Montgomery, J.A., Jr, Stratmann, R.E., Burant, J.C., Dapprich, S., Millam, J.M.,

Daniels, A.D., Kudin, K.N., Strain, M. C., Farkas, O., Tomasi, J., Barone, V., Cossi, M., Cammi, R., Mennucci, B., Pomelli, C., Adamo, C., Clifford, S., Ochterski, J., Petersson, G.A., Ayala, P. Y., Cui, Q., Morokuma, K., Salvador, P., Dannenberg, J.J., Malick, D.K., Rabuck, A.D., Raghavachari, K., Foresman, J.B., Cioslowski, J., Ortiz, J.V., Baboul, A.G., Stefanov, B.B., Liu, G., Liashenko, A., Piskorz, P., Komaromi, I., Gomperts, R., Martin, R.L., Fox, D.J., Keith, T., Al-Laham, M.A., Peng, C.Y., Nanayakkara, A., Challacombe, M., Gill, P.M.W., Johnson, B., Chen, W., Wong, M.W., Andres, J.L., Gonzalez, C., Head-Gordon, M., Replogle, E.S. and Pople, J.A. (2001) *Gaussian 98*, Revision A.11, Gaussian, Inc., Pittsburgh, PA.

75 Rezek, B., Shin, D., Nakamura, T. and Nebel, C.E. (2006) Geometric properties of covalently bonded DNA on single-crystal line diamond. *Journal of the American Chemical Society*, **128** (12), 3884–5.

76 Uetsuka, H., Shin, D., Tokuda, N., Saeki, K. and Nebel, C.E. (2007) Electrochemical grafting of boron-doped single crystalline CVD diamond. *Langmuir*, **23**, 3466–72.

77 Yang, N., Uetsuka, H., Watanabe, H., Nakamura, T. and Nebel, C.E. (2007) Photochemical amine layer formation on H-terminated single-crystalline, CVD diamond. *Chemistry of Materials* **19**, 2852.

78 Furthmüller, J., Hafner, J. and Kresse, G. (1996) Dimer reconstruction and elecronic surface states on clean and hydrogenated diamond (100) surfaces. *Physical Review B*, **53**, 7334–50.

79 Nichols, B.M., Butler, J.E., Russel, J.N., Jr and Hamers, R.J. (2005) Photochemical functionalization of hydrogen-terminated diamond surfaces: a structural and mechanistic study. *The Journal of Physical Chemistry B*, **109** (44), 20938–47.

80 Cicero, R.L., Chidsey, E.D., Lopinski, G.P., Wayner, D.D.M. and Wolkow, R. A. (2002) Olefin additions on H-Si(111): evidence for a surface chain reaction initiated at isolated dangling bonds. *Langmuir*, **18** (2), 305–7.

81 Hamers, J.R., Butler, J.E., Lasseter, T., Nichols, B.M., Russell, J.N., Jr, Tse, K.-Y. and Yang, W. (2004) Molecular and biomolecular monolayers on diamond as an interface to biology. *Diamond and Related Materials*, **14**, 661–8.

82 Kelley, S.O., Barton, J.K., Jackson, N.M., McPherson, L.D., Potter, A.B., Spain, E.M., Allen, M.J. and Hill, M.G. (1998) Orienting DNA helices on gold using applied electric fields. *Langmuir*, **14** (24), 6781–4.

83 Erts, D., Polyakov, B., Olin, H. and Tuite, E. (2003) Spatial and mechanical properties of dilute DNA monolayers on gold imaged by AFM. *The Journal of Physical Chemistry B*, **107** (15), 3591–7.

84 Poghossian, A., Cherstvy, A., Ingebrandt, S., Offenhaeusser, A. and Schoening, M. J. (2005) Possibilities and limitations of label-free detection of DNA hybridization with field-effect-based devices. *Sensors and Actuators B*, **111–112**, 470–80.

85 Takeuchi, D., Yamanaka, S., Watanabe, H. and Okushi, H. (2001) Device grade B-doped homoepitaxial diamond thin films. *physica status solidi (a)*, **186**, 269–80.

86 Pinson, J. and Podvorica, F. (2005) Attachment of organic layers to conductive or semiconductive surfaces by reduction of diazonium salts. *Chemical Society Reviews*, **34**, 429–39.

87 Allongue, P., Henry de Villeneuve, C., Cherouvrier, G., Cortes, R. and Bernard, M.-C. (2003) Phenyl layers on H-Si(111) by electrochemical reduction of diazonium salts: monolayer versus multilayer formation. *Journal of Electroanalytical Chemistry*, **550–551**, 161–74.

88 Xu, S., Miller, S., Laibinis, P.E. and Liu, G.Y. (1999) Fabrication of nanometer scale patterns within self-assembled monolayers by nanografting. *Langmuir*, **15** (21), 7244–51.

89 Zhou, D.J., Sinniah, K., Abell, C. and Rayment, T. (2002) Use of atomic force microscopy for making addresses in DNA coatings. *Langmuir*, **18** (22), 8278–81.

90 Schwartz, P.V. (2001) Meniscus force nanografting: nanoscopic patterning of DNA. *Langmuir*, **17** (19), 5971–7.

91 Crampton, N., Bonass, W.A., Kirkham, J. and Thomson, N.H. (2005) Formation of aminosilane-functionalized mica for atomic force microscopy imaging of DNA. *Langmuir*, **21** (17), 7884–91.

92 Ishige, Y., Shimoda, M. and Kamahori, M. (2006) Immobilization of DNA probes onto gold surface and its application to fully electric detection of DNA hybridization using field-effect transistor sensor. *Japanese Journal of Applied Physics*, **45** (4B), 3776–83.

93 Yang, J.-H., Song, K.-S., Kuga, S. and Kawarada, H. (2006) Characterization of direct immobilized probe DNA on partially functionalized diamond solution-gate field-effect transistors. *Japanese Journal of Applied Physics*, **45** (42), L1114–17.

94 Kelley, S.O., Boon, E.M., Barton, J.K., Jackson, N.M. and Hill, M.G. (1999) Single-base mismatch detection based on charge transduction through DNA. *Nucleic Acids Research*, **27** (24), 4830–7.

95 Chang, H.-C., Chen, K.W. and Kwok, S. (2006) Nanodiamond as a possible carrier of extended red emission. *Astrophysical Journal*, **639** (2), L63–6, Part 2.

96 Lee, J.-K., Anderson, M.W., Gray, F.A., John, P. and Lee, J.-Y. (2005) Reactions of amines with CVD diamond nanopowders. *Diamond and Related Materials*, **14**, 675–8.

97 Wu, E., Jacques, V., Treussarta, F., Zeng, H., Grangier, P. and Roch, J.-F. (2006) Single-photon emission in the near infrared from diamond color center. *Journal of Luminescence*, **119–120**, 19–23.

98 Treussart, F., Jacques, V., Wu, E., Gacoin, T., Grangier, P. and Roch, J.-F. (2006) Photoluminescence of single color defects in 50 nm diamond nanocrystals. *Physica B*, **376–377**, 926–9.

99 Wei, P.-K., Tsao, P.-H., Chang, H.-C., Fann, W., Fu, C.-C., Lee, H.-Y., Chen, K., Lim, T.-S., Wu, H.-Y. and Lin, P.-K. (2007) Characterization and application of single fluorescent nanodiamonds as cellular biomarkers. *Proceedings of the National Academy of Sciences of the United States of America*, **104** (3), 727–32.

100 Schrand, A.M., Huang, H., Carlson, C. and Schlager, J.J., Oh Sawa, E., Hussain, S.M. and Dai, L. (2007) Are diamond nanoparticles cytotoxic? *The Journal of Physical Chemistry B*, **2007** (111), 2–7.

101 Wang, J., Butler, J.E., Hsu, D.S.Y. and Nguyen, C.T.-C. (2002) *High-Q Micromechanical Resonators in CH4-Reactant-Optimized High Acoustic Velocity CVD Polydiamond*. Solid State Sensor, Actuator, and Microsystems Workshop Proceedings, Hilton Head Island, SC, pp. 61–2.

102 Wang, J., Butler, J.E., Feygelson, T. and Nguyen, C.T.-C. (2004) *1.51-GHz Polydiamond Micromechanical Disk Resonantor with Impedance Mismatch Isolating Support*. Proceedings of the 17th IEEE Micro Electro Mechanical Systems Conference, Maastrich, The Netherlands, 25–29 January, pp. 641–4.

103 Sekaric, L., Parpia, J.M., Craighead, H.G., Feygelson, T., Houston, B.H. and Butler, J.E. (2002) Nanomechanical resonant structures in nanocrystalline diamond. *Applied Physics Letters*, **81**, 4455–7.

104 Braun, T., Barwich, V., Ghatkesar, M.K., Bredekamp, A.H., Gerber, C., Hegner, M. and Lang, H.P. (2005) Micromechanical mass sensors for biomolecular detection in physiological environment. *Physical Review E*, **72**, 31907.

105 Ilic, B., Yang, Y., Aubin, K., Reichenbach, R., Krylov, S. and Craighead, H.G. (2005) Enumeration of DAN molecules bound to nanomechanical oscillator. *Nano Letters*, **5** (5), 925–9.

106 Han, S.W., Nakamura, C., Obataya, I., Nakamura, N. and Miyake, J. (2005) Gene expression using an ultrathin needle enabling accurate displacement and low invasiveness. *Biochemical and Biophysical Communications*, **332**, 633–9.

7
Diamond-Based Acoustic Devices

Vincent Mortet, Oliver Williams and Ken Haenen
Hasselt University, Institute for Materials Research (IMO), Wetenschapspark 1, B-3590 Diepenbeek, Belgium
IMEC vzw, Division IMOMEC, Wetenschapspark 1, B-3590 Diepenbeek, Belgium

7.1
Introduction

The genesis of acoustic wave devices comes at the end of the nineteenth century with the discovery of piezoelectricity [1] and the proof of elastic vibrations at the surface of solid materials [2]. The first application of piezoelectricity was to emit and receive acoustic waves under water (SONAR). Later on, piezoelectric materials have been used as oscillators in radio transmission (quartz crystals). In the 1960s, the major discovery of interdigital transducers [3] lead to the development of surface acoustic wave devices (SAW). Nowadays, SAW filters are widely used in radiofrequency applications with operating frequencies up to the gigahertz range.

Acoustic waves are highly sensitive to small perturbations; hence, acoustic devices can be operated as sensors. The first acoustic sensor was the so-called quartz crystal microbalance (QCM). QCMs were analyzed and improved by a succession of workers starting in the 1950s. They consist of a quartz crystal, initially made to stabilize the frequencies of radio transmitters, coated with a sorptive film. The next important step in acoustic sensors was made in the late 1970s when Wohltjen and Dessy made a chemical vapor sensor using a surface acoustic wave delay line [4]. More recently, acoustic plate mode (APM) and flexural plate wave (FPW) sensors were introduced. They employ similar principles, but exploit different acoustic propagation modes. In the last decade, two new types of acoustic devices have been developed for sensing applications: film bulk acoustic resonators (FBARs) and microcantilevers (MCs). FBARs were first developed for signal treatment, while MCs were first developed for scanning probe microscopy (SPM).

Despite the fact that diamond is not piezoelectric, diamond has an important role to play in the development of composite acoustic devices because of its outstanding mechanical properties. The first successful acoustic application of

Physics and Applications of CVD Diamond. Satoshi Koizumi, Christoph Nebel, and Milos Nesladek

ISBN: 978-3-527-40801-6

diamond was in the field of surface acoustic wave (SAW) filters. With the development of biotechnologies, many researchers have been working on diamond for sensor applications [5–7] due to diamond's chemical stability and bio-inertness. Thus, the combination of diamond properties and the high sensitivity of acoustic sensors is very appealing.

In this chapter, we will first consider the interest of diamond layers for acoustic applications. Then, we are going discuss the progress that has been made in R&D on diamond-based SAW filters. We will also review and compare the different types of acoustic devices and sensors. We will present the latest results on the development of composite acoustic sensors based on diamond.

7.2 Diamond Layers

Diamond is highly attractive for many different types of composite acoustic devices because of its mechanical properties. Diamond's extreme values of hardness, Young's modulus (E), stiffness and fracture strength (σ_f) make it suitable for flexural plate wave (FPW) sensors and microcantilever (MC) sensors. Those values surpass those of Si, Si_3N_4 and SiC [11] (see Table 7.1) which are commonly used for MEMS and sensors applications. The large sound velocity and high thermal conductivity of diamond have opened up possibilities to produce SAW filters operating in the GHz range since the operating frequency of SAW filters is proportional to the acoustic wave velocity. The extreme chemical stability and bio-inertness [12] of diamond make it an ideal material for sensors operating in harsh or biologic environments. Most significantly, the surface of diamond is particularly stable when functionalized with biomolecules [13]. This makes diamond of particular interest for biosensing applications.

Not withstanding diamond's exciting properties, diamond layers suffer in general from a high surface roughness due to van der Drift growth regime of CVD diamond. Both the grain size and roughness of CVD diamond layers increase with the film thickness [14]. Root mean square (RMS) roughness is in the range of 15 to 40 nm for a 1 μm thick film [15, 16] and several micrometers roughness is usual

Table 7.1 Stiffness and fracture strength of different materials compared to CVD diamond.

	Stiffness (GPa)	Fracture strength (GPa)
CVD diamond [8]	900–1000	2.2
Natural diamond [9]	1345	2.8
SiC [10]	140–450	0.1–0.3
Silicon [10]	180	0.3

for thick (>100 μm) polycrystalline CVD diamond films [14, 15]. Surfaces for acoustic applications must be flat to prevent wave scattering and propagation losses. Surface roughness must also be compatible with the photolithographic pattern resolution and deposition of piezoelectric thin films.

Several approaches have been successfully used to obtain flat diamond surfaces. Despite its extreme difficulty, diamond polishing has been the primary method. It leads to smooth and large polycrystalline diamond wafers. Bi *et al.* [17] have used nanocrystalline diamond (NCD) films with RMS roughness of 50 nm, while Mortet *et al.* [18] have used the flat nucleation side of thick freestanding diamond layer grown on silicon substrates. The roughness varies with the nucleation process, that is 30 to 50 nm between 30 to 50 nm for mechanical nucleation and ~10 nm for bias enhanced nucleation.

Growth of uniform and pinhole-free thin diamond layers, necessary for FPW, MCs, and MEMS applications, is not easy due to the diamond nucleation mechanism on foreign substrates. Nucleation density is a key parameter to control the uniformity and the roughness of thin diamond films. Diamond nucleates and grows as individual grains until they coalesce to form a continuous film. Thus, the film's coalescence thickness depends on the nucleation density. Thin coalesced films with small grains size are obtained together with a high nucleation density [19, 20]. Sub-micron thick diamond membranes of ~10 mm diameter have been produced by increasing the diamond nucleation density above 10^{10} cm^{-2} [21]. NCD films grow in the van der Drift regime, its grain size and hence roughness increase with film thickness. Above 1 μm thickness, NCD films become microcrystalline. Ultra-nanocrystalline diamond (UNCD) is of particular interest for acoustic applications compared to nanocrystalline diamond. UNCD consists of a fine grain (3–5 nm) material grown with a high renucleation rate and a low surface roughness [22, 23] independent of film thickness.

High stiffness and tensile stress are needed in order to avoid any mechanical in-plane and out-of-plane distortion of acoustic devices. Aikawa *et al.* [8] have shown that smaller grain sizes and high phase purity, which depends on the substrate pretreatment conditions and the deposition parameters, are required to obtain membranes with high fracture strength. Furthermore, diamond films often contain large residual stress, thus a tight temperature control of the whole substrate is needed during the deposition.

7.3 Acoustic Sensors

Electrochemical, optical and acoustic wave sensing devices have emerged as some of the most promising biochemical sensor technologies. Together with the development of these technologies, many researchers have been working on diamond for sensor applications due to its chemical stability and bio-inertness.

Acoustic sensors operate by detecting changes in the resonant response. They are highly sensitive. They are sensitive to variations of mechanical stress, temperature, damping and mass loading. Acoustic sensors have already been used to study physical, chemical and biological properties of gases and liquids for decades.

Acoustic sensors exist in many different configurations (see Figure 7.1), which can be classified into two main categories: bulk acoustic wave (BAW) devices and surface acoustic waves (SAW) devices. Acoustic waves are propagated through the material in BAW devices and they travel along or near the material surface in SAW devices. In addition to BAW and SAW devices, microcantilever sensors are another and new type of acoustic sensor. Both BAW and SAW sensors use longitudinal waves or shear waves (SH). Longitudinal waves have particle displacements parallel to the direction of wave propagation while shear waves have particle displacements normal to the direction of wave propagation. The particle displacements in shear waves are either normal to the sensing surface (vertical shear wave) or parallel to the sensing surface (horizontal shear wave). Acoustic devices operating with horizontal shear waves are of particular interest since there are no acoustic losses when operated in liquids compared to vertical shear waves. The nature of acoustic waves generated in piezoelectric materials is determined by the piezoelectric material orientation as well as the metal electrodes configuration employed to generate the electric field that induces acoustic waves. There are several important parameters for a sensor: the cost, the size, the robustness, the reliability, the dynamic response range, the response time, the quality factor (Q) [24], the detection limit and the sensitivity. The detection limit and the relative (S_r) mass sensitivity can be used to compare the different types of acoustic sensors (see Table 7.2):

$$S_r = \lim_{\Delta\mu \to 0} \frac{1}{f} \frac{\Delta f}{\Delta\mu} = \frac{1}{f} \frac{df}{d\mu} \tag{7.1}$$

where f is the resonant frequency, and μ is the mass/area ratio.

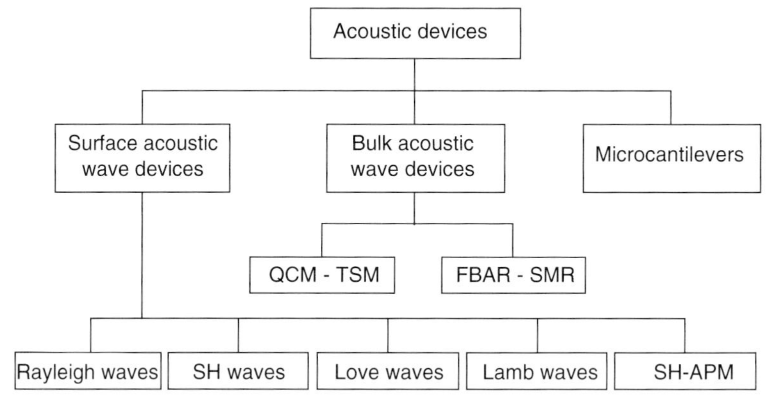

Figure 7.1 Different types of acoustic devices that can be used for sensing application.

Table 7.2 Comparison of the relative sensitivity and minimum detectable mass density of the different acoustic sensors.

Sensors	Sensitivities ($cm^2\ g^{-1}$)	Minimum detectable mass density ($ng\ cm^{-2}$)
Flexural plate wave	100–1000	~0.5
SH-SAW	65	1
FBAR	1000–10 000	0.1–0.01
Quartz microbalance	10	10
SAW	151	1.2
Micro-cantilever	1000–10 000	0.02–0.04

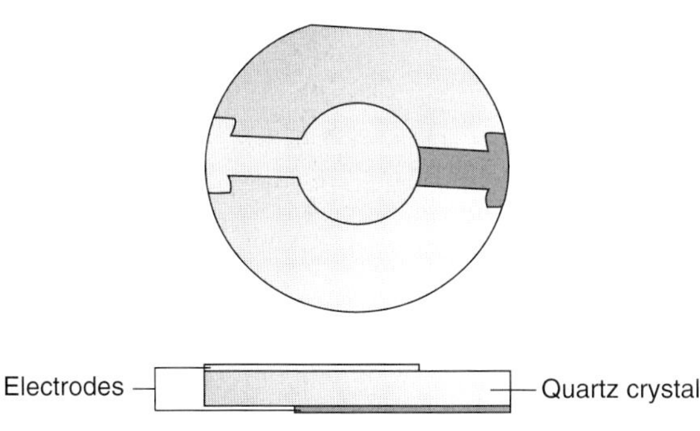

Figure 7.2 Schematic of a QCM.

7.3.1 Bulk Acoustic Wave Sensors

7.3.1.1 Quartz Crystal Microbalance

In 1959, Sauerbrey was the first to relate resonant frequency change of a piezoelectric quartz crystal to mass attachment at its surface, establishing the quartz crystal microbalance (QCM). QCMs are a particular case of thickness shear mode (TSM) resonators using quartz. QCMs are typically a few hundred micrometers thick and ~10 mm in diameter with a fundamental resonant frequency in the 3–30 MHz range. Either shear or longitudinal waves are excited by an AC voltage applied to the electrodes placed on the two sides of the quartz crystal (see Figure 7.2), depending on the crystal cut. The fundamental resonant frequency (f_0) is determined by the thickness of the piezoelectric material and its acoustic wave velocity (Equation 7.1). The relative sensitivity (S_r) is in inverse proportion with the sensor thickness (Equation 7.2):

$$f_0 = \frac{v}{2h_0} \tag{7.2}$$

$$S_r = \frac{-2 \cdot f_0}{\rho_p \cdot v} = -\frac{1}{\rho_p \cdot h_0} \tag{7.3}$$

with v, the sound velocity (longitudinal or shear); h_0, the quartz crystal thickness and ρ_p, the piezoelectric material density.

Quartz crystal microbalance sensors are simple to use and robust, but their resonant frequency is limited by the mechanical thinning process. They have a low relative sensitivity ~70 cm^2/g compared to others acoustic sensors with a mass resolution ~10 ng/cm^2 at a fundamental resonant frequency.

7.3.1.2 Film Bulk Acoustic Resonators

Basically identical to QCM, thin film bulk acoustic resonators (FBAR) consist of a thin (from 100 nm to few micrometers) piezoelectric thin film (ZnO, AlN) deposited by physical vapor deposition (PVD) and/or chemical vapor deposition (CVD) techniques and sandwiched between two metallic electrodes (see Figure 7.3). Their dimensions can be scaled down to a few tens of micrometers. FBAR were first developed for high frequency filtering applications [25, 26]. The fundamental difference is in the operating frequency. FBARs have a fundamental resonant frequency around 1GHz due to the thinner piezoelectric layer. This also means higher mass sensitivity (see Equation 7.3).

Like TSMs, FBARs need interfaces that efficiently confine waves into the piezoelectric material. Ideally, and in most of the cases, these interfaces are solid/air interfaces although it is possible to use acoustic mirrors at one interface which allow them to have solidly mounted resonators (SMRs). FBARs and SMRs generally use longitudinal waves, which make them inappropriate for operation in liquids. This is due to the difficulty of controlling the crystalline orientation of thin piezoelectric films. Recently new techniques have been developed to deposit c-axis inclined ZnO and AlN [27–30] that generate both longitudinal and shearing waves and thus, can be operated in liquids. The use of FBARs for biochemical sensing applications is recent [31]. FBARs and SMRs have a high relative sensitivity ~1000 cm^2/g at operating frequency ~1GHz.

7.3.2 Surface Acoustic Wave Devices

Even though surface acoustic waves were discovered by Lord Rayleigh in 1885, the use of metallic interdigital transducers (IDTs) to simply and efficiently convert

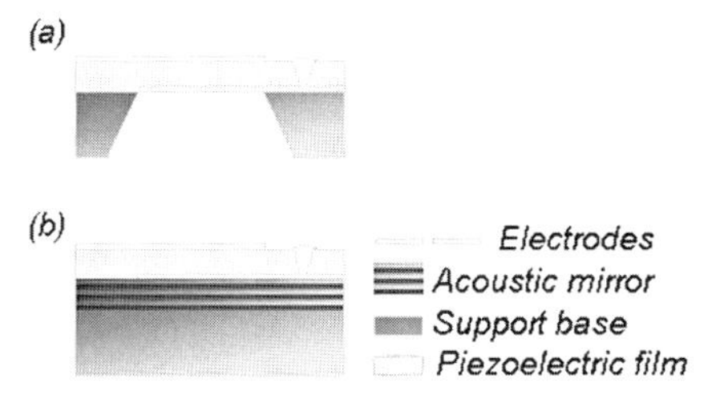

Figure 7.3 Schematic of a FBAR (a) and a SMR (b).

surface acoustic wave signals into electrical signals and vice versa was only demonstrated in 1965 [3]. A surface acoustic wave device consists of two pairs of metallic interdigital transducers deposited on a piezoelectric material, for example, quartz, $LiNbO_3$ and separated by an integer number of the spatial period λ (see Figure 7.4). Surface acoustic waves are generated by converse piezoelectric effect at the input IDT and they are converted back into an electric signal by direct piezoelectric effect. SAW devices were first used for military high frequency signal filtering applications, but nowadays they are commonly used in televisions, video recorders, and mobile telephones. There exist many different types of surface acoustic waves that are used for sensor applications.

- Named after their discoverer, Rayleigh waves have longitudinal and vertical shear components. Because of the shear component that couple with the medium in contact with the device's surface, SAW devices operating with Rayleigh waves cannot operate in liquids. They are generally used to make high frequency filters and gas sensors.
- Shear horizontal-surface acoustic waves (SH-SAW) have particle displacements perpendicular to the propagation direction and parallel to the surface. Thus, they are adequate for sensing applications in liquids.
- Also named after their discoverer [32], Love waves are a particular form of SH surface waves and they can be used for sensors operating in liquids. They are formed by the constructive interference of multiple reflections at the thin coating interface layer that has an acoustic velocity lower than the substrate. Love waves are dispersive, that is, the wave velocity is not solely determined by the material constants, but also by the ratio between the thickness of the piezoelectric layer and the wave length defined by the IDTs special periode.
- Bulk acoustic waves generated by IDTs can be confined between the upper and lower surfaces of a plate that acts as an acoustic waveguide. As a result, both sides of the plate are vibrating. Thus, IDTs can be placed on one side of the plate and the other side can be used for sensing purpose. Shear horizontal-acoustic plate mode sensors (SH-APM) can also be operated in liquids.
- If the substrate thickness is smaller than the wavelength (membrane), longitudinal and flexural waves, called Lamb waves, are generated. Devices based on such waves, also called flexural plate wave sensors (FPWs) are of

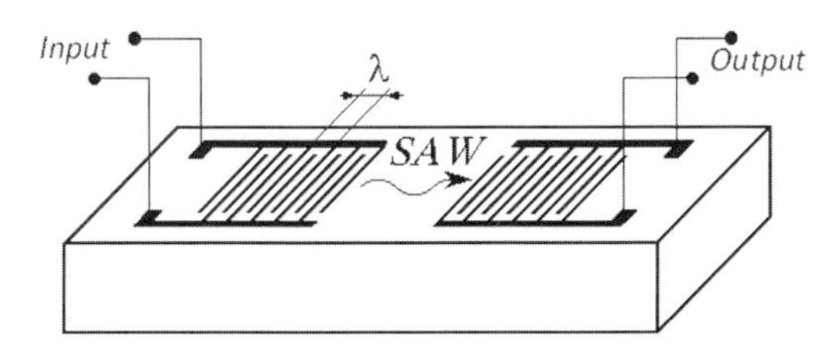

Figure 7.4 Schematic of a surface acoustic wave device.

special interest since they are more sensitive than other SAW sensors, and they can be operated in liquids. Their sensitivity is inversely proportional to density (ρ) and the thickness (t) of the membrane: $S_r = -2/\rho.t$.

7.3.3 Microcantilevers

Microcantilevers are mainly used in atomic force microscopy (AFM). They are use to transduce minute forces (from 1 nN up to 10 μN) that interact with a sharp tip at the cantilever's free end. The interest in microfabricated cantilevers has grown with the development of the atomic force microscope. MCs are simple mechanical devices. Micromachined cantilevers are extremely sensitive, miniature, mass produced, and cheap. Most microcantilevers are made of silicon. They are tiny plates of leaf spring, typically 0.2–1 μm thick, 10–100 μm wide and 100–500 μm long, which are connected at one end to an appropriate support for convenient handling. Their first use as sensor was reported in 1994 [33]. They operate either in static bending mode or in dynamic mode where the resonant frequencies of the cantilever are monitored. While there exist many different types of excitation (electrostatic, piezoelectric, resistive or optical heating, magnetic) and detection methods (capacitive, piezoresistive, magnetic) when operated in dynamic mode, the deflection of the cantilever is generally monitored using an optical lever. In this technique, visible light from a low power laser reflected on the free apex of the cantilever is displaced as the cantilever bends. This displacement is converted into an electronic signal by projecting the reflected laser beam onto a position sensitive photodetector. Microcantilevers are sensitive to external forces, temperature, damping, and mass loading. The relative sensitivity (S_r) of a microcantilever is shown in Equation 7.4:

$$S_r = \frac{1}{\rho \cdot t} \tag{7.4}$$

where ρ is the density of the cantilever material and t is the cantilever thickness. Microcantilevers have a relative sensitivity ~1000 cm^2/g at an operating frequency of few a tens to hundred kilohertz.

Except for microcantilever sensors, which have been recently developed, all the described acoustic devices have proven to be excellent gravimetric sensors. The use of diamond for these devices is new and appealing because of the excellent mechanical properties of diamond. The recent progress in diamond growth allows researchers to integrate diamond in the acoustic sensors field. FPW sensors, FBARs, and MCs seem particularly promising sensors due to their high sensitivity. FPW devices are particularly promising since they operate in liquids. FBARs and MCs are miniature and compatible with the integrated circuit technology. Composite FBAR operation in liquids has already been demonstrated [34].

7.4 Diamond Acoustic Devices

7.4.1 Surface Acoustic Wave Devices

Surface acoustic wave (SAW) filters are used in a wide range of applications such as keyless entry devices, radio frequency modems and telemetry systems, alarm systems, and so on. The increasing volume of information and communication media has produced a growing demand for high performance surface acoustic wave (SAW) devices, operating in the GHz frequency range. In filtering applications, the electrical signal to be filtered is converted to an acoustic signal at the first interdigital transducer (IDT) by the converse piezoelectric effect and is reconverted to an electrical signal at the second IDT by the direct piezoelectric effect. Filtering is performed during these conversions and it is directly related to the geometry of the IDTs and the mechanical and piezoelectric properties of the piezoelectric material. The operating frequency *(f)* of SAW devices is proportional to the acoustic wave velocity and inversely proportional to wavelength (λ), which is equal to the period of interdigital transducers, as shown in Equation 7.5.

$$f = \frac{v}{\lambda} \tag{7.5}$$

The operating frequencies of SAW filters based on standard piezoelectric materials remain under 1GHz due the limitation of optical photolithography resolution. There are several approaches to achieving higher operation frequencies on SAW filters: (i) to reduce the wavelength, that is, the period of IDTs, (ii) to use substrates with higher propagation velocity, or (iii) to use different propagation modes with high propagation velocities, for example, leaky waves [35]. The most attractive approach is to use substrates with high propagation velocities such as silicon, sapphire, SiC or diamond. Diamond is the most suitable material for acoustic parts because of its highest acoustic wave velocities (see Table 7.3). But these materials are not piezoelectric and thus they must be combined with piezoelectric films.

The use of diamond in surface acoustic wave filters has been studied after the discovery and the development of CVD method to grow diamond. Several piezoelectric materials ($KNbO_3$ [36], ZnO [17, 37–39], PZT, $LiTaO_3$ [39], $LiNbO_3$ [39], AlN [40, 41]) in combination with diamond substrates and structures, including SiO_2 and shorting metal layers have been investigated theoretically and experimentally. Theoretical calculation of the different Rayleigh wave propagation modes is difficult due to the multilayer structure. The problem is overcome by numerical calculation methods. Diamond-based SAW devices are dispersive, that is, the wave velocity is not solely determined by the material constants but also by the ratio between the thickness of the piezoelectric layer and the wave length defined by the IDTs. For the ZnO/diamond structure, wave velocities from 6400 to

Table 7.3 Longitudinal acoustic wave velocity: $V_L = (C_{11}/\rho)^{1/2}$ and transversal acoustic wave velocity: $V_T = (C_{44}/\rho)^{1/2}$ sound velocities along the (100) direction in different materials.

Materials	V_L (km/s)	V_T (km/s)
Diamond	17.52	12.82
Cubic boron nitride	15.4	11.8
4H-SiC and 6H-SiC	12.5	7.1
3C-SiC	9.5	4.1
Silicon	8.43	5.84
AT cut quartz	5.96	3.31
PZT	4.5	2.2
AlN	11.37	6.09
ZnO	6.33	2.88

10 800 m s^{-1} and electromechanical coupling coefficient K^2 from 3 to 7% were calculated. Wave velocity of 11 090 m s^{-1} and K^2 of 1.4% were calculated for the AlN/diamond structure. The $LiNbO_3$/diamond structure is very attractive since it is theoretically possible to obtain waves velocities of 12 000 m s^{-1} and a K^2 of 9% but it also suffers from the high temperature coefficient of frequency (TCF) of $LiNbO_3$, that is, the drift of the central frequency with temperature variation. The SAW velocities of these structures are twice as large as those of conventional SAW materials (see Table 7.3) and their K^2 values are large enough for practical applications [42].

The most intensively and experimentally studied structure is the ZnO/diamond structure. This is probably due to the relative ease to deposit these materials with good piezoelectric properties, usually by sputtering techniques. The use of a SiO_2 layer was introduced to reduce/cancel the temperature coefficient of frequency (TCF) of the ZnO/diamond SAW filters. Since the 1990s, Sumitomo Electronic Ltd has developed and commercialized ZnO/diamond SAW filters with an operating frequency from 1.8 GHz to 3.8 GHz, on polished CVD diamond films. These filters operate on the second Sezawa mode with $\lambda = 4$ μm, have a high electromechanical coupling coefficient $K^2 = 1.1\%$, a high phase velocity v = 9500 m/s, low insertion loss (7–9 dB) with 50 dB rejection, zero temperature deviation and a quality factor Q ~630–750 (Sezawa wave: second acoustic mode of the Rayleigh type in layered solid structures). The power durability of ZnO/diamond SAW filters and $LiTaO_3$ SAW filters has been compared. Diamond SAW filters have superior high power durability at even 3.5 times higher frequency [43].

AlN/diamond SAW filters have also been experimentally studied by different authors [18, 44]. Compared to ZnO/diamond structrure, AlN/diamond structure has higher phase velocities, from 6000 to 12 000 m/s. Mortet *et al.* have used the flat nucleation side of a thick freestanding polycrystalline CVD diamond to study AlN/diamond SAW filters (see Figure 7.5). This method presents the advantage

Figure 7.5 Picture of AlN/diamond SAW filter made on a freestanding CVD diamond layer.

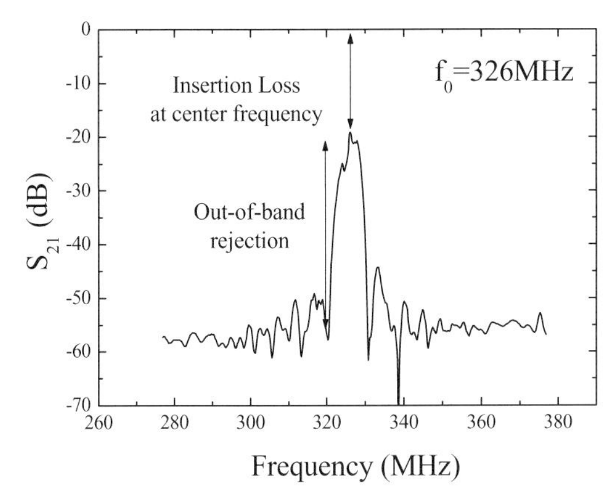

Figure 7.6 Frequency response of AlN/diamond SAW filter at the first Sezawa propagation mode with IDT periode of 32 mm and a normalized AlN film thickness $2\pi.h_{AlN}/\lambda = 1.92$, h_{AlN}: the AlN film thickness.

that it does not need a difficult diamond polishing step at the expenses of longer deposition time to obtain a mechanically stable substrate. Figure 7.6 shows the frequency response of such SAW filter with an operating frequency at 326 MHz for a wavelength of 32 μm. This filter operates on the first Sezawa mode with a phase velocity of 10400 m/s, it has low insertion loss (<20 dB) and a high band rejection (~25 dB). Higher electromechanical coupling coefficients and lower TCF values than ZnO/diamond structures have also been measured on this structure [45]. The operating frequencies of Rayleigh SAW devices on diamond substrates remain under 2–3 GHz due the limitation of optical photolithography resolution. The combination of the high velocity of diamond with the fine resolution of the electron beam lithography allows the realization of SAW devices with higher operating frequency (4.6 GHz) as it has been shown by Kirsch *et al.* on an AlN/diamond structure with spatial IDT periodicity of $\lambda = 2$ μm [46]. This device, which operates at the first Sezawa mode, has 34 dB insertion loss and more than 20 dB band rejection.

On the other hand, Benetti *et al.* have studied the propagation of a pseudo-surface-acoustic-wave on AlN/diamond. They reported a high value of phase velocity (16 000 m/s) operating at ~2 GHz, using an IDT with a line width resolution of 2 μm. However this device suffers from high insertion loss [47].

Insertion losses are the sum propagation loss and the IDT bidirectional loss. The bidirectional loss can be reduced using unidirectional IDTs. The contribution of the diamond substrate to propagation loss has been studied as a function of the diamond grain size by Fujii *et al* [48]. Smaller grain size, narrower grain distribution, and preferential grain orientation reduce the propagation loss of SAW on polycrystalline diamond films without influencing the TCF, K^2 and phase velocity of the devices [48, 49]. These results are consistent with the results of Elmasria *et al.* who observed that the propagation losses on the nucleation side of freestanding CVD diamond are three times lower in the case of bias enhanced nucleation compared to mechanically seed layers [50], since bias enhanced nucleation leads to high nucleation density, smaller grain size and lower surface roughness without the need of a polishing step.

Despite the interest in using diamond, there is, to our knowledge, little research on SAW sensors using diamond substrates. Benetti *et al.* have reported that AlN/diamond SAW filters can operate as CO and ethanol sensors in the gas phase using Co-tetra-phenyl-porphyrin as sorptive film [51]. This fact might be due to the fact that SAW devices based on Rayleigh or Sezawa waves are significantly attenuated and can not be operated in liquids. It can also be related to the problem to deposit piezoelectric layer with appropriate crystalline orientation to generate shear waves SAW devices and the competition with well established piezoelectric materials used for SAW sensors.

7.4.2
Flexural Plate Wave Sensors

Flexural plate wave (FPW) sensors are promising devices for chemical and biological sensing. The basic FPW device consists of a rectangular diaphragm (a few micrometers thick) coated with a piezoelectric layer. The piezoelectric material's thickness is generally 0.2–1 μm with a pair of ITD electrodes as any other type of SAW device. Conducting interdigitated electrodes are placed on the piezoelectric material. In the FPW, as opposed to the surface acoustic wave (SAW), the diaphragm is assumed to be thin compared to the vibrating mode's wavelength so that the two surfaces are strongly coupled and a single wave propagates along the diaphragm [52]. Several materials combinations have already been studied for FPW sensors [53–55] but none using a diamond layer. Diamond, because of its high Young's modulus and high fracture strength, allows the fabrication of less fragile and thinner membranes, and thus FPW sensors with higher sensitivities than standard materials. L. A. Francis *et al.* have characterized nanocrystalline membranes for FPW sensors applications and they demonstrate better detection limit for diamond membranes [56].

7.4.3 Bulk Acoustic Wave Devices

The quartz crystal microbalance (QCM) has evolved over the last 20 years from simple vacuum based deposition monitoring to sophisticated biosensing in liquid and gaseous environments. As these experiments increase in complexity there is a real need for the standardization of the surface. For example, the stability of gold–thiol functionalization strategies is questionable for long term monitoring of pathogens or other critical areas where a false negative could be a real risk. The stability of the diamond surface is unrivalled [13], and diamond also offers the widest electrochemical window and lowest background noise for electrochemistry [57].

The idea of coating the top electrode of a QCM with diamond is not new, but unfortunately the curie point of quartz is below the temperature of conventional diamond growth processes [58]. Even low temperature deposition techniques have failed to produce a working device, presumably because the quartz crystal starts to lose its piezoelectric properties considerably below the curie point. Some success has been already obtained by bonding a free standing diamond layer to a QCM, but these results in a greatly reduced Q due to the thickness of the diamond layer and are not commercially viable [59].

One way to circumvent the problems of the low curie point of quartz is to use another piezoelectric crystal. Several novel high temperature piezoelectric materials have been discovered and commercialized in the last 10 years, such as langasite and gallium phosphate [60]. A prototype device is shown in Figure 7.7, where a SiO_2 layer was deposited on the electrode of a commercial langasite thickness shear mode (TSM) resonator. This layer was then seeded with diamond powder and nanocrystalline diamond was grown on top. Figure 7.8 shows the phase/frequency plot of this device [61]. There is a clear resonance at 5MHz with small spurious anharmonic resonances that were already present before diamond growth. Despite the fact that TSMs have a lower sensitivity than other acoustics sensors, these devices are simple, robust, easy to use and they can operate in liquid as shown in Figure 7.8 where the resonant frequency is slightly lowered by the liquid's load.

Such miniature TSM sensors, that is, composite FBAR, with operation frequency in the GHz range and higher mass sensitivity (Equation 7.3) are very

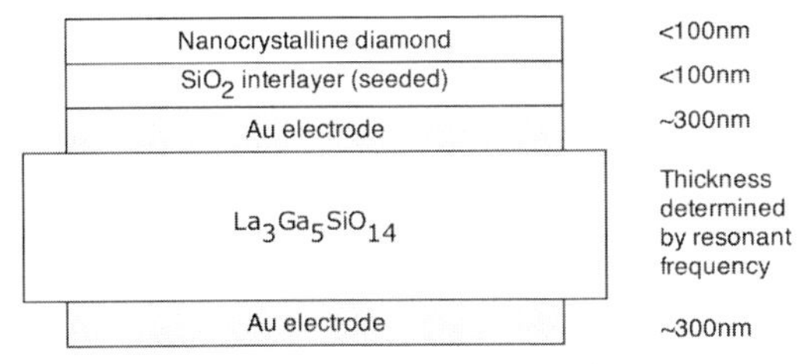

Figure 7.7 Schematic of a diamond coated TSM resonator.

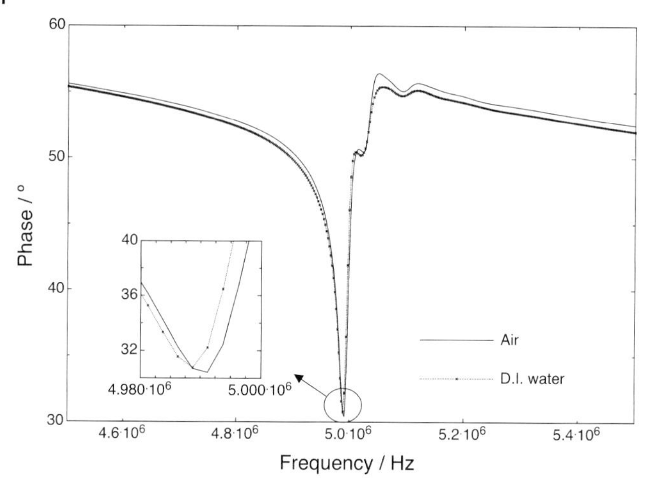

Figure 7.8 Frequency response diamond coated TSM resonator in air and in DI water.

appealing. Composite FBAR using a SiO_2 layer that operates in liquids have already been demonstrated for temperature compensation purposes. FBAR are compatible with IC technology, they are mass produced and they can be easily integrated into arrays. Most certainly, diamond has an important role to play in this technology for biological and chemical sensors.

7.4.4 Microcantilevers

Because of its extreme hardness and its low wear coefficient, diamond has been already used on tips [62] or as a protection layer for silicon tips [63] of microcantilever for scanning probe microscopy. Monolithic diamond cantilevers with integrated tip for SPM applications have also been made [64, 65]. Malave *et al.* have reported the fabrication of highly boron doped diamond tips with a resistivity of ~10^{-3} Ω cm and a tip radius of curvature ~20 nm on diamond cantilevers for applications in scanning spreading resistance microscopy, scanning capacitance microscopy and nanopotentiometry measurements [66].

While diamond possesses high fracture toughness, the high Young's modulus of diamond gives to diamond cantilevers a higher spring force constant (k), ~10 times higher than silicon cantilevers. Diamond cantilevers have also slightly higher resonant frequencies (f_i) than silicon cantilevers for the same geometry of the cantilever. Shibata *et al.* have developed diamond AFM probes integrated with piezoelectric thin films (ZnO, PZT) [67, 68]. Both side clamped diamond cantilevers, that is, diamond bridges have been used as actuators for microwave micro relays [69]. The use of diamond for this application provides large switching forces, mechanical stability and chemical inertness [70].

In 1994, it was found for the first time that a standard AFM cantilever could operate as a microcalorimeter with femto-joule (10^{-15} J) sensitivity [33, 71]. In other words, microcantilevers are excellent micromechanical sensors that can be used not only to characterized surface morphology and surface properties using SPM techniques. Microcantilevers operate by detecting changes either in resonance frequency, amplitude, Q-factor caused by mass loading and/or damping conditions or deflection caused by surface stress variation. Many applications of microcantilever sensors have already been demonstrated such as humidity sensor [72], ethanol vapor, alkanes vapor, perfume oils vapor [73], Pb^{2+} in water [74], ethanol in water and antibody/antigen recognition [75], trinitrotoluene [76], PH, albumin [77], mercury vapor, mercaptan, IR radiation, DNA [78], and so on, using standard detection techniques.

Common cantilever sensor systems are piezoelectrically, photothermally or magnetically actuated [79] and they use either optical [80] or piezoresistive detection. Electrical methods have some advantages compared with optical techniques such as no need for optical components, no laser alignment, and the read-out electronics can be integrated on the same chip and they are not affected by the optical properties of the surrounding medium. Another possibility is to use piezoelectric bimorph cantilevers. The piezoelectric layer is used as both actuator and detector at the resonance frequencies. This system is very attractive since it avoids the use of external magnetic fields or delicate adjustments of optical systems, which limit practical applications. Figure 7.9 shows a schematic of a piezoelectric bimorph cantilever. It consists of a cantilever made of two materials, one "substrate" and piezoelectric film sandwiched between two electrodes.

Vibrations of cantilevers [81] and bimorph piezoelectric cantilevers [82] have already been studied. The electromechanical coupling coefficient (k) of a piezoelectric bimorph microcantilever has been calculated analytically [83]. It is composed of two parts: the electromechanical coupling coefficient of the piezoelectric material (K_f^2) and a form factor (F). The electromechanical coupling factor of the piezoelectric material is a function of the piezoelectric coefficient at constant field (e_{31}^2), the elastic constant (c_{11}) and the permittivity at constant strain (ε_{33}) as shown in Equation 7.6:

$$K_f^2 = e_{31}^2/(c_{11} \cdot \varepsilon_{33}) \tag{7.6}$$

The most suitable piezoelectric material is PZT for piezoelectric bimorph microcantilevers (see Table 7.4). The form factor is a function of the Young's modulus (Y_s) the thickness (h_s) of the substrate, the elastic constant (C_{11}) and the thickness

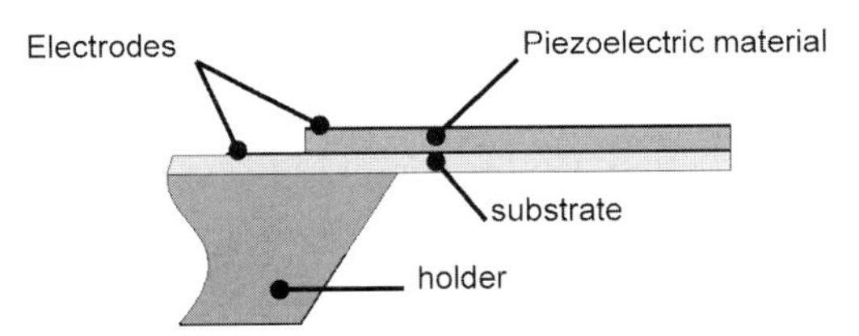

Figure 7.9 Schematic of a piezoelectric bimorph cantilever.

Table 7.4 Electromechanical coupling coefficient (K_f^2) of various piezoelectric materials.

Piezoelectric material	K_f^2
AlN	1.03
ZnO	1.71
$BaTiO_3$	2.63
PZT4	3.47

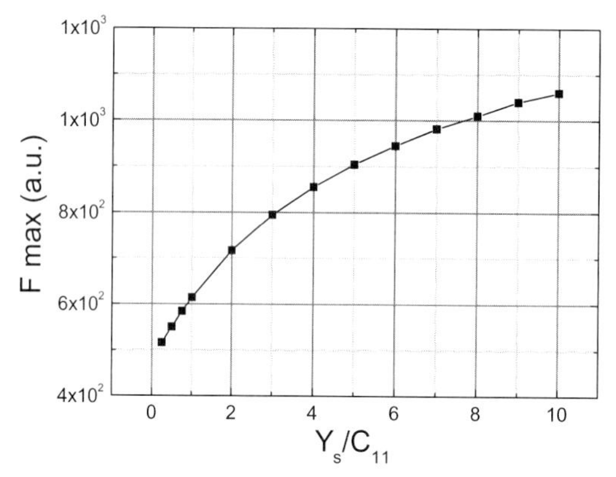

Figure 7.10 Variation of the maximal value of the form factor versus the ratio the substrate's Young's modulus (Y_s) and the piezoelectric elastic constant (C_{11}).

(h_c) of the piezoelectric material. It is inversely proportional with the cantilever length (L). For each ratio Y_s/C_{11}, it has a maximum value with the normalized thickness. The maximum value increases with the ratio Y_s/C_{11} (see Figure 7.10). In another words, the stiffer the substrate, the higher the form factor and the electromechanical coupling coefficient. Thus, diamond, which has the highest known Young's modulus, is the most suitable material for this type of cantilever. To our knowledge, the first piezoelectric bimorph microcantilever that has been studied are the Microactuated silicon probes (DMASP-Micro-Actuated Silicon Probes) provided by Veeco© [84]. It consists of a Si/ZnO cantilever with two Ti/Au electrodes that sandwich the piezoelectric layer. Modeling and characterization of the frequency response and the impedance of this cantilever have been reported recently [85–87]. In a previous work, we have shown that this microcantilever can be use as a as gas pressure sensors [88]. It operates by monitoring the frequency shift of the resonant modes of the cantilever, which acts as a driven and damped oscillator. The change in the resonant frequencies is due to the variation of drag force with the pressure of the surrounding gas. The sensitivity of the piezoelectric bimorph cantilever varies with the vibration mode and the nature of the gas (see

Table 7.5 Pressure sensitivity of a Si/ZnO microcantilever in different gases for the first and the third resonant vibration mode [88].

	1st resonant vibration mode (ppm/mbar)	3rd resonant vibration mode (ppm/mbar)
Argon	5.7	4.5
Nitrogen	4.2	3.4
Helium	0.7	0.6

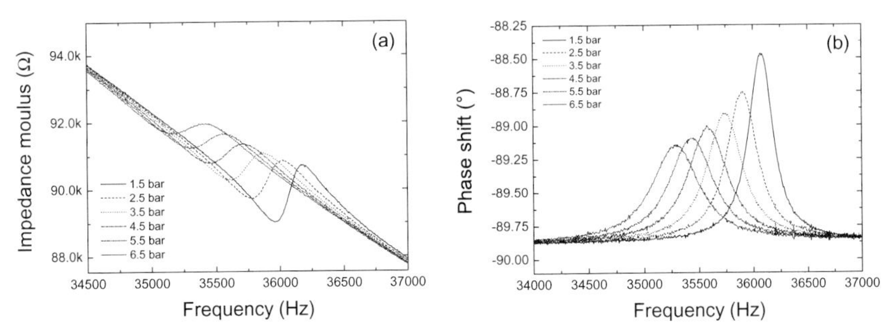

Figure 7.11 Impedance variation of a diamond/AlN microcantilever at the first resonant mode as function of nitrogen pressure.(cantilever length ~200 μm, cantilever wide ~50 μm, diamond thickness ~650 vm and AlN thickness ~1 μm).

Table 7.5) [88]. The bimorph cantilevers are also sensitive to temperature. It has been shown that the determination of the frequency shift of two resonant modes can be used to determine simultaneously both pressure and temperature of a gas [89]. We have also made and operated diamond/AlN microcantilevers as gas pressure sensors [83]. The variation of the impedance of the first resonant frequency of the diamond/AlN as a function of nitrogen gas pressure is plotted on Figure 7.11. The resonant frequency decreases nearly linearly with the increasing pressure of nitrogen (see Figure 7.12) with a sensitivity of ~4.2 ppm/mbar for nitrogen pressure up to 7 bars and at center frequency f~36.5 kHz.

7.5 Conclusion

In this chapter, we have briefly reviewed the different types of acoustic devices with special attention to sensor applications. Despite the fact that acoustic sensors have been already used to study physical and chemical properties of gases and liquids for decades, the use of diamond in this field is new. We have discussed

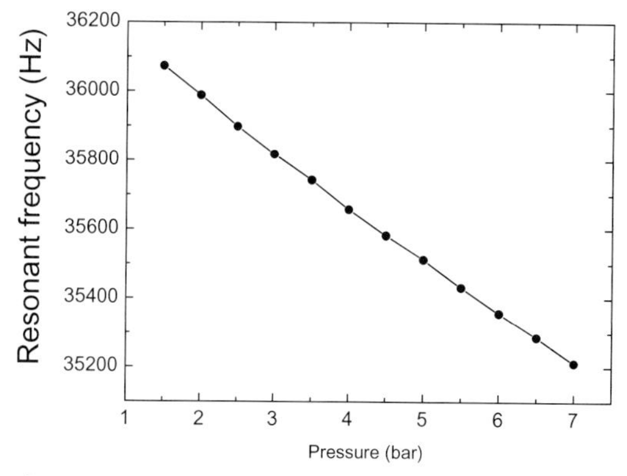

Figure 7.12 Variation of the first mode's resonant frequency as a function of the nitrogen pressure for a diamond/AlN micro-cantilever.

and shown the interest to use diamond in composite acoustic devices: SAW filters, PFW sensors, TSM sensors, FBAR sensor and microcantilever sensors. While diamond-based SAW filter technology is mature, diamond-based acoustic sensors are still in their infancy and should not be neglected. The specific combination of exceptional properties of diamond: high mechanical properties, chemical stability, bio-inertness are highly desirable for the new generation of chemical and biological sensors.

References

1 Curie, J. and Curie, P. (1880) *Bulletin de la Société française de Minéralogie*, **3**, 90.
2 Lord Rayleigh (1885) Proceedings of the London Mathematical Society, **s1-17**, 4.
3 White, R.M. and Voltmer, F.W. (1965) *Applied Physics Letters*, **7**, 314.
4 Wohltjen, H. and Dessy, R. (1979) *Analytical Chemistry*, **51**, 1458.
5 Wenmackers, S., Christiaens, P., Daenen, M., Haenen, K., Nesládek, M., van deVen, M., Vermeeren, V., Michiels, L., Ameloot, M. and Wagner, P. (2005) *physica status solidi (a)*, **202**, 2212.
6 Hernando, J., Pourrostami, T., Garrido, J.A., Williams, O.A., Gruen, D.M., Romka, A., Teinmüller, D. and Tutzmann, M. (2007) *Diamond and Related Materials*, **16**, 138.
7 Wang, J. and Carlisle, J.A. (2006) *Diamond and Related Materials*, **15**, 279.
8 Aikawa, Y. and Baba, K. (1993) *Japanese Journal of Applied Physics*, **32**, 4680.
9 Field, J.E. (ed.) (1979) *The Properties of Diamonds*, Academic Press, London.
10 Peterson, K.E. (1982) *IEEE Proceedings*, **70**, 5.
11 Kohn, E., Gluche, P. and Adamschik, M. (1999) *Diamond and Related Materials*, **8**, 934.
12 Specht, C.G., Williams, O.A., Jackman, R.B. and Schoepfer, R. (2004) *Biomaterials*, **25**, 4073.
13 Yang, W., Auciello, O., Butler, J.E., Cai, W., Carlisle, J.A., Gerbi, J., Gruen, D.M., Knickerbocker, T., Lasseter, T.L., Russell, J.N., Smith, L.M. and Hamers, R.J. (2002) *Nature Materials*, **1**, 253.
14 Assouar, M.B. (2001) *Etude del dispositifs a ondes acoustiques de surface (SAW) a structure multicouche nitrure d'aluminium/ diamant: croissance de materiaux en couches*

minces et technologie de réalisation, PhD Thesis, Nancy.
15 Ravet, M.F. and Rousseaux, F. (1996) *Diamond and Related Materials*, **5**, 812.
16 Mortet, V., D'Haen, J., Potmesil, J., Kravets, R., Drbohlava, I., Vorlicek, V., Rosa, J. and Vanecek, M. (2005) *Diamond and Related Materials*, **14**, 393.
17 Bi, B., Huang, W.-S., Asmussen, J. and Golding, B. (2002) *Diamond and Related Materials*, **11**, 677.
18 Mortet, V., Elmazria, O., Nesladek, M., Assouar, M.B., Vanhoyland, G., D'Haen, J., D'Olieslaeger, M. and Alnot, P. (2002) *Applied Physics Letters*, **81**, 1720.
19 Philip, J., Hess, P., Feygelson, T., Butler, J.E., Chattopadhyay, S., Chen, K.H. and Chen, L.C. (2003) *Journal of Applied Physics*, **93**, 2164.
20 Williams, O.A., Daenen, M., D'Haen, J., Haenen, K., Maes, J., Moshchalkov, V.V., Nesladek, M. and Gruen, D.M. (2006) *Diamond and Related Materials*, **15**, 654.
21 Michaelson, Sh., Akhvlediani, R. and Hoffman, A. (2002) *Diamond and Related Materials*, **11**, 721.
22 Espinosa, H.D., Peng, B., Prorok, B.C., Moldovan, N., Auciello, O., Carlisle, J.A., Gruen, D.M. and Mancini, D.C. (2003) *Journal of Applied Physics*, **94**, 6076.
23 Hernández, F.J., Janischowsky, K., Kusterer, J., Ebert, W. and Kohn, E. (2005) *Diamond and Related Materials*, **14**, 411.
24 Sepulveda, N., Aslam, D. and Sullivan, J.P. (2006) *Diamond and Related Materials*, **15**, 398.
25 Lee, J.B., Kim, H.J., Kim, S.G., Hwang, C.S., Hong, S.-H., Shin, Y.H. and Lee, N.H. (2003) *Thin Solid Films*, **435**, 179.
26 Agilent FBAR appears in wireless products (2002) *III-Vs Review*, **15**, 8.
27 Link, M., Weber, J., Schreiter, M., Wersing, W., Elmazria, O. and Alnot, P. (2007) *Sensors and Actuators B: Chemical*, **121**, 372.
28 Wingqvist, G., Bjurström, J., Liljeholm, L., Yantchev, V. and Katardjiev, I. (2007) *Sensors and Actuators B: Chemical*, **123**, 466.
29 Krishnaswamy, S.V., McAvoy, B.R., Takei, W.J. and Moore, R.A. (1982) IEEE Ultrasonics Symposium Proceedings, 476.
30 Wang, J.S., Lakin, K.M. and Landin, A.R. (1983) Frequency Control, 37th Annual Symposium, 144.
31 Gabl, R., Schreiter, M., Green, E., Feucht, H.D., Zeininger, H., Runck, J., Reichl, W., Primig, R., Pitzer, D., Eckstein, G. and Wersing, W. (2003) Proceedings IEEE Sensors, 1184.
32 Love, A.E.H. (1911) *Some Problems of Geodynamics*, Cambridge (England), University Press, 1911.
33 Thundat, T., Warmack, R.J., Chen, G.Y. and Allison, D.P. (1994) *Applied Physics Letters*, **64**, 2894.
34 Bjurström, J., Wingqvist, G., Yantchev, V. and Katardjiev, I. (2007) *Journal of Micromechanics and Microengineering*, **17**, 651.
35 Yamanouchi, K., Sakurai, N. and Satoh, T. (1989) IEEE Ultrasonics Symposium, 351.
36 Shikata, S.-I., Nakahata, H., Hachigo, A. and Narita, M. (2005) *Diamond and Related Materials*, **14**, 167.
37 Chen, J.J., Zeng, F., Li, D.M., Niu, J.B. and Pan, F. (2005) *Thin Solid Films*, **485**, 257.
38 Tang, I.-T., Chen, H.-J., Hwang, W.C., Wang, Y.C., Houng, M.-P. and Wang, Y.-H. (2004) *Journal of Crystal Growth*, **262**, 461.
39 Shikata, S.-I., Nakahata, H. and Hachigo, A. (1999) *New Diamond and Frontier Carbon Technology*, **9**, 75.
40 Mortet, V., Elmazria, O., Nesladek, M., D'Haen, J., Vanhoyland, G., Elhakiki, M., Tajani, A., Bustarret, E., Gheeraert, E., D'Olieslaeger, M. and Alnot, P. (2003) *Diamond and Related Materials*, **12**, 723.
41 Ishihara, M., Manabe, T., Kumagai, T., Nakamura, T., Fujiwara, S., Ebata, Y., Shikata, S.-I., Nakahata, H., Hachigo, A. and Koga, Y. (2001) *Japanese Journal of Applied Physics*, **40**, 5065.
42 Nakahata, H., Hachigo, A., Higaki, K., Fujii, S., Shikata, S.-I. and Fujimori, N. (1995) *IEEE Transactions on Ultrasonics, Ferroelectrics, and Frequency Control*, **42**, 362.
43 Higaki, K., Nakahata, H., Kitabayashi, H., Fujii, S., Tanabe, K., Seki, Y. and Shikata, S.-I. (1997) *IEEE Transactions on Ultrasonics, Ferroelectrics, and Frequency Control*, **44**, 1395.
44 Ishihara, M., Nakamura, T., Kokai, F. and Koga, Y. (2002) *Diamond and Related Materials*, **11**, 408.

45 Elmazria, O., Mortet, V., Hakiki, M. El, Nesladek, M. and Alnot, P. (2003) *IEEE transactions on Ultrasonics, Ferroelectrics, and Frequency Control*, **50**, 710.
46 Kirsch, P., Assouar, M.B., Elmazria, O., Mortet, V. and Alnot, P. (2006) *Applied Physics Letters*, **88**, 22350.
47 Benetti, M., Cannatà, D., Di Pietrantonio, F., Fedosov, V.I. and Verona, E. (2005) *Applied Physics Letters*, **87**, 033504.
48 Fujii, S., Shikata, S., Member, T. Uemura, Nakahata, H. and Harima, H. (2005) *IEEE Transactions on Ultrasonics, Ferroelectrics, and Frequency Control*, **52**, 1817.
49 Uemura, T., Fujii, S., Kitabayashi, H., Itakura, K., Hachigo, A., Nakahata, H., Shikata, S., Ishibashi, K. and Imai, T. (2002) Ultrasonics Symposium Proceedings, 431.
50 Elmazria, O., Hakiki, M. El, Mortet, V., Assouar, B.M., Nesladek, M., Vanecek, M., Bergonzo, P. and Alnot, P. (2004) *IEEE Transactions on Ultrasonics, Ferroelectrics, and Frequency Control*, **51**, 1704.
51 Benetti, M., Cannata, D., D'Amico, A., Di Pietrantonio, F., Macagnano, A. and Verona, E. (2004) Sensors, Proceedings of IEEE, 753.
52 Weinberg, M.S., Cunningham, B.T. and Clapp, C.W. (2000) *Journal of Microelectromechanical Systems*, **9**, 370.
53 Luginbuhl, Ph., Collins, S.D., Racine, G.-A., Gretillat, M.-A., de Rooij, N.F., Brooks, K.G. and Setter, N. (1998) *Sensors and Actuators A*, **64**, 41.
54 Laurent, T., Bastien, F.O., Pommier, J.-C., Cachard, A., Remiens, D. and Cattan, E. (2000) *Sensors and Actuators*, **87**, 26.
55 Choujaa, A., Tirole, N., Bonjour, C., Martin, G., Hauden, D., Blind, P., Cachard, A. and Pommier, C. (1995) *Sensor and Actuators A*, **46**, 179.
56 Francis, L.A., Kromka, A., Steinmuller-Nethl, D., Bertrand, P. and Van Hoof, C. (2006) *IEEE Sensors Journal*, **6**, 916.
57 Hupert, M., Muck, A., Wang, R., Stotter, J., Cvackova, Z., Haymond, S., Show, Y. and Swain, G.M. (2003) *Diamond and Related Materials*, **12**, 1940.
58 Bragg, W.L. and Gibbs, R.E. (1925) *Proceedings of the Royal Society A*, **109**, 405.
59 Zhang, Y.R., Asahina, S., Yoshihara, S. and Shirakashi, T. (2002) *Journal of the Electrochemical Society*, **149**, H179.
60 Damjanovic, D. (1998) *Current Opinion in Solid State and Materials Science*, **3**, 469.
61 Williams, O.A., Mortet, V., Daenen, M. and Haenen, K. (2007) *Applied Physics Letters*, **90**, 063514.
62 Álvarez, D., Fouchier, M., Kretz, J., Hartwich, J., Schoemann, S. and Vandervorst, W. (2004) *Microelectronic Engineering*, **73/74**, 910.
63 Givargizov, E.I., Stepanova, A.N., Mashkova, E.S., Molchanov, V.A., Shi, F., Hudek, P. and Rangelow, I.W. (1998) *Microelectronic Engineering*, **41/42**, 499.
64 Kulisch, W., Malave, A., Lippold, G., Scholz, W., Mihalcea, C. and Oesterschulze, E. (1997) *Diamond and Related Materials*, **6**, 906.
65 Oesterschulze, E., Malave, A., Keyser, U.F., Paesler, M. and Haug, R.J. (2002) *Diamond and Related Materials*, **11**, 667.
66 Malave, A., Oesterschulze, E., Kulisch, W., Trenkler, T., Hantschel, T. and Vandervorst, W. (1999) *Diamond and Related Materials*, **8**, 283.
67 Shibata, T., Unno, K., Makino, E. and Shimada, S. (2004) *Sensors and Actuators A*, **114**, 398.
68 Shibata, T., Unno, K., Makino, E., Ito, Y. and Shimada, S. (2002) *Sensors and Actuators A*, **102**, 106.
69 Adamschik, M., Kusterer, J., Schmid, P., Schad, K.B., Grobe, D., Floter, A. and Kohn, E. (2002) *Diamond and Related Materials*, **11**, 672.
70 Kusterer, J., Hernandez, F.J., Haroon, S., Schmid, P., Munding, A., Müller, R. and Kohn, E. (2006) *Diamond and Related Materials*, **15**, 773.
71 Barnes, J.R., Stephenson, R.J., Welland, M.E., Gerber, C. and Gimzewski, J.K. (1994) *Nature*, **372**, 79.
72 Domanski, K., Grabiec, P., Marczewski, J., Gotszalk, T., Ivanov, T., Abedinov, N. and Rangelow, I.W. (2003) *Journal of Vacuum Science and Technology B*, **21**, 48.
73 Battiston, F.M., Ramseyer, J.-P., Lang, H.P., Baller, M.K., Gerber, C., Gimzewski, J.K., Meyer, E. and Guntherodt, H.-J. (2001) *Sensors and Actuators B*, **77**, 122.
74 Liu, K. and Ji, H.-F. (2004) *Analytical Sciences*, **20**, 9.

75 Tamayo, J., Humphris, A.D.L., Malloy, A.M. and Miles, M.J. (2001) *Ultramicroscopy*, **86**, 167.

76 Pinnaduwage, L.A., Gehl, A., Hedden, D.L., Muralidharan, G., Thundat, T., Lareau, R.T., Sulchek, T., Mannings, L., Rogers, B., Jones, M. and Adams, J.D. (2003) *Nature*, **425**, 474.

77 Butt, H.-J. (1996) *Journal of Colloid and Interface Science*, **180**, 251–60.

78 Thundat, T., Oden, P.I. and Warmack, R.J. (1997) *Microscale Thermophysical Engineering*, **1**, 185–99.

79 Brown, K.B., Allegretto, W., Vermeulen, F.E. and Robinson, M. (2002) *Journal of Micromechanics and Microengineering*, **12**, 204.

80 Raiteri, R., Grattarola, M., Butt, H.-J. and Skládal, P. (2001) *Sensors and Actuators B*, **79**, 115.

81 Landau, L.D. and Lifshitz, E.M. (1986) *Theory of Elasticity*, 3rd edn, Vol. 7 Butterworth-Heinemann, Oxford.

82 Brissaud, M., Ledren, S. and Gonnard, P. (2003) *Journal of Micromechanics and Microengineering*, **13**, 832.

83 Mortet, V., Haenen, K., Potmesil, J., Vanecek, M. and D'Olieslaeger, M. (2006) *physica status solidi (a)*, **203**, 3185.

84 Veeco Probe Catalog (2005) 86.

85 Sanz, P., Hernando, J., Vazquez, J. and Sanchez-Rojas, J.L. (2007) *Journal of Micromechanics and Microengineering*, **17**, 931.

86 Vazquez, J., Sanz, P. and Sanchez-Rojas, J.L. (2007) *Sensors and Actuators A*, **136**, 417.

87 Mahmoodi, S.N., Jalili, N. and Daqaq, M.F. (2008) *Non-linear Mechanics*, **13**, 58.

88 Mortet, V., Petersen, R., Haenen, K. and D'Olieslaeger, M. (2005) *Ultrasonics Symposium*, **3**, 1456.

89 Mortet, V., Petersen, R., Haenen, K. and D'Olieslaeger, M. (2006) *Applied Physics Letters*, **88**, 133511.

8
Theoretical Models for Doping Diamond for Semiconductor Applications

Jonathan P. Goss Richard J. Eyre, and Patrick R. Briddon
University of Newcastle, School of Natural Science, Newcastle upon Tyne NE1 7RU, UK

8.1
Introduction – The Doping Problem

Diamond, with a room temperature band gap of 5.50 eV [1], is an electrical insulator, but many of the properties of the material, such as its extraordinary intrinsic carrier mobilities [2], high thermal conductivity [3], radiation hardness [4], chemical inertness, and high break-down field characteristics [5] make it highly desirable for the production of conductive material by doping.

In contrast to the case for silicon and other common semiconductor materials, the extremely compact diamond lattice affords little room for substitutional impurities. Consequently few dopant species are routinely taken up in as-grown material, those that are taken up are principally B, N, Si, P, Ni, and S. Other chemical species may be introduced by ion implantation techniques, but, as reviewed in Section 8.5, this often leads to resistive material, probably as a consequence of compensation arising from persistent radiation damage centers forming mid-gap between acceptor and donor states.

Of the chemical species that *can* be incorporated in sufficient concentrations, the most conspicuously successful is boron. Diamond may be doped with boron using gas-phase growth techniques such as chemical vapor deposition (CVD), or the high-pressure, high-temperature (HPHT) conversion of graphite into diamond using transition metal catalyst–solvents. Such material, as well as some natural diamonds, behave as a p-type semiconductors. Indeed, deliberate B-doping of diamond can yield metallic [6] and even superconducting material [7], provided the concentration is above a threshold value around $10^{21}\,cm^{-3}$. The nature of B-doping will be explored in Section 8.3.

Valence-band, hole-mediated conduction may also be achieved with diamond without boron doping, albeit within a conductive surface-layer. This property relates to an apparent exchange of charge between the valence band and physisorbed molecular species produced in aqueous solution on the diamond surface. This so-called transfer doping leads to a highly conductive layer within a few

Physics and Applications of CVD Diamond. Satoshi Koizumi, Christoph Nebel, and Milos Nesladek

ISBN: 978-3-527-40801-6

nanometres of the surface that exhibits a thermal stability consistent with the removal of a surface aqueous layer. Deliberate addition of electronegative species to diamond surfaces has also been used to induce surface valence band holes, allowing for a more deliberate engineered effect. This mode of doping is beyond the scope of this chapter.

At the other end of the band gap there is considerable difficulty. The highly soluble nitrogen donor undergoes a chemical reconstruction with one carbon neighbor leading to a hyper-deep level at E_c – 1.7 eV [8], which is of no practical use for room-temperature semiconductor devices. Phosphorus can also be incorporated in relatively high concentrations, but, as with N, leads to a deep donor level, albeit rather shallower at around E_c – 0.6 eV [9]. Although many examples of working devices employing phosphorus doping have been demonstrated [10–13], there remains a clear incentive to develop shallower n-type dopants, and much of the effort in determining likely candidates has been based in atomistic modeling. In Section 8.4 a range of these systems is reviewed. Many systems suffer from one of two obstacles. The first is solubility, but this may be less of an issue than one would think since it has been established that, for example, the uptake of phosphorus is a highly non-equilibrium event, and may be related to surface processes [14]. Indeed the very process of CVD is non-equilibrium. The second is complexity. Most suggested shallow donors are examples of co-doping, where two or more defect centers, often containing different chemical species, are combined to manufacture the desired electrical properties. In so doing, and in contrast to P_s doping, one has to ensure that all of the component parts are brought together in the correct structure, otherwise one produces ionized impurities that reduce mobility, or worst still, compensation centers that reduce the free carrier concentration.

Underlying all of the theoretically based shallow donor systems are the methodologies employed in obtaining both the structures and their electrical characteristics. Despite advances in computational sciences due to improved methods and rapidly increasing computer speeds, there remain significant issues in the accurate prediction of donor and acceptor levels in doping solutions. Indeed, the most commonly adopted computational framework, density functional theory (DFT) within either the local-density (LDA) or generalized-gradient (GGA) approximations, suffers from a number of issues, as outlined in Section 8.2.

Nevertheless, quantum-chemical computational methods are key in the design of future doping schemes, both as tools to aid interpretation of complex experimental data, and as predictive models, rapidly exploring the wide chemical spectrum of possible dopants, guiding future materials production.

8.2
Modeling of Doping Characteristics

It is generally accepted that in order to understand the electrical properties of impurities and other defects in crystalline materials purely from a computational basis, one is required to use a quantum-mechanically based method. The broad range of methods termed quantum-chemical are those which include the interactions of the electrons and nuclei, taking into account their fundamental quantum nature, but even so there are various levels of sophistication.

It is clearly beyond the scope of this review to give a detailed discussion of the benefits and validity of these various approached. Nevertheless, since it is important to have an appreciation of the basis for the calculated properties in order to provide a context for judging the predictions we present in this section, we outline the band-structure methods most typically used. We then review the common techniques for calculating the donor and acceptor levels.

There are many quantum-mechanically based computational frameworks that are currently being used to model collections of atoms. These include: the tight-binding limit which has computational simplicity and can be used for collections of large numbers of atoms, but lacks rigor; a panoply of methods centered upon DFT; and the accurate, but computational intensive first-principles Hartree–Fock methods with configuration interactions. There are many books that detail the fundamentals [15–19], and we shall not present them here. From this spectrum of methods, by a wide margin the most commonly used for solid-state quantum-chemical simulations is currently density functional theory. Density functional methods are based upon the result that the charge density of a system is bound by a one-to-one relation with the ground state potential, and therefore a complete knowledge of the system [20]. The practical application of this method generally involves expressing the charge density in one-particle like functions obtained from a set of single-particle equations [21, 22]. These are the so-called Kohn–Sham levels and states. The meaning of the functions and eigenvalues of these one-particle-like equations is a matter for debate, but they are commonly employed as approximate one-electron functions and energies. The interpretation of these quantities is expanded below.

8.2.1
Interpretation of Band Structures

The interpretation of band structure is key in terms of understanding the location of a donor or acceptor level relative to the valence band top or conduction band minimum.

The eigenvalues and eigenfunctions of the solutions to the Kohn–Sham equations are often interpreted as a one-electron description of a system, in analogy to a Hartree–Fock approach where the one-electron states are obtained by construction. Whilst interpreting Kohn–Sham spectra in this way is formally incorrect, they are related to an electronic structure, and much insight into the properties of a collection of atoms may be gleaned from them. However, in so doing, one has to take care.

In particular, one might wish to use knowledge of the highest occupied and lowest unoccupied Kohn–Sham orbitals to understand donor or acceptor properties, respectively. This has a firm footing in Janak's theorem [23] which related the energy of the highest occupied Kohn–Sham eigenvalue to the ionization energy of the system. The lowest empty state strictly cannot be identified with the electron affinity, but the presence of empty Kohn–Sham levels is highly indicative. The main problems with using the Kohn–Sham band structures for quantitative analysis of the donor and acceptor levels lie in the error in the band gap energy, and the "vertical" nature of the transitions in this approach. The latter can be understood in that the ionization energy suggested by the Kohn–Sham orbital relates to the structure for a specific charge state. In real systems the structure can and does change with charge state, sometimes dramatically, such as in the negative-U (where, for these systems, the neutral charge state is only ever metastable) interstitial hydrogen systems in many materials [24]. The former, that is the serious underestimation of the band gap, leads to questions of which band edge one should reference the level to, or even if the energies should be scaled in some fashion to regain the experimental band gap. This issue is addressed further below.

Notwithstanding the problems associated with interpreting the Kohn–Sham orbitals as one-electron energies, their location within the gap may be used as an indication of the potential for a defect to yield deep or shallow donor or acceptor levels. It is questionable whether using their energies as a quantitative measure is of much worth.

8.2.2 Electron Affinities and Ionization Energies

Electrical levels, that is the location of donor or acceptor levels within the band gap, when being viewed as a thermally activated process cannot be obtained simply from the band structure. This is because these electrical transitions relate to the thermodynamic probability of a system occurring in a given charge state, which in turn depends upon the properties of a dopant in all the charge states concerned. A contrasting view may be taken when examining optically excited changes in charge state (for example) which are not constrained to thermodynamic equilibrium states, but this is of less relevance to the question of doping.

The donor and acceptor levels of a defect can be viewed very simply as the ionization energies and electron affinities relative to those of the bulk system, which can be calculated within density-functional and other computational approaches using the difference in the total energies of the neutral and charged system. Now, this procedure involves a number of assumptions which lead to significant problems of interpretation. In this chapter we review the major issues as found for typical density-functional implementations, but they are not exclusive to such approaches.

The first problem we shall discuss relates to the choice of boundary conditions. Most calculations are performed using periodic boundary conditions (PBC) and this immediately causes a fundamental problem. PBCs cannot be used to simulate charged systems: the electrostatic energy for an infinitely repeated net charge is divergent. In practice, there is a mechanism that is used to allow for an approximate treatment of charged systems with PBCs. One assembles the nuclei and electrons one wishes to model, and the difference between the nuclear and electronic charge is offset by a uniformly distributed counter-charge.

In addition to this expedient, the modeling of a periodic array of charges associated with localized defect orbitals leads to further problems. The interaction of the charges represents a component in the total energy that is purely an artifact of the calculation, shifting the total energy of the charged system relative to that of the neutral, and thereby introducing an error into the ionization energy. The methods adopted to estimate and remove this artifact, as well as other multi-polar terms present even for neutral systems, has lead to considerable debate [25–31], and there is no clear consensus on how to proceed. The best solution appears to be to use very large atomic systems so that the interactions between dopants and their periodic images are minimal. However, despite the increasing availability of computational resources, modeling as few as 1000 atoms is a challenging calculation. Even then substitutional centers are separated by just 5 lattice constants, and the simulations represent a concentration of impurities which is high in comparison to typical values for real materials.

For cluster calculations, where defects are simulated by embedding them in large molecules, typically terminated by a passivant such as hydrogen, there is no issue regarding charging, but two other issues arise. These originate from the interactions of the defect states with the surface species, and the related artificial confinement of the electron states within the volume of the molecule. Again, there are *post-hoc* adaptations to calculations to account for the "quantum confinement", which sometimes are as simple as the addition of a constant related to the kinetic energy of a particle in an appropriately sized three-dimensional potential well (particle in a box) to the ionization energies [32].

Independent of the atomic systems used, it is common to calculate the electrical levels of a system by one of two approaches, which have been reviewed previously by some of the current authors [33, 34]. The first is to calculate a charge-dependent formation energy and determine the critical values of the electronic chemical potential for which the most stable charge state changes. For LDA-DFT or GGA-DFT calculations, the interpretation of this data is problematic since these methods significantly underestimate the band gap. Where a donor level is calculated to be (say) 4 eV above the valence band top, comparison with the experimental band gap suggests a very deep donor at around $E_c - 1.5\,\text{eV}$. Conversely, if compared to the theoretical band gap of around 4.2 eV [35], this represents a very shallow donor. It is therefore preferable to compare the donor levels of similar defects to one another, and use the experimental electrical levels of one of these systems to reference the others to the band edges. This is the so-called marker method [34].

8.3 p-type Diamond

Boron occurs in the rare type IIb natural diamond, and can be readily incorporated in synthetic diamond, both via CVD and HPHT methods. For gas-phase growth, boron is typically added in the form of diborane gas. In addition, boron is possibly the only proved example of implantation doped diamond [36–42]. In order to recover the crystalline material high temperature annealing is required. For example, Tsubouchi reports 1450 °C [40, 42], and Wu employs annealing at 1700 °C [41]. Such temperatures can also be employed for in-diffusion of boron [43].

Acceptor concentrations over a wide range can be achieved, with the characteristics of the films varying from relatively resistive p-type samples to highly conductive but opaque material, which exhibits metallic conduction.

8.3.1
Substitutional Boron

Isolated substitutional boron (B_s) theoretically lies close to or on a host site [44–49], and slightly dilates the surrounding lattice [47].

The measured activation energy for exciting a hole bound to B_s into the valence band is 0.37 eV [50], and this relatively deep value has also been obtained using computational methods [33, 51, 52].

The presence of neutral substitutional boron in diamond films leads to electronic transitions in the infrared with prominent peaks at 2450 and 2820 cm^{-1} (304 and 350 meV) [53, 54]. The presence or otherwise of this optical characteristic may be used to monitor the concentration of neutral acceptors, such as during hydrogen-passivation experiments [54, 55]. Note, the electronic transition has been erroneously linked in one theoretical study with transitions involving hydrogen–boron pairs [56].

As the concentration of B_s increases the activation energy for conduction diminishes, and at $[B_s] \sim 5 \times 10^{20}$ cm^3, the material undergoes a transition from a semiconductor to a metal [6, 57]. There are two distinct mechanisms that might lead to this effect, as illustrated in Figure 8.1. The first limiting case is the formation of an impurity band within the diamond band gap, so that there is a nonzero density of states at the Fermi energy, as represented in Figure 8.1(a). The alternative extreme is that the boron-related hole states lie below the valence-band maximum, and then the conduction path is within a depopulated host band, as shown schematically in Figure 8.1(c). Of course, there is a range of possible scenarios that bridge the gap between these limits, with a boron-related impurity band being made up from a combination of the valence band top and states arising from the boron acceptors, as depicted in Figure 8.1(b).

Of the limiting cases, both models have some support from experiment. The impurity band model has been argued to be the appropriate explanation for cathodoluminescence spectra [58], whereas X-ray angle-resolved photoemission spectroscopy appears to reveal a Fermi-level lying below the host valence band top [59]. Much of the computational analysis appears to support the latter [49, 60–64].

At low temperatures, B-doped material exhibiting metallic conduction superconducts. This is a subject discussed elsewhere in this volume, and we shall not cover it here.

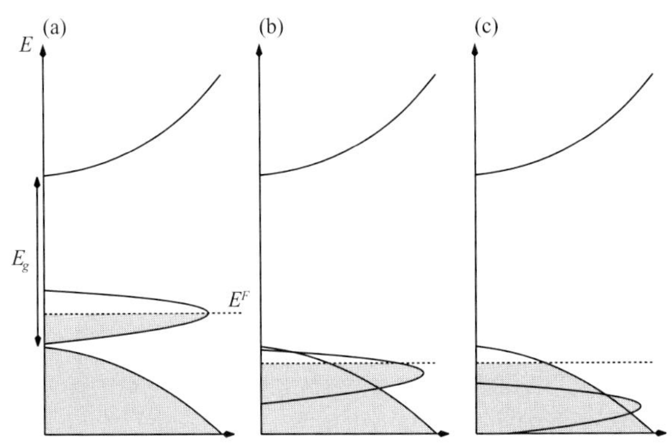

Figure 8.1 A schematic representation of three simplified densities of states for heavily B-doped diamond. (a) Boron forms an impurity band wholly within the diamond band gap. (b) A boron impurity band lies at the top of the valence band. (c) The boron impurity band lies deeper in the valence band and depopulates the valence band top. The band gap, E_g, is indicated, the occupied densities of states are shown as shaded, and the Fermi levels (E^F) are shown by dashed horizontal lines.

8.3.2
H-passivation of Boron

The injection of hydrogen or deuterium into boron doped films, typically using a plasma source, has been shown to lead to profound modifications of the materials properties.

A key result is that the addition of hydrogen (or deuterium) to the material results in closely correlated impurity concentration profiles as a function of depth as measured by secondary ion mass spectroscopy (SIMS) [54]. These measurements may be interpreted most simply as the formation of close-by B_s—H pairs. This is quite typical of hydrogen passivation of shallow dopants in other materials, such as silicon [65, 66] and GaAs [67]. As one would expect, the resultant material does not exhibit the electronic infrared transitions [54, 55], and the electrical conductivity is lost [68–70].

Theoretically, the pairing of B_s and interstitial hydrogen has been examined using a range of techniques. Approximate molecular orbital based methods suggest that hydrogen lies close to a B_s—H bond center, with a < B—H—C of 113°, this structure being 0.2 eV more stable than H lying in a C—B bond [71]. Both cluster and PBCs using first-principles approaches also support a puckered bond-centered geometry [44, 48, 51, 56, 72], perhaps indicative of a three-center bonding configuration [72]. The structure described in references [44, 45, 72] with a smaller bond angle is shown schematically in Figure 8.2.

In all the theoretical studies except reference [56], the B_s—H pairing is found to be electrically passive, and therefore support the view taken from experimental investigations.

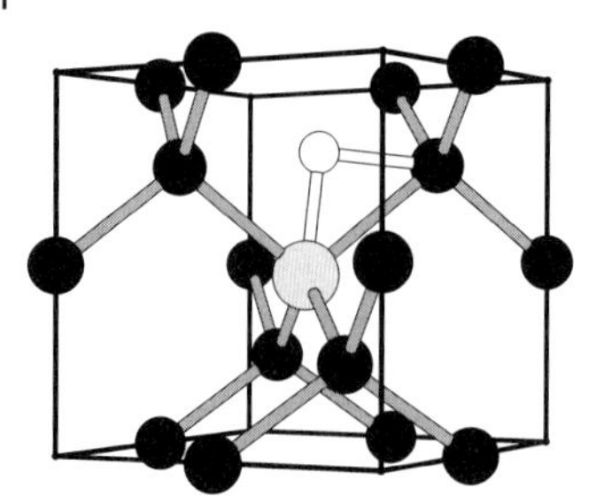

Figure 8.2 Schematic of the B_s–H complex in diamond. White, dark, and light gray atoms are H, C, and B, respectively. The box indicates the underlying cubic axes of the diamond lattice.

The passivation of the boron acceptors can be reversed by annealing in vacuum, with an activation energy of 2.5 eV inferred for the trap-limited diffusion of deuterium [45, 73]. Theoretical estimates for the binding energy have been published at 2.4–2.9 eV [45], 3.8 eV [48], and 3.24 eV [51]. It seems reasonable to conclude that the model of nearby B_s–H pairs forming a bound, electrically passive system is correct.

Interestingly, under excitation by an electron beam, that is where non-equilibrium conditions apply, the dissociation rate is enhanced [74]. The observations have been interpreted as hot electrons exciting the B_s–H pairs vibrationally, lowering the effective dissociation barrier. Such a mechanism has also been suggested for enhanced mobility of self-interstitials in low temperature electron-irradiated diamond.

There have been investigations of further hydrogen trapping at B_s, related to the apparent conversion of p-type diamond to n-type material [55, 75–77]. We review these complexes and their electronic properties in Section 8.4.4.4.

8.3.3 Boron Aggregates

Since boron has been incorporated in CVD diamond films in very high concentrations, it would not be surprising to find a proportion of these impurities being taken up in an aggregated form if such configurations are energetically favorable. There is some indirect experimental evidence for the presence of such complexes, with a vibrational band around 500 cm^{-1} proposed to arise from a nearest-neighbor pair of boron acceptors [45, 49, 78].

This defect pair has an analogue in the aggregation of nitrogen, and as with the N-pair (labeled an A-center), theory suggests that placing substitutional impurities on neighboring sites is favored in part due to bond-counting: the "bond" lying between the two B_s centers is depleted of charge, resulting in a relaxation and the formation of a defect state in the band gap. A schematic of the structure of the boron pair is shown in Figure 8.3(a). The displacement of the B atoms away from

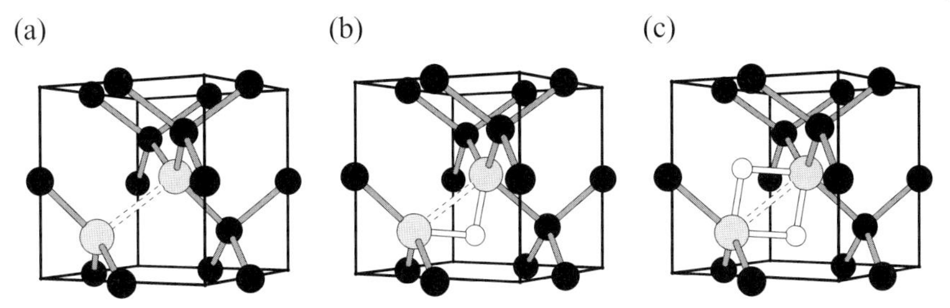

Figure 8.3 Schematic of the B—B complex with zero, one, or two hydrogen. White, dark, and light gray atoms are H, C, and B, respectively. The box indicates the underlying cubic axes of the diamond lattice.

each other was calculated to be around 0.56 Å in cluster-based simulations, with a lower value of 0.41 Å (B—B separations of 1.94–1.95 Å) in two independent studies employing PBCs [45, 49].

A range of structures made up from boron in forms other than simple substitutional acceptors have been modeled using computational techniques [45, 49, 51, 79]. Support for an energetic preference for the formation of nearest-neighbor substitutional pairs is obtained, with the pair being bound with respect to two isolated B_s centers. The magnitude of the binding appears to vary with different boundary conditions: calculations using PBCs yield values of 0.8 eV [79], whereas cluster calculations suggest a much larger value of 2.09 eV [51]. Although the binding energy is not quoted, an independent PBC calculation also presented the energy differences between nearest-neighbor pairs and pairs with intervening carbon atoms to be ≤0.2 eV [49], consistent with a modest binding. However, the electronic structure of the pairs is qualitatively independent of the supercell/cluster model, with the pairs leading to very deep acceptor levels.

The implication is that material containing only substitutional boron pairs would not be able to conduct in the same manner as material containing only B_s. However, Cai *et al.* [51] suggest that the presence of such structures may lead to impurity-band related conduction, which we review in Section 8.4.6.

Substitutional boron pairs may also be responsible for additional trapping of hydrogen [45, 51]. In these structures H lies in between the two B_s, as shown schematically in Figure 8.3(b). These complexes have an electrical level close to that of a simple B_s center [45, 51], and would therefore, in contrast to B_s—H and B—B complexes, contribute to p-type conduction. B_s—H—B centers are more strongly bound than B_s—H, by around 0.6 eV, with the trapping of a second H_i to form the structure depicted in Figure 8.3(c) exhibiting a binding energy close to B_s—H [45].

More extensive aggregation of substitutional boron has been investigated [51]. Trimers may also act as acceptors and trap hydrogen. In contrast to dimers, there is no experimental evidence at this time to suggest boron trimers may be present in as-grown diamond. This is most probably due to their low stability, with PBC calculations suggesting that the dissociation into a dimer and B_s is slightly exothermic [49].

Boron theoretically also favors incorporation in tandem with a lattice vacancy, shown schematically in Figure 8.4 [79]. Perhaps of most interest is the electronic structure of a complex made up from a lattice vacancy completely surrounded by boron atoms. This defect, analogous to the B-center in nitrogen-containing material, is theoretically a shallower acceptor than B_s, but is much less efficient in terms of the number of conduction holes per boron atom incorporated in the lattice. The electrical levels of boron–vacancy complexes [79] are plotted in Figure 8.4(e), and indicate that for three or four B species neighboring a lattice vacancy,

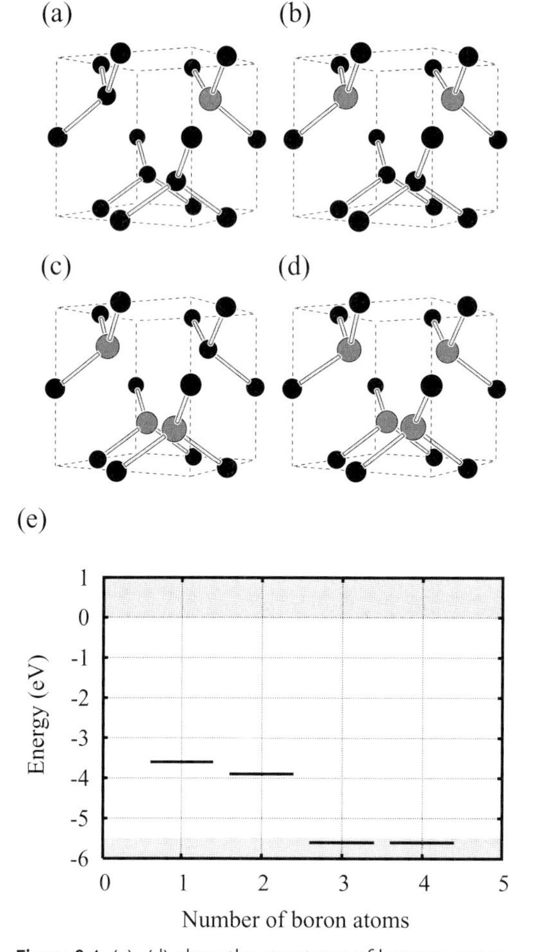

Figure 8.4 (a)–(d) show the structures of boron-vacancy complexes with one to four boron atoms, with (e) showing the calculated acceptor levels of these systems relative to E_c at zero, and the conduction and valence bands shown as shaded regions. For the atomic structures, light and dark gray atoms are boron and carbon, respectively, and the dashed boxes indicate the underlying cubic axes of the diamond lattice.

the acceptor level is close to the valence band top. Perhaps significantly, the formation energy of these latter systems are comparable to or lower than B_s, and thus, provided a formation mechanism exists, in thermal equilibrium there will be a tendency for the formation of these centers.

The presence of boron-pairs and boron–vacancy complexes may contribute to the reported reduction in doping efficiency in heavily doped diamond [80]. For example, although VB_3 may have a shallow acceptor level, it will only contribute one hole for three boron impurities, that is a third that of B_s doping.

In addition to this effect, *V*B and other interstitial related boron centers, act as deep donors, which can then compensate shallow acceptors [79].

There is no quantum-chemical evidence that interstitial boron, or interstitial boron pairs can give rise to shallow levels, which has been postulated as a contribution to superconductive material [81]. Indeed, the opposite is true, and the dumbbell structures proposed in reference [81] have been shown to be unstable [79].

8.3.4 Other Group III Acceptors

Originally, the acceptor level at $E_v + 0.37\,\text{eV}$ had been assigned to substitutional aluminum [82]. Although this view has been revised, and despite the conspicuous success of boron-doping, it is useful to determine the scope of Al and other group-III substitutional species to act as shallow acceptors.

Theory predicts that as the atomic number increases, so the acceptor level associated with the on-site substitutional impurity moves deeper into the band gap [47]. The values from LDA-DFT calculations using cubic supercells are 1.4 eV, 1.4 eV, and 1.8 eV above the valence band top for Al, Ga, and In, respectively. Such deep levels are clearly of no practical importance.

8.4 The Problem of n-Type Diamond

8.4.1 Phosphorus and Other Pnictogen Substitutional Doping

Traditional doping techniques generally supplant a host species with dopants with one more valence electron, and for diamond and other group-IV materials this implies the use of pnictogens. In silicon, nitrogen is relatively insoluble, but phosphorus may be easily incorporated and leads to a shallow donor level at around 44 meV [83], and an even shallower one at 12 meV for germanium [84]. In contrast, nitrogen is highly soluble in diamond, and is commonly present in natural crystals where it is present in various aggregated forms. N is also soluble in SiC, where it substitutes for carbon and leads to a shallow donor level at 50 meV [85]. The presence of N in SiC leads to the common n-type conductivity of as-grown materials.

However, in diamond, nitrogen does not lead to a shallow donor level. This is at least in part due to a rather dramatic localization of the donor electron on a neighboring carbon atom concurrent with the formation of a lone-pair on the nitrogen atom. The relaxed structure, and its unpaired electron wave function is shown schematically in Figure 8.5. The dramatic structural relaxation is sometimes referred [86] to as a Jahn–Teller or pseudo-Jahn–Teller effect, but this is probably not the case [87]. A Jahn–Teller effect occurs for systems with an orbitally degenerate ground state of non-zero effective spin, but on-site substitutional N has an orbitally non-degenerate electronic state [87]. The relaxation should therefore be considered a chemical rebonding.

The localization of the donor electron for N_s is accompanied by the formation of a very deep donor level, usually quoted at around $E_c - 1.7\,eV$ [8]. Clearly such a deep level is of little value for the manufacture of n-type semiconducting material, with a room-temperature ionization fraction of practically nil. Phosphorus, on the other hand, as a large impurity is not thought to relax in an analogous fashion to N_s, and leads to a substantially shallower donor level. Experimentally, the donor level is reported to be at $E_c - 0.6\,eV$ [9], which is still rather deep for room-temperature ionization.

Although P_s does not form a hyper-deep level, it does undergo a Jahn–Teller distortion [88, 89]. However, theory suggests that this relaxation does not lead to a significant deepening of the donor level: the Jahn–Teller energy in the neutral charge state obtained for relaxation to a tetragonal structure is less than 10% of the donor level energy [88, 89].

Calculations have also suggested that other pnictogens also give rise to potentially shallow donor levels. In particular, arsenic may be around 0.1–0.2 eV shallower than P_s, and substitutional antimony may be shallower still [90, 91]. Indeed, although P and As have similar donor levels in Si and Ge, the donor level of Sb is shallower in both materials [84].

The key issue with these alternative, unproven donor species appears to relate to the practicalities of incorporation in diamond from growth or implantation. To

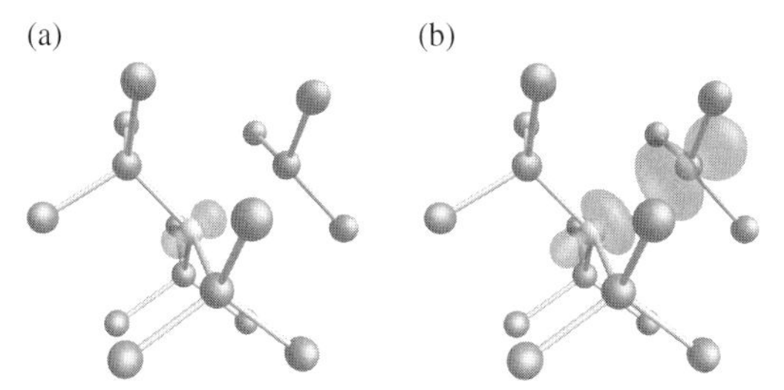

Figure 8.5 Schematics of the neutral substitutional nitrogen center in diamond. (a) Shows the charge density of the lone-pair orbital close the the valence band, and (b) the partially occupied antibonding combination of sp^3 orbitals localized upon the N atom and one of its four carbon neighbors resulting in the dilation of the bond and the trigonal symmetry. The nitrogen atom is at the center.

date, attempts to grow arsenic into diamond during CVD have apparently resulted in no measurable arsenic content in the diamond films as determined by SIMS [May *et al.*, private communication]. This may relate to the low solubility of As_s in diamond, but such arguments must be tempered with the knowledge that P_s also theoretically has a very low solubility.

To circumvent solubility or growth related obstacles, one may also examine As-implanted diamond. Studies on ion implanted material have not shown any clear indication of n-type conductivity due to As_s, although resistive n-type material with a donor activation of 0.42 eV has been reported [92]. Nuclear experiments performed on As-implanted diamond suggests that only around half of the impurities lie on the substitutional sites [93, 94]. The reported n-type conduction was similar to that seen with other implantation species, and it was concluded that the conduction related to the damage modified by the impurities, rather than formation predominantly of As_s. We shall return to implantation-related damage and the consequences for doping later in Section 8.5.

8.4.2 Chalcogen Donors

Chalcogens, and particularly sulfur, have also been used widely in attempts to dope diamond. Chalcogens, with two additional valence electrons per atom, represent double-donors. Substitutional S, Se, and Te in silicon lead to relatively deep donor levels [95] in comparison to pnictogens. For example, the activation energies of S and P are measured at 0.29 eV [96] and 44 meV [83] below the conduction band respectively. It is worthy of note that, at least in the case of Si, the trend is to become a more shallow donor with increasing atomic number, so that Se and Te are 0.29 eV and 0.20 eV, respectively [97].

For diamond, sulfur is typically added to the growth gas in the form of H_2S, leading to some intriguing observations related to n-type doping. For example, devices such as p–n junctions, detectors and thermionic emitters have been demonstrated [98–102]. However, there is a lack of clarity in the role of sulfur in these materials. For example, it seems clear that early reports [103, 104] of shallow donors with an activation energy of 0.38 eV actually relate to p-type conduction from the acceptor level of substitutional boron unintentionally incorporated due to contamination from the CVD reactor [105].

Analysis of the gas-phase chemistry where sulfur is present suggests its role may be one of modification of the key growth species [106–109]. Nevertheless, analysis such as SIMS [110] and particle induced X-ray emission spectroscopy (PIXE) [101, 111] clearly shows that sulfur is incorporated during growth, the mechanisms for which have some theoretical support [112, 113].

Implantation studies have also indicated relatively shallow donor levels, and in such cases contamination by boron or other impurities can be assumed to be minimal. The reported activation energy for conduction in these samples varies considerably. Hasegawa *et al.* [99] interpret the temperature dependence of the resistance to indicate a donor level between 0.19 eV and 0.33 eV for samples with [S] of $8 \times 10^{19} - 3 \times 10^{20}\,cm^{-3}$. Material prepared in a similar way yielded a level in the 0.28–0.42 eV range, determined by deep-level transient spectroscopy [114], and a third implantation study reported an activation energy of 0.32 eV [115].

Activation energies from material where the sulfur is incorporated during growth also vary. Besides the initial reports of 0.38 eV [103, 104] values in the 0.5–0.75 eV have also been reported, where boron is intentionally included and may be directly involved in the n-type conduction [116].

However, the origin of the donor levels in the n-type conductive material is far from clear. One possibility would be substitutional sulfur. A number of computational approaches have been employed to estimate the donor level of such a defect. Early, semi-empirical molecular orbital cluster methods [117] and later first-principles plane-wave calculations [118, 119] have suggested shallow donor levels, quoted as 0.37 eV [117], 0.375 eV [118] and 0.15 eV [119]. However, the weight of evidence appears to be that substitutional sulfur has a deep donor level: $\sim E_c - 1$ eV [120], $E_c - 1.63$ eV [121–123], $E_c - 0.77$ eV [124], $E_v + 3.7$ eV [125], $E_c - 1.4$ eV [90, 126] $E_c - 1.12$ eV [48], $E_c - 1.2$ eV [33] and $E_c - 1.02$ eV [51]. Where the donor level of substitutional sulfur is found to be deep, it appears to be at least partly due to a large structural relaxation which lowers the total energy relative to the on-site center by 0.4–0.5 eV [90, 125]. The resultant structure, at least in the neutral charge state, is calculated to prefer a trigonal geometry with a single dilated S—C bond [48, 90, 123, 125].

The practical upshot is that simple substitutional doping with S, be it via growth or implantation, is unlikely to be the origin of any measured n-type conduction.

Recent interpretation is suggestive of a combined role for S with boron and/or hydrogen in the donor activity of sulfur-containing material [51], and the suggestion from implantation studies is that the doping is either wholly [127] or partly damage-related [114]. The interaction of sulfur with other impurities or lattice vacancies may be important either in dopant compensation [47], or even the formation of impurity bands [51], and we shall return to such systems in Sections 8.4.4, 8.4.5, and 8.5.

Other chalcogens have also been analyzed, with Se and Te also resulting in rather deep levels, albeit potentially shallower than substitutional sulfur: the donor levels are reported to lie at $E_c - 1.4$ eV and $E_c - 1.2$ eV, for Se and Te, respectively [90]. The deep levels appear to be in response to structural relaxations, as with sulfur, and it seems likely that substitutional Se and Te are not good candidates for n-type diamond.

A remaining possibility lies with oxygen. In a similar fashion to the case of sulfur, n-type diamond appears to result from implantation of oxygen [37, 128, 129]. The activation energy 0.323 eV appears very promising, but as with the n-type doping via sulfur implantation, it seems likely that the conduction mechanism is not simply via a substitutional dopant [128]. First principles calculations indicate that substitutional oxygen, as a divalent species with a covalent radius similar to that of carbon, leads to very deep donor and acceptor levels. The (0/+) level has been variably predicted to lie at $E_v + 1.97$ eV [130], $E_c - 3.1$ eV [32], and $E_c - 2.8$ eV

[33], and (−/0) at $E_v + 2.87$ eV [130] and $E_c - 1.9$ eV [33]. The likely acceptor property of substitutional oxygen would mean that these centers would compensate oxygen-related shallow donors, so that the conduction mechanism for O-implanted diamond is more probably a result of the lattice disorder, possibly involving the oxygen species.

8.4.3
Interstitial Dopants: Alkali Metals

Another plausible class of dopants for n-type conductivity in diamond arises with interstitial monovalent species: this has lead to investigation of the alkali metals, primarily lithium (Li_i) and sodium (Na_i).

Early theoretical simulations involving small systems suggest that for these impurities lying at the T-site, the donor levels are relatively shallow [131–133], and indeed, working devices based on Li doping have been demonstrated [134–137]. The simulations yield donor levels relative to E_c at 0.1 eV [131, 132] and 0.4 eV [133] for Li_i and 0.3 eV for Na_i [131, 132]. Where Li-doping has resulted in conductive material and electrical activation energies have been measured, they generally lie within 0.2 eV of the conduction band minimum [134, 138–140], close to the theoretical values for Li_i. This may be viewed as some support for the possibility that the conduction is proceeding via Li_i dissolved within the diamond lattice, but one has to be cautious. An activation energy for conduction is not spectroscopically linked to any system conclusively Li-related, and may result from the mechanisms involved in introducing Li to the diamond, such as lattice damage in the case of ion implantation.

Indeed, as with other impurities, it is very important to consider the likely influence of the means by which the dopants are introduced into the diamond, with variations shedding some light on the likely underlying processes that may generate conduction electrons. Li has been introduced via a range of methods: implantation [135, 138, 139, 141–144], in-diffusion [145–147], and during growth [134, 148, 149]. Different methods yield various levels of success in terms of the resultant conductivity of the doped diamond.

Ion implantation often yields n-type material, but annealing to the relatively modest temperature of 600 °C dramatically reduces conductivity [136]. The loss of carriers has been interpreted as passivation of Li_i donors due to complex formation with residual lattice damage [139], since self-interstitials become mobile above around 400 °C and neutral vacancies at 600 °C, corresponding to activation energies of diffusion of 1.68 ± 0.15 eV and 2.4 ± 0.3 eV, respectively [150, 151].

In reality, the true nature of the donor system in n-type Li-implanted diamond is far from obvious. Indeed, since conduction is in the hopping regime, it seems a plausible conclusion that conduction proceeds via implantation damage rather than Li [140]. Nuclear techniques involving the radioactive decay of ^{8}Li [152], show that just 40% and 17% of implanted Li lie on interstitial and substitutional sites, respectively. The sizable contribution from *substitutional* lithium is very important: since substitutional Li (Li_s) is (at least theoretically) a deep acceptor [133], the presence of substantial concentrations of Li at this site will inevitably lead to significant levels of self-compensation.

Where lithium is in-diffused into diamond, the material appears generally to be characterized as insulating [145–147, 153, 154], which is also the case for material where lithium is grown into the diamond with a Li-source such as Li_2O or C_4H_9OLi [134, 148, 149].

There is some evidence that de-activation of the potential Li interstitial donors is a result of impurity clustering, and there are concomitant estimates of the diffusion energy of Li. These vary rather widely: 0.26 eV [154], 0.9 ± 0.3 eV [153] and >1.25 eV [152]. In comparison, the *calculated* diffusion barrier for Li_i at 0.85 eV is more in line with the higher experimental estimates [131]. However, again, there is nothing in the experimental studies that uniquely identify the nature of the atomic processes that are activated by the various measured amounts: in short, it is not obvious what is diffusing in the experiments.

Photothermal defect spectroscopy and photoconductivity measurements [155–157] indicate the presence of gap levels yielding absorption thresholds around 0.9 and 1.5 eV in Li-doped diamond. However, the correlation of these features directly with Li point defects, or even the aggregated lithium clusters is unclear [157].

The assignment of optical and electrical effects to lithium is further complicated as it is thought that Li is incorporated in diamond with other impurities such as oxygen and boron [134, 147, 149]. Such complexes are inevitably going to have electronic structures quite different from isolated Li.

Relative to lithium, much less experimental data is available for sodium doped diamond. However, Na-implanted material is reported to yield n-type diamond with a small activation energy for conduction (0.13–0.42 eV), but high resistivities [92, 140]. As with Li, theoretically interstitial Na is a candidate n-type dopant, albeit with a deeper donor level than Li, reported at $E_c - 0.3$ eV [131, 132]. Nevertheless, such a level would represent an improvement over P_s.

Recent calculations, chiefly examining the potential for Li as an n-type dopant, have proved rather negative. There appear to be two properties of Li in diamond that, when combined, suggest that Li may have a reduced efficacy in terms of n-type doping: the first relates to the mobility of the interstitial atom within the diamond lattice, and the second concerns the formation of a range of clusters which are either electrically inactive or, worse still, act as acceptors and will compensate donors that remain in solution.

We first return to the activation barrier for diffusion of the simple interstitial lithium donor. Most recently the barrier has been estimated using first principles density functional simulations to be 1.09 eV [158] and 1.2 eV [159] for the neutral donor. These come from simulations involving 72 and 64 atom supercells, respectively, and the activation energy is around 30–40% higher than the values obtained using 32-atom cells [131].

In the recent simulations, and in contrast to earlier calculations, the donor electron is relatively localized in a single dilated C—C bond, in the cage enclosing the interstitial ion [159]. The structure and wave function are shown schematically in Figure 8.6(a). This localization does not happen spontaneously in simulations, but lowers the total energy by around 0.2 eV.

So, what is the significance of an activation energy, E_a, for diffusion between 1.09 and 1.2 eV? Assuming a simple Boltzmann process, the temperature at which the rate of diffusion steps is 1 Hz is given by the expression $E_a/k_B \ln(\omega)$, where k_B is Boltzmann's constant and ω is the attempt frequency. We may further make the assumption that the attempt frequency is of the order of the localized vibrational frequencies of Li_i. These have been calculated [159] to be in the vicinity of 1400–1500 cm^{-1}. The interplay between attempt frequency and activation energy can be seen by looking at the isotherms that satisfy the above hopping condition, as shown in Figure 8.7. The attempt frequency is plotted on a logarithmic scale, showing that for the calculated range of activation energies, Li_i would be mobile

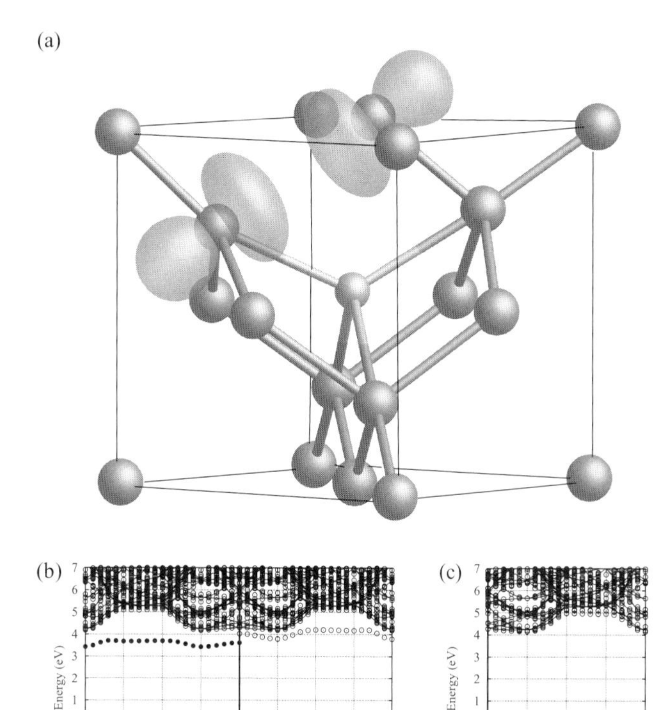

Figure 8.6 (a) An iso-surface showing the localization of the charge density associated with the donor electron wave function for Li_i in diamond. (b) and (c) show the band structures of Li_i^0 and Li_i^+ in a simple cubic supercell of side length 3 a_0 along high-symmetry directions in the Brillouin zone in the vicinity of the band gap. Filled circles show filled bands, empty circles, empty bands. The left and right panels of (b) are spin up and spin down spectra, respectively. Solid lines are for defect-free material and the zero of energy is the bulk E_v.

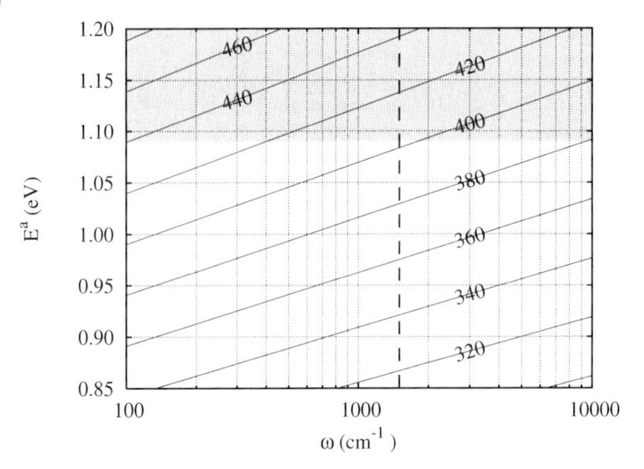

Figure 8.7 Plot of the isotherms (Kelvin) satisfying the migration of Li_i in diamond as a function of activation energy (E_a) and attempt frequency (ω). The shaded area indicates the range of calculated activation energies, and the vertical, dashed line shows the the attempt frequency corresponding to the calculated local mode of Li_i.

in the range 400–440 K over a reasonable range of ω. If the earlier estimate of 0.85 eV is correct, then for the same attempt frequency the temperature for diffusion is lowered to the order of 50 °C. Even the higher temperatures are low in comparison to that experienced during growth and those typical of in-diffusion experiments, meaning that during preparation Li_i would be expected to be lost to traps in the material.

The diffusion of Li_i would permit the formation of complexes, either with a lattice vacancy in implanted material, or perhaps with other Li_i donors. The former case would generate a substitutional Li acceptor which represents a very deep trap for additional mobile Li_i, thus forming a stable acceptor or even electrically passive complexes [159].

The pairing of Li_i to form interstitial dimers may not be an obvious result, but theoretically such structures were independently obtained using different DFT based simulation methods [158, 159]. The presence of a dilated C—C bond for the isolated neutral Li_i donor can then be accommodated for a pair where they coincide, leading to a substantial binding energy for the dimers of around 1.7 eV per Li_i [159]. Aggregates of three Li_i are also bound [158], with these small clusters leading to deep donor levels. The aggregation of Li_i is expected to remove any shallow donor properties [158, 159].

Despite their energetic favorability, Li_i-pairs can diffuse through the diamond with a similar diffusion barrier to that of Li_i calculated at around 1.4 eV [159]. This means that not only will Li_i shallow donors be able to form pairs at the temperatures typical of in-diffusion, growth, and annealing, they will in turn be able to diffuse to other traps.

The diffusion of Li_i and Li_i-pairs may result in significant clustering. Indeed, large clusters have been inferred from optical experiments [153]. Prototype planar precipitates have been shown theoretically to be strongly bound relative to Li_i in solution [160]. The formation of such extended defects is consistent with the accounts of Li in carbon composites used for ion storage applications [161–163]. In such cases there appears to be a greater uptake of Li in non-diamond bonded regions. The application of Li (and other alkali metal interstitial dopants) for thermally stable n-type doping therefore seems doubtful on the basis of the calculations.

8.4.4 Hydrogen-Modification for n-Type Doping

CVD of diamond typically results in an abundance of H-containing species in the growth mixture, and there is evidence that a small fraction of this may be taken up in the material. Defect systems, typically also involving carbon dangling bonds related to vacancies or vacancy-like environments, are seen in electron paramagnetic resonance experiments involving as-grown CVD material [164–172]. Some of these centers involve impurities other than H and, although the mechanism for the formation of these complexes is not well understood, this may suggest that other H-containing complexes are present in as-grown material.

Furthermore, subsequent exposure of diamond to hydrogen or deuterium plasmas is a relatively common method for addition of these species to diamond films, as reviewed in Section 8.3.2 for the determination of hydrogen passivation of boron acceptors.

Hydrogen has appeared as a component of a number of suggested shallow donor systems, some of which are known to perform well in other materials.

8.4.4.1 Chalcogen–Hydrogen Complexes

For silicon, it has been established that a relatively deep donor level associated with the simple substitutional chalcogen species can be significantly shifted upwards in energy by the partial passivation with a single interstitial hydrogen [173]. For example, the donor level of substitutional sulfur in silicon is 0.29 eV below E_c [96], but this is theoretically reduced to just 10 meV for S_s—H where the hydrogen atom lies anti-bonding to a neighboring Si atom [173].

The conversion of deep to shallow donor levels for simple substitutional chalcogen centers has also been examined for the case of diamond [33, 48, 90, 120, 121, 123, 125, 126, 174, 175].

The majority of analysis has been applied to sulfur–hydrogen interactions. The consensus is that simple S—H pairs are lowest in energy when the hydrogen atom lies at the AB1 site [90, 121, 123, 175] (Figure 8.8), which is also found for phosphorus–hydrogen complexes [48, 72, 124]. Antibonding directly to the chalcogen impurity is in contrast to the same centers in silicon [173], where the H lies either in the bond center but more closely associated with a Si atom, or at the antibonding site labeled AB3 in Figure 8.8.

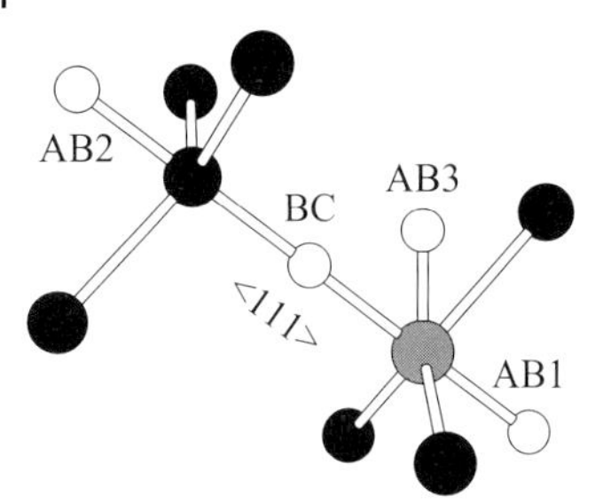

Figure 8.8 A schematic representation of four possible locations of hydrogen in impurity–hydrogen complexes. BC and AB refer to bond-centered and anti-bonding sites respectively. The black, white, and gray atoms represent carbon, hydrogen, and impurity sites, respectively.

For the bonding center, AB2, and AB3 sites, S—H complexes are predicted to be less stable by around 0.8–2.0 eV [90, 123]. One report [48] based on cluster calculations indicated that AB2 is the more stable conformation. However, in all sites, the binding energy relative to dissociation into the deep S_s double-donor and an interstitial H_i is relatively small in the context of growth temperatures: for neutral products 2.56 eV from LDA-DFT supercell calculations [121, 123], and 1.57 eV using LDA-DFT cluster calculations [48].

The precise location of the donor level of a S—H_{AB} pair is unclear, ranging from very deep to around the same as that of P_s: E_c − 1.07 eV [121], E_c − 0.61 eV [48], 0.3 eV deeper than P_s [33], E_c − 1.0 eV [90], and 0.5 eV–0.4 eV below E_c [175]. However, it may be that sulfur–hydrogen complexes of this type are responsible for the electrical activity seen in as-grown S-doped diamond which exhibits levels in the 0.75–0.5 eV range [116].

A general trend appears to be that hydrogenation of substitutional chalcogens raises the donor level in the band gap. The levels for O—H, Se—H and Te—H complexes from one study [33] are plotted in Figure 8.9.

This suggests that, at least theoretically, it would be preferable to dope with chalcogen–hydrogen complexes made up using species from further down the periodic table. However, it is likely that the uptake of large impurities would be difficult on the basis of what is known to be grown into diamond, and on the calculated solubilities. We shall return to the chemical trend in Section 8.4.5.

8.4.4.2 Sulfur–Vacancy–Hydrogen

Using a GGA-DFT approach it was found [174, 175] that there are combinations of sulfur, hydrogen, and a lattice vacancy that lead to donor levels within 0.5–0.6 eV of the conduction band minimum. Specifically, SVH_3 and SVH_4H_{AB} were proposed as shallow donors, and schematics of the atomic geometries are shown in Figure 8.10. It is clear that S—H_{AB} and SVH_4H_{AB} are iso-electronic, and one would expect them to have qualitatively similar properties.

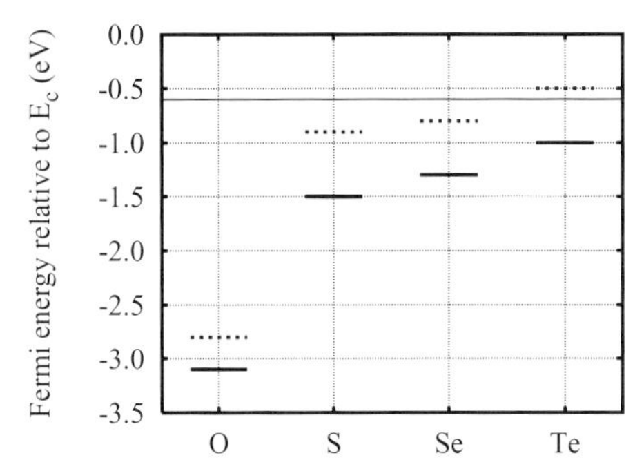

Figure 8.9 Donor levels for substitutional chalcogens (solid lines) and chalcogen–hydrogen pairs (dotted lines) relative to E_c calculated using 216-atom cubic supercells and LDA-DFT [33]. Levels are evaluated using the marker-method, aligned to the phosphorus donor at $E_c - 0.6\,\text{eV}$.

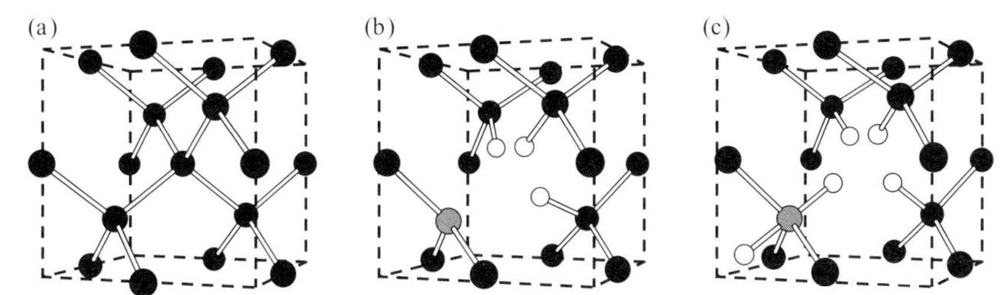

Figure 8.10 Atomic structures of S–H–V complexes proposed to produce shallow donor levels [174, 175]. Black, gray, and white atoms represent C, S, and H respectively. (b) SVH_3 and (c) SVH_4H_{AB}. (a) Shows the corresponding section of defect-free material for comparison.

One issue that mitigates against the use of such complex structures as donors is that the calculated (zero-temperature) formation energy (4 eV and greater) suggest that they would not occur in high concentrations under equilibrium conditions. More importantly, alternative arrangements of the constituent species, including simple vacancy–hydrogen complexes or even isolated hydrogen, are both thermodynamically more stable over a wide range of hydrogen chemical potentials, and give rise to *deep* levels, leading to electronic compensation. We return to this later in the chapter in Section 8.5.

8.4.4.3 **Nitrogen–Hydrogen Complexes**

Since nitrogen is highly soluble in diamond, and the aggregated forms are prevalent in natural materials, the potential for the modification of such centers to form shallow donors has obvious benefits.

Miyazaki *et al.* [176] performed density-functional simulations within the supercell approach for nitrogen pairs trapping an interstitial hydrogen species, forming a relatively shallow donor. The resultant donor wave function was found to be relatively delocalized [176], and is shown schematically in Figure 8.11(a). The initial calculated donor level lying around $E_c - 0.4\,\text{eV}$ was close to that predicted using the same methods for P_s, and it is therefore not clear if the use of N—H—N donors could result in a superior doping level to the simpler P_s approach.

Significantly, there are two effects not shown in the initial study which might exclude it as a candidate donor. The first is that recent calculations [Goss *et al.*, Submitted] suggest the bond-centered structure proposed in reference [176] is unstable relative to the a symmetry lowering reconstruction (Figure 8.11(b)). The relaxation localizes the highest occupied orbital onto a dilated nitrogen–carbon bond, pushes the donor level deeper into the band gap, and thereby renders N—H—N a deeper donor than P_s.

The same calculations also show that the center may act as an acceptor involving a spontaneous displacement of the hydrogen atom to a neighboring bond. This

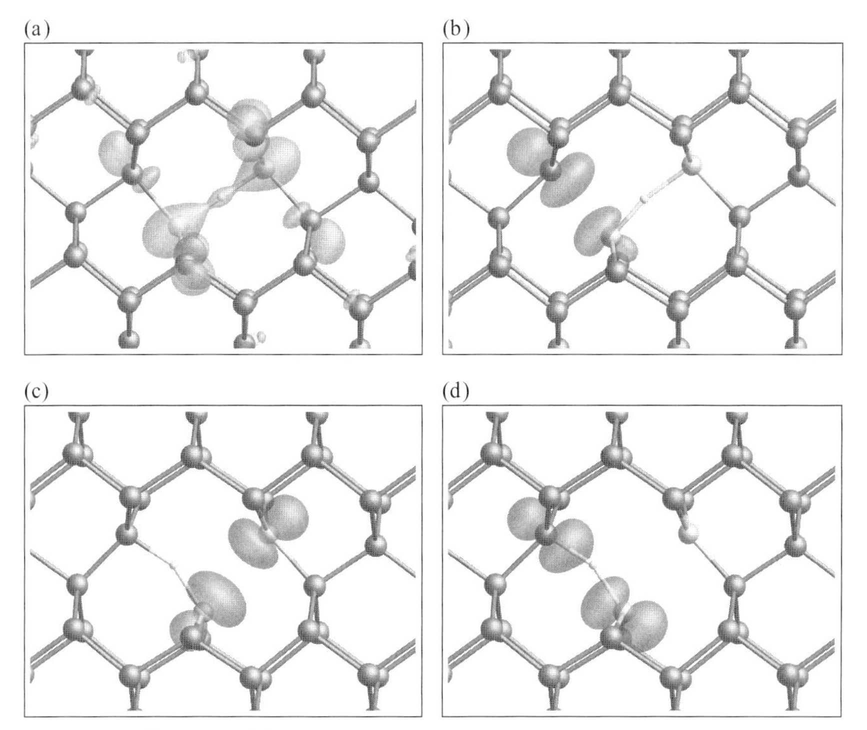

Figure 8.11 Schematics of the structures and gap-state localization for the N—H—N complex is diamond. In each panel the N-pair is at the center, and hydrogen is shown as a small white atom. The iso-surfaces are of the charge density for (a) the highest occupied orbital (HOMO) of axially symmetric N—H—N, (b) the HOMO of the perturbed N—H—N complex [177], (c) the second highest occupied orbital of $(\text{N—H—N})^-$, and (d) the HOMO of $(\text{N—H—N})^-$. The structures are viewed projected onto a (110) plane, with the vertical direction being [001].

renders the center negative-U in nature (i.e. the neutral charge state is never thermodynamically favored), and the (−/+) level of the complex is naturally deeper in the band gap than the donor level [Goss *et al.*, Submitted]. The negatively charged system is effectively an A-center neighbored by a bond-centered H_i, as can be seen from the lone-pair orbitals (Figure 8.11(c)) and mid-gap H_i state (Figure 8.11(d)).

8.4.4.4 **Boron–Hydrogen Complexes**

Recent experimental evidence may suggest the formation of shallow donors made up from B_s and multiple hydrogen or deuterium species [55, 75–77]. The experiment, as with the passivation of boron acceptors with single hydrogen or deuterium atoms, does not directly probe the nature of the electrically active defects, but there are several points of circumstantial evidence consistent with the model. First, Hall measurements point to a shallow level associated with electron transport. The location of the donor level depends upon the boron concentration and the resultant n-type carrier concentration matches the original hole concentration prior to exposure to the hydrogen/deuterium source.

In addition, SIMS shows a deuterium profile that matches the boron, but exceeds it by a factor of two or more. The modification of the acceptor to become a donor through the interaction with interstitial hydrogen is broadly supported by its theoretical amphoteric nature: H_i has mid-gap donor and acceptor levels [72, 124].

Since the original co-doping model for type inversion in this material was presented, several independent computational studies into the structure and properties of boron–hydrogen complexes have been published.

The simplest system, made up from the addition of a single H_i to a passive B_s—H complex, has been shown to be a deep donor, and indeed a deep acceptor [46, 48, 51]. The structure is strongly charge-state dependent, and a possible low-energy configuration is shown in Figure 8.12(a). The electrical properties of B_s—H_2 are related to those of H_i, since the passive B_s—H complex plays a limited role. It seems unlikely on the basis of current first-principles modeling that the electrical conduction observed experimentally arises from B_s—H_2 complexes.

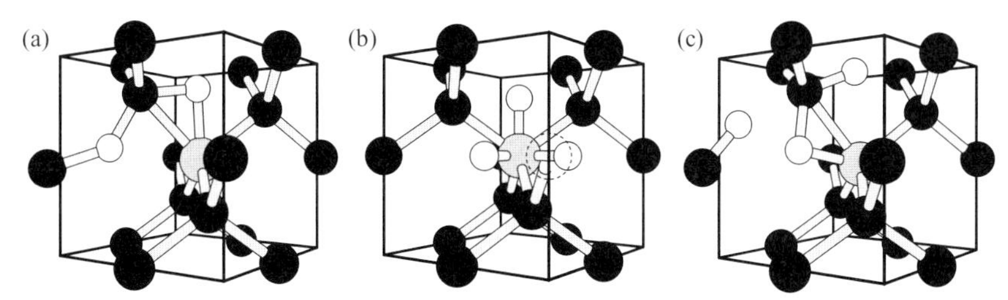

Figure 8.12 Schematic representations of B_s—H_2 and B_s—H_3 complexes, as obtained using density-functional methods. (a) Shows the stable form of B_s—H_2 from reference [46]. (b) the proposed BH_3 donor system from reference [56], and (c) the ground state structure for B_s—H_3 from reference [177].

In contrast, presented as theoretical support for the shallow donor codoping model, Dai *et al.* [56] reported the results of cluster-based density-functional calculations. They suggested that structures made up from B_s and *three* interstitial H atoms would be a shallow donor. Their proposed structure, shown in Figure 8.12(b), gives rise to an occupied Kohn–Sham level at around E_c – 1.0 eV in comparison with a defect-free cluster band gap. Even ignoring the use of Kohn–Sham eigenvalues as a quantitative measure of donor levels, this represents a deep donor level, and cannot explain the 0.34 and 0.24 eV activation energies reported from experiment [55, 76]. More seriously, this configuration of B_s—H_3 has since been shown to be unstable [51, 177].

One of the more stable structures than that in Figure 8.12(b) is shown in Figure 8.12(c). This form of B_s—H_3 is passive, being made up from a passive B_s—H complex and a hydrogen pair in a H_2^* arrangement, which is composed of colinear bond-centered and anti-bonding hydrogen, which although observed in silicon [178], was first predicted for hydrogen pairs in diamond [179].

In summary, the weight of evidence from density-functional modeling of complexes made up from B_s and more than one hydrogen atom therefore has provided no substantiated support for a shallow donor species made up from these components. An alternative model involving impurity band conduction has been presented in the literature [51], and is reviewed here in Section 8.4.6.

8.4.5
Co-doping: Multi-Substitutional Species as Shallow Donors

The modification of defects by the addition of hydrogen, such as the type-inversion of B-doped diamond, can be thought of as a post-growth processing stage. However, the migration energy typical for substitutional impurities in diamond render such mechanisms costly due to the high temperatures and pressures required. However, under favorable circumstances substitutional impurities might be introduced together during growth, or, provided that their radiation damage can be removed, via implantation and annealing. The complexes of impurities that may result *could* give rise to levels closer to the the band edges than the constituent species, and the drive for a shallow donor in diamond has lead to investigation of this effect for the production of n-type material.

There are many schemes that might give rise to the elusive donor system, and we present an overview of each type in turn.

8.4.5.1 Donor–Acceptor–Donor Complexes

The first candidate multi-impurity donor system which we shall discuss is made up from two donors and an acceptor, resulting in a defect with a single donor level. Yu *et al.* suggested a surfactant-mediated incorporation of N—Al—N structures during growth [180]. Such a mechanism overcomes one challenge of co-doping, namely the way one achieves the incorporation of a number of impurities in a specific structure. The initial calculations were promising, with an estimation of the donor level at E_c – 0.4 eV [180]. However, subsequent calculations have indicated rather deeper levels [176, 181].

N–Al–N is an example of a broad class of defects made up from two single-donors and an acceptor, for which one may adopt the short-hand notation "DAD". The basis of DAD centers giving rise to a shallow donor level may be thought of as having electronic and structural components. The latter is especially important for N-containing complexes, since suppressing the localization of the donor electron into a carbon dangling bond will immediately result in a level significantly shallower than the donor on its own. The former component is more subtle [122]. For DAD centers one can consider a simple model made up from a combination of three states, as indicated in Figure 8.13(a).

A wide range of combinations have been modeled [181]. As highlighted in Section 8.4.4.1, there appears to be a clear trend in the donor level being shallower for systems involving "larger" impurity species, as shown in Figure 8.14(a), but conversely the energetic stability for these systems is rather poor, as shown in Figure 8.14(b).

For those involving combinations of nitrogen and boron which have previously been reported as good candidates [122], there are serious issues regarding the order in which the three impurity atoms are incorporated, and the equivalence or otherwise of the two N donors. For an orthorhombic center, N–B–N theoretically yields a donor level significantly above that of N_s [122].

However, N and B are of a similar atomic size to the host atoms. Then, in contrast to the N–Al–N system, one of the nitrogen atoms may relax in a fashion similar to that of isolated N_s. This in turn would be expected to yield deep donor levels. The necessary step is to allow the nitrogen atoms to be equivalent. The reconstructed center can be described as (N–B)–N*, where the (N–B) component is iso-electronic with the host, and N* closely resembles the hyper-deep N_s donor:

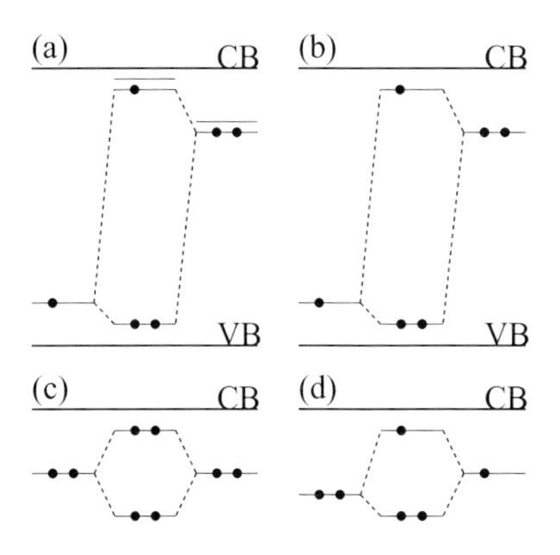

Figure 8.13 Schematics of the interaction between donor and acceptors that might lead to shallow levels in co-doping. (a) DAD, (b) double-donor–acceptor, (c) double-donor pairs, and (d) donor–double-donor complexes. Black circles indicate occupancy.

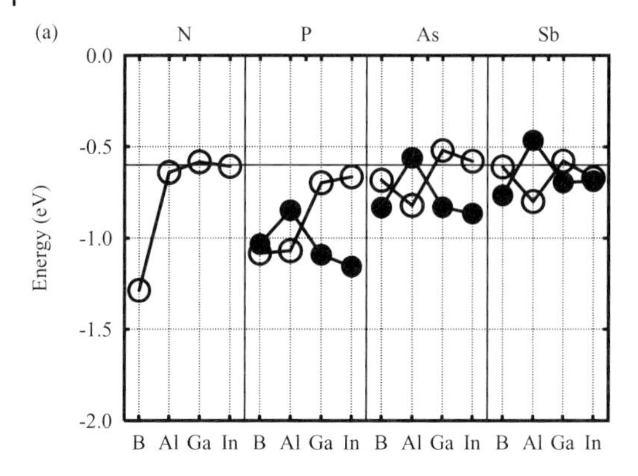

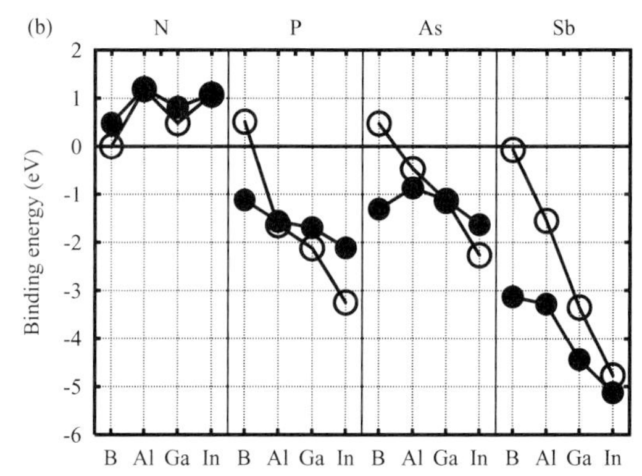

Figure 8.14 (a) Donor levels of D—A—D and D—D—A systems calculated using the marker method. The level of the phosphorus marker is denoted by the horizontal line at −0.6 eV, with E_c defined at zero energy. Results are taken from 216 atom systems, and are plotted for the D—A—D (empty circles) and D—D—A (filled circles) conformers. (b) Shows a plot of the corresponding binding energies (eV). Binding energies above zero represent bound systems. In each plot the acceptor species is indicated along the bottom and the donor species along the top.

we find a donor level of $E_c - 1.3$ eV (Figure 8.14(a)). The shallow donor, orthorhombic structure [122] where the nitrogen atoms are equivalent is at best metastable, and not a strong candidate for producing n-type diamond.

More seriously, reordering the components into N—N—B produces an acceptor. This can be understood in terms of B_s neighboring an A-center (nearest neighbor N atoms). The latter center has a donor level close to the valence band top [33, 182]. In the neutral charge state N—N—B can be thought of $(N{-}N)^{+}{-}B^{-}$, which then has an acceptor level around the same energy as the donor level of the A-center.

If co-doping leads to a mixture of N—B—N and N—N—B conformations, the latter would compensate the former. To avoid this, one would desire the donor system to be more stable, but this is not the case, with the total energy of N—N—B is 0.5 eV lower than N—B—N (Figure 8.14(a)). This conformational issue is not unique to

N/B co-doping, and we note the similarity to the difficulties with sulfur–vacancy–hydrogen complexes where partially hydrogenated systems will compensate the shallow donor systems (Section 8.4.4.2).

Another system suggested, also DAD in nature, is an S—B—S complex which has been calculated to be quite shallow at around $E_c - 0.5$ eV [121, 125], although recent cluster-based calculations predicts that it yields a deep donor level close to that of the B—S pair [51].

Perhaps the main problem with the shallower centers, such as Sb—Sb—Al (Figure 8.14(a)) however, is that they are thermodynamically unstable. In particular, they are unbound relative to their component parts (Figure 8.14(b), and combined with entropic considerations [183] would therefore only be able to be produced if grown into the lattice intact. There is very limited evidence for aggregated impurities being incorporated in this way, even for highly soluble species such as boron and nitrogen.

8.4.5.2 **Substitutional Pairs**

Alternative structures to three-impurity DAD systems involving modification of double donors have also been examined. A simplistic electronic structure argument, shown in Figure 8.13(b) may lead one to believe that double-donor–acceptor complexes, being relatively simple, are an improvement over the DAD systems.

Of particular interest are those made up from sulfur and boron since experimentally there is some evidence that the simultaneous incorporation of these species may be the root cause of the apparent n-type activity in CVD material grown in the presence of sulfur [110]. Here one can envisage a repulsive interaction between the states on the B and S, and indeed early, molecular orbital calculations suggested this to be the case [117]. However, more accurate calculations show that a simple nearest-neighbor S—B pair yields very deep donor levels [32, 51, 181]. The problem is that the picture shown in Figure 8.13 for DAD, double-donor–acceptor complexes, double-donor pairs, and donor–double-donor complexes is too simplistic. There are many more states involved in the electronic structure than shown, and rehybridization, structural relaxation, strain, and other effects complicate the picture considerably. For example, for systems involving states close to the valence band top, the energy difference between these levels and those closer to the conduction bands is likely to lead to much smaller quantum-mechanical repulsive interactions than shown schematically in Figure 8.14 and previous publications [52, 122].

We note that a model whereby separate boron- and sulfur-related defects are responsible for the conduction via mid-gap impurity bands has been suggested [51]. We return to this later in Section 8.4.6.

Of course, as with DAD complexes, one can calculate the donor activity for a wide range of combinations of pairs of substitutional dopants. A recent study examined a broad spectrum [181] but found that only the largest of species were even plausible as shallow donors. It is highly unlikely that such combinations would form. For example, phosphorus pairs at nearest neighbor sites are shallower than isolated P_s due to the increased strain of having two large impurities in close proximity [181]. However, at other separations the P—P interactions lead to rather deep donors.

Double-donor–acceptor complexes made up from Te—Al pairs also appear relatively shallow [181], but there is no compelling evidence that either species can be incorporated in diamond during growth.

8.4.5.3 Si—N Co-doping – Iso-electronic Effects

In line with the constraining effect of Al preventing structural relaxation of the N donors in N—Al—N complexes, Segev and Wei suggested that encapsulation of N_s by silicon within the diamond matrix may lead to a *very* shallow donor [52]. The N_sSi_4 complex, shown schematically in Figure 8.15, has tetrahedral symmetry with N_s on a lattice site surrounded by four substitutional Si impurities. Using an approach based upon the formation energies and 128 atom supercells, the donor level was estimated to lie around 90 meV from the conduction band minimum, which, if true, would result in an improved room temperature ionization fraction relative to B_s acceptors. Despite the silicon centers being rather large, the complex is theoretically bound relative to dissociated components. Indeed the binding energy is rather large, being estimated at 3.17–3.7 eV [52, 184]. Under such conditions, one might therefore expect N_sSi_4 doping to be thermally stable.

The mechanism proposed for the shallow nitrogen donor level involves the repulsive interaction between occupied, strain-induced states associated with the iso-electronic Si_s centers as well as the constraining effect they have on the potential relaxation of N_s into the hyper-deep configuration. The combination of these effects was presented as the origin of a shallow donor system that might be obtained experimentally by co-implantation of the dopant species.

Indeed, part of the background that may be used as a justification for supposing N_sSi_4 systems to be a realistic candidate for doping diamond comes from experimental studies regarding Si implantation studies of diamond. In this case N is present as an accidental contaminant during implantation, with the subsequent material *measured* to have n-type characteristics [185–187]. Since nitrogen substi-

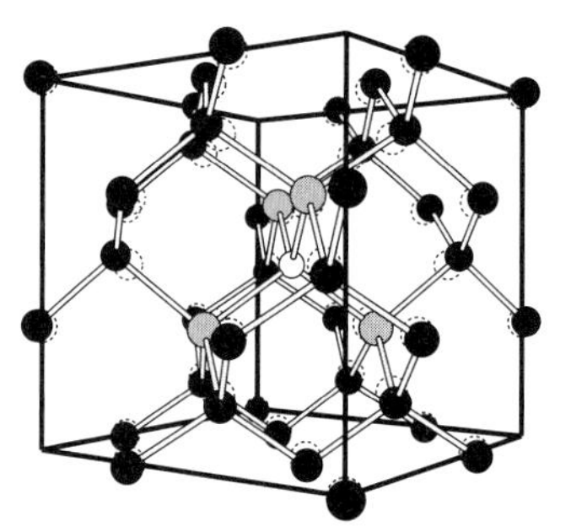

Figure 8.15 Schematic representation of the atomic structure for the N_sSi_4 complex [52, 91]. White, dark gray, and light gray circles represent nitrogen, carbon, and silicon, respectively. Dashed circles indicate unrelaxed sites. The box is aligned along cubic axes.

tuting for silicon in SiC is a shallow donor [85], one may take the view that the N_sSi_4 center in diamond is effectively N-doped SiC embedded within diamond.

Later modeling using much larger supercells revealed that the donor level of the complex may be artificially enhanced in simulations of such large defects employing small atomic cells, corresponding to high impurity concentrations [91]. Indeed, in 1000 atom supercells it was shown that the donor levels of P_s and As_s are comparable or shallower than N_sSi_4 [91].

It should also be noted that the favored interpretation of the silicon implantation doping of diamond involves hopping conduction via unannealed damage [186, 187], and cannot easily be connected to N-doped SiC inclusions.

8.4.6 Impurity Band Conduction

In the case of boron, p-type doping, one model for the conversion from semiconductor to metal involves the formation of an impurity band within the diamond band gap. A similar suggestion has been made for the conduction by electrons in boron–hydrogen or boron–sulfur–hydrogen co-doped materials [51]. The model involves the formation of mid-gap states due to combinations of these impurities, where the band gap wave functions are delocalized. Under this assumption, conduction through a band formed by the overlap of these defect states may occur.

In the proposal of Cai *et al.* [51], the defects involved are, in the case of boron–sulfur co-doping, substitutional boron aggregates with mid-gap acceptor levels, and S—V—S complexes for which they assign a donor level in the lower half of the band gap. The conduction would then proceed by activating the donor electrons thermally from the S—V—S complexes to the boron aggregates.

A closely related model has been proposed for the type inversion of boron-doped and hydrogenated diamond [51]. In this case the acceptor species are also boron aggregates but with varying levels of hydrogenation. The donor system is made up from interstitial hydrogen. There are a number of problems with this model. First, the positively charged interstitial hydrogen is mobile well below room temperature and would be expected to move to passivate the acceptors. Second, there is no evidence for the large concentration of boron aggregates that would be required for the number of carriers in post-hydrogenated material to be of the same order of magnitude as the hole concentration in the p-type, as-grown diamond, as observed experimentally [55]. In addition, the donor and acceptor states would have to be highly delocalized, since the concentration of boron in experiment is of the order of $10^{18}\,cm^{-3}$, well below the semiconductor–metal transition. For a mid-gap state to be as delocalized as a shallow donor or acceptor is unexpected: for the conduction mechanism of this type to be correct, the defect levels have to be close in energy and overlap spatially.

In addition to the model for mid-gap band-like conduction, an impurity band conduction model has been proposed for P—H and S—H complexes. The hydrogen in P—H complexes acts as an acceptor and partially passivates the phosphorus donor, with the complex possessing a filled state in the band gap [72, 121, 124]. The incomplete electrical passivation of phosphorus can be traced to the deep levels of hydrogen in diamond, and the P—H complex has been calculated to lead to a donor level at $E_c - 3.0\,\text{eV}$ [72] from LDA-DFT and $E_c - 3.29\,\text{eV}$ from GGA-DFT [124]. These values relate to thermodynamic levels involving both charge states. In contrast, it has been proposed using the Kohn–Sham spectra that the P—H defect also gives rise to an *empty* state very close to the conduction band minimum. Such a state would be expected to be rather diffuse, and lend itself to conduction. The model is that thermal population of this empty state from unpassivated P_s donors is easier than activation into the conduction band, which may result in enhanced conduction in material containing both P_s and P—H centers [121]. A similar model was also proposed for sulfur–hydrogen co-doping [121, 123]. However, the use of the Kohn–Sham spectra as evidence for such a mechanism must be viewed with caution in the light of the fact that thermodynamically there is no acceptor level associated with P—H [72].

8.5 Limiting Factors in Doping: Compensation and Solubility

With many of the proposed dopants for diamond electronics, there are significant issues arising from the formation of the desired structures.

For implantation of simple dopants, the main problem to overcome is to ensure that the vast majority of implanted ions reside on the correct lattice site: for example, substitution of a host atom for phosphorus or arsenic, or insertion into an interstitial site for lithium or sodium. However, implantation of atoms into the diamond lattice *always* results in damage to the diamond crystal structure, with the formation of vacancies and self-interstititals. Therefore, subsequent to implantation the diamond is annealed to high temperatures to remove implantation damage. In particular, the isolated lattice vacancy migrates with an activation energy of $2.4 \pm 0.3\,\text{eV}$ [151], for which annealing temperatures above around 600 °C are required.

In the case of boron-implanted materials, there has been some success in obtaining p-type material [36–42]. In such experiments annealing temperatures above 1130 °C have been employed.

However, boron appears to be somewhat of an exception in terms of achieving high levels of on-site dopants. For other substitutional dopant impurities the annealed material generally remains resistive. Indeed, nuclear measurements performed on arsenic-implanted material suggests just 50% of the dopants lie on the preferred substitutional site.

So, where are the remaining implanted dopants? One may be guided here by knowledge of structures adopted by other "large" impurities during growth. Silicon, has a covalent radius of 1.11 Å, which is 44% greater than that of the host at 77 Å, but rather close to the pnictogen dopants phosphorus (1.06 Å) and arsenic (1.19 Å). Now, silicon, when incorporated into diamond during CVD growth, commonly gives rise to an optical center around 1.682 eV which exhibits fine structure consistent with the natural abundance of Si isotopes for a transition between orbitally degenerate ground and excited states [188]. The electronic structure of the negatively charged complex of silicon with a lattice vacancy is consistent [189] with the experimental observations. The complex theoretically adopts a "split vacancy" structure where the impurity resides at the center of a divacancy, and it turns out that theoretically [47] this is also the structure of other complexes of a lattice vacancy with large impurity species (i.e. those of atomic number greater than 10).

Once formed, the very large (4.1–15.1 eV) calculated binding energies of these vacancy–impurity pairs from groups III, IV, V, and VI between atomic numbers 13 and 53 renders their dissolution improbable. This is in contrast to the case of boron where the binding energy of a boron–vacancy complex is just 1.8 eV, and therefore can be overcome at the annealing temperatures employed [36–42]. The practical implication is that one will not easily remove implantation damage from material containing impurities such as P and As so that only substitutional impurities remain.

Perhaps more serious than the trapping of potential shallow donors, the majority of vacancy–impurity complexes are also deep acceptors. They will therefore act as compensating centers in implanted material even if the implantation results in 50% of the impurities being the desired substitutional species.

In-diffusion of substitutional dopants has been achieved for boron, but the temperatures required are rather high: 1600 °C [43]. For in-diffusion of large donors such as P, it seems entirely likely that the temperatures required would be even higher. The arguments for this view are two-fold. The first is that the in-diffusion depends upon the solubility of the dopant, which is high for boron, but very low for phosphorus. (P_s has a calculated formation energy of around 6.6 eV [91, 124].) The second issue relates to kinetics. The activation energy for the diffusion of B or N by concerted exchange is calculated at 7–8 eV [79]. Diffusion with lower activation energies maybe mediated by vacancies [190] or other species, but if this were true for large impurities, due to the relatively larger binding energies for X—V complexes, one would most likely produce large concentrations of compensating centers. Assuming that the activation energy in the absence of catalyzing species such as vacancies is similar to that calculated for N or B, then the in-diffusion of P would require extremely high temperatures, and would have to overcome the very low solubility implied by the very high formation energy.

In contrast, the mobility of hydrogen in diamond can be very high, especially in the positive charge state [72]. Therefore, modification of implantation species might seem an attractive proposition: for example, it is likely that sulfur–vacancy complexes will be formed subsequent to implantation and annealing, and addition of hydrogen might convert these deep acceptors into the potentially shallow SVH_3 and SVH_4H_{AB} complexes (see Section 8.4.4.2). However, one has to examine the properties of other species that may be generated from these constituents. In particular, complexes made up from vacancies and fewer than four hydrogen atoms are strongly bound and result in deep acceptor levels [33, 51]. These complexes are thermodynamically favored over those containing sulfur [174] which might introduce significant compensation in thermal preparation of sulfur-implanted diamond.

In fact, although the application of thermodynamics is questionable for the incorporation of impurities during CVD, the solubilities as estimated from formation energies tend to support the view that it is difficult to incorporate impurities in diamond. For instance, the formation energies calculated for substitutional phosphorus, arsenic, and sulfur are all very high: P_s quoted at 6.67 eV [124] and 6.5 eV [91], arsenic at 10.4eV [91], and S_s at 10.67 eV [124].

The calculation of these values is both computationally difficult and requires the definition of reference states. Typically the atomic chemical potentials are taken from the respective solids, but in diamond growth this may be inappropriate. For example, it might be more reasonable to take phosphorus to be in dynamic equilibrium with phosphine, methane, and diamond. This then leads to formation energies being lowered by around 0.8 eV [91]. This scale of variation would have no qualitative effect upon the solubility of phosphorus which is effectively zero. However, it is not possible to exclude such effects having a more dramatic impact, and calculated solubilities from first-principles calculations must always be viewed with care.

The formation energies also do not take into account excitation and non-equilibrium conditions prevalent during growth. If one accepts the qualitative result that the solubility of phosphorus is very low, this suggests that its observed incorporation in high concentrations is a non-equilibrium process and the high formation energies of substitutional As and other “large” impurity species should not be taken as an absolute barrier to their use.

8.6 Concluding Remarks

A wide range of dopant species have been introduced into diamond during growth and via ion implantation in order to modify the electrical conductivity for use in semiconductor devices.

The success of doping using boron is something of an exception, and as a rule diamond resists the insertion of foreign species that lead to shallow levels. Even then, dilute solutions of boron in diamond, with an activation energy of 0.37 eV leads to low room-temperature ionization fractions and it is the incorporation of high concentrations of boron that lead to metallic behavior.

The deep donor levels associated with nitrogen and phosphorus have lead to the exploration of many species and complex doping techniques in the attempt to produce n-type diamond. In the vanguard of this research is the application of quantum-chemical computational techniques which are now beginning to be able to treat realistic concentrations of impurities. However, despite the proposal of many imaginative schemes for the generation of shallow donors, such as surfactant-mediated incorporation of multi-impurity complexes, the best candidates appear to be simple pnictogens: substitutional arsenic and antimony. Moreover, these, along with many other centers, suffer from extremely low solubilities and to date attempts to incorporate them during growth have been unsuccessful.

Nevertheless, the potential impact of the fabrication of n-type diamond during growth with an activation energy less than that of P_s would compliment the current state of p-type production, and in turn allow for the exploitation of a wide range of electronic applications currently difficult to achieve. Indeed, devices have been demonstrated using dopant such as S and Li. The nature of the donors in these systems, the microscopic origin for the type-inversion of boron-doped diamond, and in general the engineering of electronic grade diamond demands further investigation.

References

1 Singh, J. (1993) *Physics of Semiconductors and Their Heterostructures*, McGraw-Hill, New York.
2 Isberg, J., Hammersberg, J., Johansson, E., Wikström, T., Twitchen, D.J., Whitehead, A.J., Coe, S.E. and Scarsbrook, G.A. (2002) *Science*, **297**, 1670–2.
3 Windischmann, H. (2001) *Properties, Growth and Applications of Diamond* (eds M.H. Nazaré and A.J. Neves), number 26 in EMIS Datareviews Series, Chapter C2.2. INSPEC, Institute of Electrical Engineers, London, pp. 410–15.
4 Kagan, H. (2005) *Nuclear Instruments and Methods in Physics Research Section A*, **546**, 222–7.
5 Butler, J.E., Geis, M.W., Krohn, K.E., Lawless, J., Jr, Deneault, S., Lyszczarz, T.M., Flechtner, D. and Wright, R. (2003) *Semiconductor Science and Technology*, **18**, S67–71.
6 Thonke, K. (2003) *Semiconductor Science and Technology*, **18**, S20–6.
7 Ekimov, E.A., Sidorov, V.A., Bauer, E.D., Mel'nik, N.N., Curro, N.J., Thompson, J.D. and Stishov, S.M. (2004) *Nature*, **428**, 542–5.
8 Farrer, R. (1969) *Solid State Communications*, **7**, 685.
9 Gheeraert, E., Koizumi, S., Teraji, T. and Kanda, H. (2000) *Solid State Communications*, **113**, 577–80.
10 Koizumi, S., Watanabe, K., Hasegawa, M. and Kanda, H. (2001) *Science*, **292**, 1899–901.
11 Suzuki, M., Yoshida, H., Sakuma, N., Ono, T., Sakai, T. and Koizumi, S. (2004) *Applied Physics Letters*, **84**, 2349–51.
12 Tajani, A., Tavares, C., Wade, M., Baron, C., Gheeraert, E., Bustarret, E., Koizumi, S. and Araujo, D. (2004) *Physica Status Solidi (a)*, **201**, 2462–6.
13 Tavares, C., Tajania, A., Baron, C., Jomard, F., Koizumi, S., Gheeraert, E. and Bustarret, E. (2005) *Diamond and Related Materials*, **14**, 522–5.
14 Yan, Y., Zhang, S.B. and Al-Jassim, M.M. (2002) *Physical Review B*, **66**, 201401(R).
15 Kohanoff, J. (2006) *Electronic Structure Calculations for Solids and Molecules*, Cambridge University Press, Cambridge, UK.
16 Singleton, J. (2001) *Band Theory and Electronic Properties of Solids. Oxford Master Series in Condensed Matter Physics*, Oxford University Press, Oxford, UK.
17 Szabo, A. and Ostlund, N.S. (1989) *Modern Quantum Chemistry. Introduction to Advanced Electronic Structure Theory*, 1st edn, McGraw-Hill, New York, USA.
18 Levine, I.N. (2000) *Quantum Chemistry*, 5th edn, Prentice Hall, New Jersey, USA.
19 Thijssen, J.M. (1999) *Computational Physics*, 1st edn, Cambridge University Press, Cambridge, UK.
20 Hohenberg, P. and Kohn, W. (1964) *Physics Review*, **136**, 864–71.
21 Kohn, W. and Sham, L.J. (1965) *Physics Review*, **137**, A1697–705.
22 Kohn, W. and Sham, L.J. (1965) *Physics Review*, **140**, A1133–8.

23 Janak, J.F. (1978) *Physical Review B*, **18**, 7165–8.
24 van de Walle, C.G. and Neugebauer, J. (2003) *Nature*, **423**, 626–8.
25 Makov, G. and Payne, M.C. (1995) *Physical Review B*, **51**, 4014–122.
26 Nozaki, H. and Itoh, S. (2000) *Physical Review E*, **62**, 1390–6.
27 Gerstmann, U., Deák, P., Rurali, R., Aradi, B., Frauenheim, T. and Overhof, H. (2003) *Physica B*, **340–342**, 190–4.
28 Castleton, C.W.M. and Mirbt, S. (2004) *Physical Review B*, **70**, 195202.
29 Shim, J., Lee, E.K., Lee, Y.J. and Nieminen, R.M. (2005) *Physical Review B*, **71**, 035206.
30 Castleton, C.W.M., Höglund, A. and Mirbt, S. (2006) *Physical Review B*, **73**, 035215.
31 Erhart, P., Albe, K. and Klein, A. (2006) *Physical Review B*, **73**, 205203.
32 Albu, T.V., Anderson, A.B. and Angus, J.C. (2002) *Journal of the Electrochemical Society*, **149** E143–7.
33 Goss, J.P., Briddon, P.R., Sque, S.J. and Jones, R. (2004) *Diamond and Related Materials*, **13**, 684–90.
34 Goss, J.P., Shaw, M.J. and Briddon, P.R. (2007) *Theory of Defects in Semiconductors*, Vol. **104** of *Topics in Applied Physics* (eds D.A. Drabold and S.K. Estreicher), Springer, Berlin/Heidelberg, pp. 69–94.
35 Liberman, D.A. (2000) *Physical Review B*, **62**, 6851–3.
36 Fontaine, F., Uzan-Saguy, C., Philosoph, B. and Kalish, R. (1996) *Applied Physics Letters*, **68**, 2264–6.
37 Prins, J.F. (2002) *Diamond and Related Materials*, **11**, 612–17.
38 Tshepe, T., Kasl, C., Prins, J.F. and Hoch, M.J.R. (2004) *Physical Review B*, **70**, 245107.
39 Vogel, T., Meijer, J. and Zaitsev, A. (2004) *Diamond and Related Materials*, **13**, 1822–5.
40 Tsubouchi, N., Ogura, M., Kato, H., Ri, S.G., Watanabe, H., Horino, Y. and Okushi, H. (2005) *Diamond and Related Materials*, **14**, 1969–72.
41 Wu, J., Tshepe, T., Butler, J.E. and Hoch, M.J.R. (2005) *Physical Review B*, **71**, 113108.
42 Tsubouchi, N., Ogura, M., Kato, H., Ri, S.G., Watanabe, H., Horino, Y. and Okushi, H. (2006) *Diamond and Related Materials*, **15**, 157–9.
43 Krutko, O.B., Kosel, P.B., Wu, R.L.C., Fries-Carr, S.J., Heidger, S. and Weimer, J. (2000) *Applied Physics Letters*, **76**, 849–51.
44 Breuer, S.J. and Briddon, P.R. (1994) *Physical Review B*, **49**, 10332–6.
45 Goss, J.P., Briddon, P.R., Jones, R., Teukam, Z., Ballutaud, D., Jomard, F., Chevallier, J., Bernard, M. and Deneuville, A. (2003) *Physical Review B*, **68**, 235209.
46 Goss, J.P., Briddon, P.R., Sque, S.J. and Jones, R. (2004) *Physical Review B*, **69**, 165215.
47 Goss, J.P., Briddon, P.R., Rayson, M.J., Sque, S.J. and Jones, R. (2005) *Physical Review B*, **72**, 035214.
48 Lombardi, E.B., Mainwood, A. and Osuch, K. (2004) *Physical Review B*, **70**, 205201.
49 Bourgeois, E., Bustarret, E., Achatz, P., Omnés, F. and Blase, X. (2006) *Physical Review B*, **74**, 094509.
50 Crowther, P.A., Dean, P.J. and Sherman, W.F. (1967) *Physics Review*, **154**, 772–85.
51 Cai, Y., Zhang, T., Anderson, A.B., Angus, J.C., Kostadinov, L.N. and Albu, T.V. (2006) *Diamond and Related Materials*, **15**, 1868–77.
52 Segev, D. and Wei, S.H. (2003) *Physical Review Letters*, **91**, 126406.
53 Smith, S.D. and Taylor, W. (1962) *Proceedings of the Physical Society*, **79**, 1142–53.
54 Chevallier, J., Theys, B., Lusson, A., Grattepain, C., Deneuville, A. and Gheeraert, E. (1998) *Physical Review B*, **58**, 7966–9.
55 Teukam, Z., Chevallier, J., Saguy, C., Kalish, R., Ballutaud, D., Barbé, M., Jomard, F., Tromson-Carli, A., Cytermann, C., Butler, J.E., Bernard, M., Baron, C. and Deneuville, A. (2003) *Nature Materials*, **2**, 482–6.
56 Dai, Y., Dai, D., Liu, D., Han, S. and Huang, B. (2004) *Applied Physics Letters*, **84**, 1895–7.
57 Johnston, C., Crossley, A., Werner, M. and Chalker, P.R. (2001) *Properties, Growth and Applications of Diamond* (eds M.H. Nazaré and A.J. Neves), number 26 in EMIS Datareviews Series, Chapter

B3.3. INSPEC, Institute of Electrical Engineers, London, pp. 337–47.

58 Baron, C., Wade, M., Deneuville, A., Jomard, F. and Chevallier, J. (2006) *Diamond and Related Materials*, **15**, 597–601.

59 Yokoya, T., Nakamura, T., Matsushita, T., Muro, T., Takano, Y., Nagao, M., Takenouchi, T., Kawarada, H. and Oguchi, T. (2005) *Nature*, **348**, 647–50.

60 Xiang, H.J., Li, Z., Yang, J., Hou, J.G. and Zhu, Q. (2004) *Physical Review B*, **70**, 212504.

61 Boeri, L., Kortus, J. and Andersen, O.K. (2004) *Physical Review Letters*, **93**, 237002.

62 Lee, K.W. and Pickett, W.E. (2004) *Physical Review Letters*, **93**, 237003.

63 Blase, X., Adessi, C. and Connetable, D. (2004) *Physical Review Letters*, **93**, 237004.

64 Lee, K.W. and Pickett, W.E. (2006) *Physical Review B*, **73**, 075105.

65 Pearton, S.J., Corbett, J.W. and Stavola, M. (1992) *Hydrogen in Crystalline Semiconductors*, Chapter 3, Springer-Verlag, Berlin.

66 Stavola, M. (1999) *Properties of Crystalline Silicon* (ed. R. Hull), number 20 in EMIS Datareviews Series, Chapter 9.9. INSPEC, Institute of Electrical Engineers, London, pp. 522–37.

67 Nickel, N.H. (ed.) (1999) *Hydrogen in Semiconductors II*, Vol. **61** of *Semiconductors and Semimetals*, Academic Press, San Diego, USA.

68 Landstrass, M.I. and Ravi, K.V. (1989) *Applied Physics Letters*, **55**, 1391–3.

69 Ogura, M., Mizuochi, N., Yamasaki, S. and Okushi, H. (2005) *Diamond and Related Materials*, **14**, 2023–6.

70 Zeisel, R., Nebel, C.E. and Stutzmann, M. (1999) *Applied Physics Letters*, **74**, 1875–6.

71 Mehandru, S.P. and Anderson, A.B. (1994) *Journal of Materials Research*, **9**, 383–95.

72 Goss, J.P., Jones, R., Heggie, M.I., Ewels, C.P., Briddon, P.R. and Öberg, S. (2002) *Physical Review B*, **65**, 115207.

73 Teukam, Z., Ballutaud, D., Jomard, F., Chevallier, J., Bernard, M. and Deneuville, A. (2003) *Diamond and Related Materials*, **12**, 647–51.

74 Barjon, J., Chevallier, J., Jomard, F., Baron, C. and Deneuville, A. (2006) *Applied Physics Letters*, **89**, 232111.

75 Chevallier, J., Teukam, Z., Saguy, C., Kalish, R., Cytermann, C., Jomard, F., Barbé, M., Kociniewski, T., Butler, J.E., Baron, C. and Deneuville, A. (2004) *physica status solidi (a)*, **201**, 2444–50.

76 Kalish, R., Saguy, C., Cytermann, C., Chevallier, J., Teukam, Z., Jomard, F., Kociniewski, T., Ballutaud, D., Butler, J. E., Baron, C. and Deneuville, A. (2004) *Journal of Applied Physics*, **96**, 7060–5.

77 Saguy, C., Kalish, R., Cytermann, C., Teukam, Z., Chevallier, J., Jomard, F., Tromson-Carli, A., Butler, J.E., Baron, C. and Deneuville, A. (2004) *Diamond and Related Materials*, **13**, 700–4.

78 Bernard, M., Baron, C. and Deneuville, A. (2004) *Diamond and Related Materials*, **13**, 896–9.

79 Goss, J.P. and Briddon, P.R. (2006) *Physical Review B*, **73**, 085204.

80 Chen, Y.H., Hu, C.T. and Lin, I.N. (1999) *Applied Physics Letters*, **75**, 2857–9.

81 Pogorelov, Y.G. and Loktev, V.M. (2005) *Physical Review B*, **72**, 075213.

82 Dean, P.J., Lightowlers, E.C. and Wight, D.R. (1965) *Physics Review*, **140**, A352–68.

83 Long, D. and Myers, J. (1959) *Physics Review*, **115**, 1119–21.

84 Kittel, C. (2005) *Introduction to Solid State Physics*, 8th edn, John Wiley & Sons, Inc., Chichester.

85 Götz, W., Schöner, A., Pensl, G., Suttrop, W., Choyke, W.J., Stein, R. and Leibenzeder, S. (1993) *Journal of Applied Physics*, **73**, 3332–8.

86 Messmer, R.P. and Watkins, G.D. (1973) *Physical Review B*, **7**, 2568–90.

87 Lombardi, E.B., Mainwood, A., Osuch, K. and Reynhardt, E.C. (2003) *Journal of Physics: Condensed Matter*, **15**, 3135–49.

88 Eyre, R.J., Goss, J.P., Briddon, P.R. and Hagon, J.P. (2005) *Journal of Physics: Condensed Matter*, **17**, 5831–7.

89 Orita, N., Nishimatsu, T. and Katayama-Yoshida, H. (2007) *Japanese Journal of Applied Physics Part 1*, **46**, 315–17.

90 Sque, S.J., Jones, R., Goss, J.P. and Briddon, P.R. (2004) *Physical Review Letters*, **92**, 017402.

91 Goss, J.P., Briddon, P.R. and Eyre, R.J. (2006) *Physical Review B*, **74**, 245217.

92 Hunn, J.D., Parikh, N.R., Swanson, M.L. and Zuhr, R.A. (1993) *Diamond and Related Materials*, **2**, 847–51.

93 Bharuth-Ram, K., Quintel, H., Hofsäss, H., Restle, M. and Ronning, C. (1996) *Nuclear Instruments and Methods in Physics Research Section B*, **118**, 72–5.

94 Correia, J.G., Marques, J.G., Alves, E., Forkel-Wirth, D., Jahn, S.G., Restle, M., Dalmer, M., Hofsäss, H., Bharuth-Ram, K. and Collaboration, I. (1997) *Nuclear Instruments and Methods in Physics Research Section B*, **127–128**, 723–6.

95 Wagner, P. (1999) *Properties of crystalline Silicon* (ed. R. Hull), number 20 in EMIS Datareviews Series, Chapter 9.13. INSPEC, Institute of Electrical Engineers, London, pp. 573–82.

96 Pettersson, H. and Grimmeiss, H.G. (1990) *Journal of Applied Physics B*, **42**, 1381–7.

97 Grimmeiss, H.G. and Janzém, E. (1996) *Deep Centers in Semiconductors*, 2nd edn (ed. S.T. Pantelides), Gordon and Breach, Switzerland, p. 97.

98 Okushi, H. (2001) *Diamond and Related Materials*, **10**, 281–8.

99 Hasegawa, M., Takeuchi, D., Yamanaka, S., Ogura, M., Watanabe, H., Kobayashi, N., Okushi, H. and Kajimura, K. (1999) *Japanese Journal of Applied Physics Part 2*, **38**, L1519–22.

100 Horiuchi, K., Kawamura, A., Ide, T., Ishikura, T., Takamura, K. and Yamashita, S. (2001) *Japanese Journal of Applied Physics Part 2*, **40**, L275–8

101 Trajkov, E., Prawer, S., Butler, J.E. and Hearne, S.M. (2005) *Journal of Applied Physics*, **98**, 023704.

102 Koeck, F.A.M. and Nemanich, R.J. (2005) *Diamond and Related Materials*, **14**, 2051–4.

103 Sakaguchi, I., Nishitani-Gamo, M., Kikuchi, Y., Yasu, E., Haneda, H., Suzuki, T. and Ando, T. (1999) *Journal of Applied Physics B*, **60**, R2139–41

104 Nishitani-Gamo, M., Yasu, E., Xiao, C.Y., Kikuchi, Y., Ushizawa, K., Sakaguchi, I., Suzuki, T. and Ando, T. (2000) *Diamond and Related Materials*, **9**, 941–7.

105 Kalish, R., Reznik, A., Uzan-Saguy, C. and Cytermann, C. (2000) *Applied Physics Letters*, **76**, 757–9.

106 Sternschulte, H., Schreck, M. and Stritzker, B. (2002) *Diamond and Related Materials*, **11**, 296–300. Sp. Iss. SI.

107 Petheridge, J.R., May, P.W., Crichton, E.J., Rosser, K.N. and Ashfold, M.N.R. (2002) *Physical Chemistry Chemical Physics*, **4**, 5199–206.

108 Petheridge, J.R., May, P.W., Fuge, G.M., Robertson, G.F., Rosser, K.N. and Ashfold, M.N.R. (2002) *Journal of Applied Physics*, **91**, 3605–13.

109 Petheridge, J.R., May, P.W., Fuge, G.M., Rosser, K.N. and Ashfold, M.N.R. (2002) *Diamond and Related Materials*, **11**, 301–6.

110 Eaton, S.C., Anderson, A.B., Angus, J.C., Evstefeeva, Y.E. and Pleskov, Y.V. (2003) *Diamond and Related Materials*, **12**, 1627–32.

111 Nishitani-Gamo, M., Xiao, C.Y., Zhang, Y.F., Yasu, E., Kikuchi, Y., Sakaguchi, I., Suzuki, T., Sato, Y. and Ando, T. (2001) *Thin Solid Films* **382**, 113–23.

112 Dandy, D.S. (2001) *Thin Solid Films*, **381**, 1–5.

113 Haubner, R. (2004) *Diamond and Related Materials*, **13**, 648–55.

114 Troupis, D.K., Gaudin, O., Whitfield, M.D. and Jackman, R.B. (2002) *Diamond and Related Materials*, **11**, 342–6.

115 Prins, J.F. (2001) *Diamond and Related Materials*, **10**, 1756–64.

116 Nakazawa, K., Tachiki, M., Kawarada, H., Kawamura, A., Horiuchi, K. and Ishikura, T. (2003) *Applied Physics Letters*, **82**, 2074–6.

117 Anderson, A.B., Grantscharova, E.J. and Angus, J.C. (1996) *Physical Review B*, **54**, 14341–8.

118 Zhou, H., Yokoi, Y., Tamura, H., Takami, S., Kubo, M., Miyamoto, A., Nishitani-Gamo, M. and Ando, T. (2001) *Japanese Journal of Applied Physics Part 1*, **40**, 2830–2.

119 Saada, D., Adler, J. and Kalish, R. (2000) *Applied Physics Letters*, **77**, 878–9.

120 Miyazaki, T. and Okushi, H. (2001) *Diamond and Related Materials*, **10**, 449–52.

121 Nishimatsu, T., Katayama-Yoshida, H. and Orita, N. (2001) *Physica B*, **302–303**, 149–54.

122 Katayama-Yoshida, H., Nishimatsu, T., Yamamoto, T. and Orita, N. (2001) *Journal of Physics: Condensed Matter*, **13**, 8901–14.

123 Nishimatsu, T., Katayama-Yoshida, H. and Orita, N. (2002) *Japanese Journal of Applied Physics Part 1*, **41**, 1952–62.

124 Wang, L.G. and Zunger, A. (2002) *Physical Review B*, **66**, 161202(R).

125 Miyazaki, T. (2002) *physica status solidi (a)*, **193**, 395–408.

126 Sque, S.J., Jones, R., Goss, J.P. and Briddon, P.R. (2003) *Physica B*, **340–342**, 80–3.

127 Baskin, E., Reznik, A., Saada, D., Adler, J. and Kalish, R. (2001) *Physical Review B*, **64**, 224110.

128 Prins, J.F. (2000) *Diamond and Related Materials*, **9**, 1275–81.

129 Prins, J.F. (2000) *Physical Review B*, **61**, 7191–4.

130 Gali, A., Lowther, J.E. and Deák, P. (2001) *Journal of Physics: Condensed Matter*, **13**, 11607–13.

131 Kajihara, S.A., Antonelli, A., Bernholc, J. and Car, R. (1991) *Physical Review Letters*, **66**, 2010–13.

132 Bernholc, J., Kajihara, S.A., Wang, C., Antonelli, A. and Davis, R.F. (1992) *Materials Science and Engineering B*, **11**, 265–72.

133 Anderson, A.B. and Mehandru, S.P. (1993) *Physical Review B*, **48**, 4423–7.

134 Borst, T.H. and Weis, O. (1995) *Diamond and Related Materials*, **4**, 948–53.

135 Melnikov, A.A., Denisenko, A.V., Zaitsev, A.M., Shulenkov, A., Varichenko, V.S., Filipp, A.R., Dravin, V.A., Kanda, H. and Fahrner, W.R. (1998) *Journal of Applied Physics*, **84**, 6127–34.

136 Weima, J.A., von Borany, J., Meusinger, K., Horstmann, J. and Fahrner, W.R. (2003) *Diamond and Related Materials*, **12**, 1307–14.

137 Zaitsev, A.M., Bergman, A.A., Gorokhovsky, A.A. and Huang, M.B. (2006) *physica status solidi (a)*, **203**, 638–42.

138 Vavilov, V.S., Gukasyan, M.A., Guseva, M.I. and Konorova, E.A. (1972) *Soviet Physics–Semiconductors*, **6**, 741–6.

139 Job, R., Werner, M., Denisenko, A., Zaitsev, A. and Fahrner, W.R. (1996) *Diamond and Related Materials*, **5**, 757–60.

140 Prawer, S., Uzan-Saguy, C., Braunstein, G. and Kalish, R. (1993) *Applied Physics Letters*, **63**, 2502–4.

141 Popovici, G., Melnikov, A., Varichenko, V.V., Sung, T., Prelas, M.A., Wilson, R.G. and Loyalka, S.K. (1997) *Journal of Applied Physics*, **81**, 2429–31.

142 Galkin, V.V., Krasnopevtsev, V.V. and Milyutin, Y.V. (1970) *Soviet Physics–Semiconductors*, **4**, 709–16.

143 Vavilov, V.S., Konorova, E.A., Stepanova, E.B. and Trukhan, É.M. (1979) *Soviet Physics–Semiconductors*, **13**, 604–6.

144 Vavilov, V.S., Konorova, E.A., Stepanova, E.B. and Trukhan, É.M. (1979) *Soviet Physics–Semiconductors*, **13**, 635–8.

145 Okumura, K., Mort, J. and Machonkin, M. (1990) *Applied Physics Letters*, **57**, 1907–9.

146 Popovici, G., Prelas, M.A., Sung, T., Khasawinah, S., Melnikov, A.A., Varichenko, V.S., Zaitsev, A.M., Denisenko, A.V. and Fahrner, W.R. (1995) *Diamond and Related Materials*, **4**, 877–81.

147 Popovici, G., Sung, T., Khasawinah, S., Prelas, M.A. and Wilson, R.G. (1995) *Journal of Applied Physics*, **77**, 5625–9.

148 Sachdev, H., Haubner, R. and Lux, B. (1997) *Diamond and Related Materials*, **6**, 494–500.

149 Sternschulte, H., Schreck, M., Stritzker, B., Bergmaier, A. and Dollinger, G. (2000) *Diamond and Related Materials*, **9**, 1046–50.

150 Allers, L., Collins, A.T. and Hiscock, J. (1998) *Diamond and Related Materials*, **7**, 228–32.

151 Davies, G., Lawson, S.C., Collins, A.T., Mainwood, A. and Sharp, S.J. (1992) *Physical Review B*, **46**, 13157–70.

152 Restle, M., Bharuth-Ram, K., Quintel, H., Ronning, C., Jahn, S.G., ISOLDE-Collaboration and Wahl, U. (1995) *Applied Physics Letters*, **66**, 2733–5.

153 te Nijenhuis, J., Cao, G.Z., Smits, P.C.H.J., van Enckevort, W.J.P., Giling, L. J., Alkemade, P.F.A., Nesládek, M. and Remes, Z. (1997) *Diamond and Related Materials*, **6**, 1726–32.

154 Uzan-Saguy, C., Cytermann, C., Fizgeer, B., Richter, V., Brener, R. and Kalish, R. (2002) *physica status solidi (a)*, **193**, 508–16.

155 Nesládek, M., Van ek, M. and Stals, L.M. (1996) *physica status solidi (a)*, **154**, 283–303.

156 Neslàdek, M., Meykens, K., Stals, L.M., Quaeyhaegens, C., D'Olieslaeger, M., Wu, T.D., Van ek, M. and Rosa, J. (1996) *Diamond and Related Materials*, **5**, 1006–11.

157 Zeisel, R., Nebel, C.E., Stutzmann, M., Sternschulte, H., Schreck, M. and Stritzker, B. (2000) *physica status solidi (a)*, **181**, 45–50.

158 Yilmax, H., Weiner, B.R. and Morell, G. (2007) *Diamond and Related Materials*, **16**, 840–4.

159 Goss, J.P. and Briddon, P.R. (2007) *Physical Review B*, **75**, 075202.

160 Goss, J.P., Eyre, R.J. and Briddon, P.R. (2007) *physica status solidi (a)*, **204**, 2978–84.

161 Ferreira, N.G., Mendonca, L.L., Airoldi, V.J.T. and Rosolen, J.M. (2003) *Diamond and Related Materials*, **12**, 596–600.

162 Kulova, T.L., Evstefeeva, Y.E., Pleskov, Y.V., Skundin, A.M., Ral'chenko, V.G., Korchagina, S.B. and Gordeev, S.K. (2004) *Physics of the Solid State*, **46**, 726–8.

163 Almeida, E.C., Trava-Airoldi, V.J., Ferreira, N.G. and Rosolen, J.M. (2005) *Diamond and Related Materials*, **14**, 1673–7.

164 Zhou, X., Watkins, G.D., McNamara Rutledge, K.M., Messmer, R.P. and Chawla, S. (1996) *Physical Review B*, **54**, 7881–90.

165 Glover, C., Newton, M.E., Martineau, P., Twitchen, D.J. and Baker, J.M. (2003) *Physical Review Letters*, **90**, 185507.

166 Glover, C., Newton, M.E., Martineau, P.M., Quinn, S. and Twitchen, D.J. (2004) *Physical Review Letters*, **92**, 135502.

167 Iakoubovskii, K. and Stesmans, A. (2001) *physica status solidi (a)*, **186**, 199–206.

168 Iakoubovskii, K. and Stesmans, A. (2002) *Journal of Physics: Condensed Matter*, **14**, R467–99

169 Iakoubovskii, K., Stesmans, A., Suzuki, K., Sawabe, A. and Yamada, T. (2002) *Physical Review B*, **66**, 113203.

170 Iakoubovskii, K. and Stesmans, A. (2002) *Physical Review B*, **66**, 195207.

171 Iakoubovskii, K., Stesmans, A., Nesladek, M. and Knuyt, G. (2002) *physica status solidi (a)*, **193**, 448–56.

172 Iakoubovskii, K., Stesmans, A., Suzuki, K., Kuwabara, J. and Sawabe, A. (2003) *Diamond and Related Materials*, **12**, 511–15.

173 Coutinho, J., Torres, V.J.B., Jones, R. and Briddon, P.R. (2003) *Physical Review B*, **67**, 035205.

174 Miyazaki, T. and Okushi, H. (2002) *Diamond and Related Materials*, **11**, 323–7.

175 Miyazaki, T., Okushi, H. and Uda, T. (2001) *Applied Physics Letters*, **78**, 3977–9.

176 Miyazaki, T., Okushi, H. and Uda, T. (2002) *Physical Review Letters*, **88**, 066402.

177 Goss, J.P., Briddon, P.R., Sachdeva, R., Jones, R. and Sque, S.J. (2005) *Physics of Semiconductors* (eds J. Menéndez and C.G. Van de Walle), Vol. **722A** of *AIP Conference Proceedings*, AIP, Melville, NY.

178 Holbech, J.D., Bech Nielsen, B., Jones, R., Sitch, P. and Öberg, S. (1993) *Physical Review Letters*, **71**, 875–8.

179 Briddon, P.R. and Jones, R. (1993) *Physica B*, **185**, 179–89.

180 Yu, B.D., Miyamoto, Y. and Sugino, O. (2000) *Applied Physics Letters*, **76**, 976–8.

181 Eyre, R.J., Goss, J.P., Briddon, P.R. and Wardle, M.G. (2007) *physica status solidi (a)*, **204**, 2971–7.

182 Davies, G. (1976) *Journal of Physics C*, **9**, L537–42.

183 Miyazaki, T. and Yamasaki, S. (2005) *physica status solidi (a)*, **202**, 2134–40.

184 Goss, J.P., Briddon, P.R. and Shaw, M.J. (2007) *Physical Review B*, **76**, art. no. 075204.

185 Heera, V., Fontaine, F., Skorupa, W., Pécz, B. and Barna, Á. (2000) *Applied Physics Letters*, **77**, 226–8.

186 Weishart, H., Heera, V. and Skorupa, W. (2005) *Journal of Applied Physics*, **97**, 103514.

187 Weishart, H., Eichhorn, F., Heera, V., Pécz, B., Barna, Á. and Skorupa, W. (2005) *Journal of Applied Physics*, **98**, 043503.

188 Clark, C.D., Kanda, H., Kiflawi, I. and Sittas, G. (1995) *Physical Review B*, **51**, 16681–8.

189 Goss, J.P., Jones, R., Breuer, S.J., Briddon, P.R. and Öberg, S. (1996) *Physical Review Letters*, **77**, 3041–4.

190 Collins, A.T. (1980) *Journal of Physics C*, **13**, 2641–50.

9
n-Type Doping of Diamond

Satoshi Koizumi,[1] Mariko Suzuki[2] and Julien Pernot[3]
[1]National Institute for Materials Science, 1-1 Namiki, Tsukuba 305-0044, Japan
[2]Corporate Research & Development Center, Toshiba Corporation, 1 Komukai Toshiba-cho, Saiwai-ku, Kawasaki 212-8582, Japan
[3]Institut NEEL, CNRS & Université Joseph Fourier, BP 166, F-38042 Grenoble Cedex 9, France

9.1
Introduction

n-type diamond does not exist in nature. From early 1980s, when the chemical vapor deposition (CVD) technique was introduced for diamond growth, n-type doping has been investigated mainly using phosphorus (P) as an impurity. Nitrogen does not give rise to any shallow donor levels as a result of the local lattice deformation known as the Jahn–Teller effect which is induced by the substitutional nitrogen. Theoretical calculation (effective mass approximation) predicts that the ground state of P atom will be situated at about 200 meV below the conduction band minimum in diamond [1]. Although some studies have shown electrical conductivity, the evidence of n-type conductivity was lacking because of the polycrystalline nature of the samples which gives rise to various kinds of conduction routes such as grain boundaries, defects and graphitic aggregates [2–4]. There were no reports of P atoms in diamond as an n-type dopant. To clarify the validity of P-doping for n-type diamond formation it is necessary to perform Hall effect measurements intensively for high perfection crystalline P-doped diamond layers grown epitaxially. In 1996, we succeeded in growing n-type P-doped diamond thin films on diamond substrates and we characterized the nature of the conductivity clearly by temperature dependent Hall measurements [5]. P-doped layers have been grown on synthetic Ib diamond substrates with the crystalline orientation of {111}. On {100} orientated substrates, it was impossible to grow n-type layers or flat epitaxial films [6–8]. This situation is the opposite to the conventional experimental results on undoped or boron-doped CVD diamonds for which a {100} surface can be formed that is less defective than a {111} surface. It is very interesting that such a small amount of impurity atoms introduced for doping can have such a big influence on diamond growth itself. The activation energy of the carrier concentration was 0.43 eV in the first report. However this value was

Physics and Applications of CVD Diamond. Satoshi Koizumi, Christoph Nebel, and Milos Nesladek

ISBN: 978-3-527-40801-6

underestimated because the temperature dependence of carrier concentration included the influence of hopping conductivity due to heavy doping and also crystalline defects. The activation energy has been revised by further Hall measurements of higher quality P-doped films and it is 0.57 eV [9]. By optical characterization using photocurrent spectroscopy [10], FTIR [11] and cathodoluminescence [12], it has been confirmed that this value is reliable. Particle induced X-ray emission (PIXE) and Rutherford backscattering spectrometry (RBS) measurements revealed the P atoms are located at lattice substitutional sites in diamond [13]. The Hall mobility of conducting electrons was only 23 cm^2/V s from the first report and the reproducibility of n-type samples was quite poor (less than 10%). Now the highest Hall mobility for n-type diamond is 660 cm^2/V s at room temperature and the reproducibility is 100% [9]. All this progress has been achieved by an intensive effort to grow high quality P-doped diamond layers. In 2001, we reported formation of a successful pn junction and the light emitting diode operation of it emitting deep ultraviolet (UV) light using a P-doped n-type layer and boron-doped p-type layer [14]. The wavelength that was emitted from the diamond LED was 235 nm and this is the shortest wavelength ever reported for current-induced light emitting devices. The diamond pn diode has also shown good responsivity to deep UV (VUV) light detection with complete solar-blind characteristics [15]. The diamond VUV photodetectors were chosen to be launched onboard satellite PROBA-2 for solar UV radiation analysis [16]. In another trial of n-type doping, there is a report of deuterium diffusion treatment of boron-doped p-type diamond. The study suggested the formation of a boron–deuterium complex in the diamond lattice which results in n-type conductivity [17]. However, there are no reports of successful reproduction of the experiments by other research groups and the results are controversial for the moment. In 2004, new results were reported for P-doping on {100} oriented diamond [18]. The results were convincing of successful n-type doping of diamond at {100} orientation using unique growth conditions during CVD that were different from the {111} case. Although doping control is limited to about 10^{18}~10^{19} cm^{-3}, which is still far behind {111} results, the Hall mobility is reported to be rather high and it is hoped to have better quality n-type diamond than {111} in the future.

In this chapter, the current status of n-type doping of diamond will be reviewed especially focusing on high quality {111} n-type diamond growth and electrical characterization using the Schottky diode structure.

9.2 High Mobility n-Type Diamond

The mobility of semiconductors is explained by introducing several different scattering mechanisms for conducting carriers such as scattering due to phonons, coulombic forces by ionized and neutral impurities and defects. By reducing the impurity concentration in the doped semiconductor, we can expect to have a higher mobility. However, in reality, we have a certain number of defects and impurity atoms that compensate carrier activities. In the case of n-type diamond, the numerical information related to the compensating defects was unknown. We tried to find the lower limit of n-type doping by growing very lightly P-doped diamond films aiming to get high mobility so as to search for the nature of compensating defects. For this experiment, we needed to minimize the effect of crystalline imperfections. The improvement of electrical properties is strongly related to the improvement of crystalline perfection. In order to obtain P-doped films with high crystalline perfection, optimum growth conditions such as lower methane concentration and higher substrate temperature were used. In addition, the deposition experiments were performed in an ultrahigh vacuum chamber using high-purity reactant gases in this study.

9.2.1 Growth of Lightly P-Doped Diamond Thin Films

The growth of diamond is performed using a stainless-steel-type microwave plasma-assisted CVD system with the capability of precise control of the growth conditions. The vacuum chamber can be evacuated to below 1×10^{-8} Torr. The P concentration of an unintentionally doped film was about $10^{15}\,cm^{-3}$ due to the residual P compounds in the chamber. Lightly P-doped diamond films were epitaxially grown on synthetic type Ib diamond {111} substrates with the dimensions of $2 \times 2 \times 0.5$ mm. The source gas was 0.05% CH_4 (6N: purity 99.9999%) diluted with H_2 (9N). Phosphorus doping was carried out by adding PH_3 (6N) to the source gas. The PH_3/CH_4 ratio in the source gas was varied from 10 to 1000 ppm. The gas pressure, the total gas flow rate, the substrate temperature, and the growth duration were 100 Torr, 1000 sccm, 900 °C, and 3 hours, respectively. The resultant film thickness was approximately 1 μm. Detailed information about the CVD system and the growth conditions are given in the other publications [8, 19].

To check the P-doping efficiency and the profile in the grown layers, secondary ion mass spectroscopy (SIMS) measurements were performed for a sample grown with several different doping levels sequentially changed in deposition experiments. Figure 9.1 shows the SIMS depth profiles of impurities in a P-doped film consecutively grown at 10, 100 and 200 ppm of PH_3 in CH_4. We confirmed control of the P concentration at low doping levels. The P concentrations were estimated to be approximately $1 \times 10^{16}\,cm^{-3}$ for 10 ppm, $7 \times 10^{16}\,cm^{-3}$ for 100 ppm, and $1 \times 10^{17}\,cm^{-3}$ for 200 ppm, while boron, nitrogen, and hydrogen concentrations were below the detection limits. The phosphorus doping efficiency (P-atoms in the layer vs. P-atoms in the gas phase) was about 0.3%.

9.2.2 Hall Measurements of P-Doped Films

Heavily P-doped diamond layers were selectively grown on a lightly P-doped film to form ohmic contacts. The detailed process of selective doping has been described previously [20]. Before forming metal contacts, the surface of the diamond film

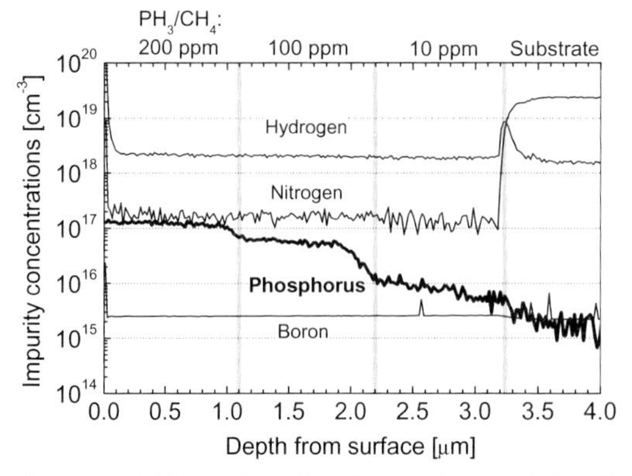

Figure 9.1 SIMS depth profile of impurities in a P-doped diamond film consecutively grown at 10, 100 and 200 ppm in the PH_3/CH_4 ratio.

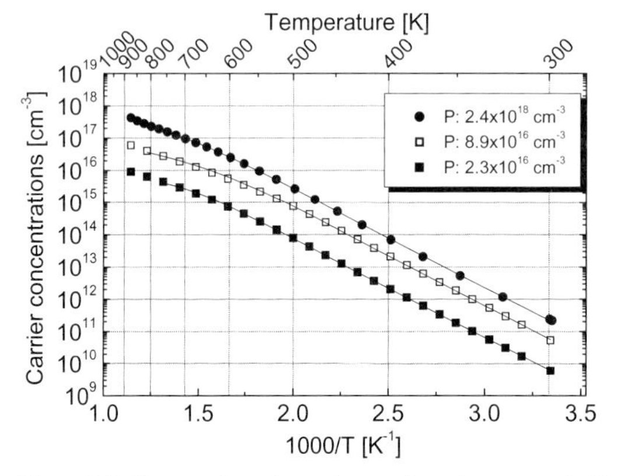

Figure 9.2 Temperature dependence of carrier concentration for P-doped diamond films with various P concentrations. Solid curves indicate the theoretical calculations using Equation 9.1.

was chemically oxidized to remove the surface conductive layer using an acid mixture of H_2SO_4 and HNO_3 at 230 °C for 1 hour. Ohmic electrodes composed of Ti capped with Au were formed on heavily P-doped diamond layers by e-beam vacuum deposition. Hall-effect measurements using the van der Pauw method were carried out in the temperature range between 300 and 873 K at an AC magnetic field of 0.6 T with a frequency of 0.1 Hz.

P-doped films with phosphorus concentrations above $1 \times 10^{16}\,cm^{-3}$ show negative Hall coefficients in the temperature range from 300 to 873 K, which indicates n-type conduction of the films. In the case of P-doped films with a P concentration of $1 \times 10^{16}\,cm^{-3}$ or less, although n-type conductivity of the films is observed at higher temperatures, negative Hall coefficients around room temperature (RT) have not been convincingly detected due to the high resistivity of the films. Figure 9.2 shows the temperature dependence of the carrier concentrations of the

P-doped diamond films with various P concentrations. The carrier concentrations are directly proportional to the P concentrations. The carrier concentration and resistivity of a P-doped film with a P concentration of $9 \times 10^{16}\,cm^{-3}$ were $5 \times 10^{10}\,cm^{-3}$ and $2 \times 10^{5}\,\Omega$ cm at RT, $6 \times 10^{16}\,cm^{-3}$ and $2\,\Omega$ cm at 873 K, respectively. For a non-degenerate semiconductor, the activation energy and the compensation ratio can be calculated using the following equation as usually given in conventional semiconductor physics textbooks:

$$\frac{n(n+N_A)}{N_D-N_A-n}=\frac{N_C}{g}\exp\left(-\frac{E_D}{kT}\right), \tag{9.1}$$

where n is the carrier concentration, N_D and N_A are the donor and acceptor concentrations, N_C is the effective density of states in the conduction band, g is the degeneracy factor for the donor, E_D is the activation energy of the donor, k is the Boltzmann constant and T is the temperature. For the P-doped films grown at PH_3/CH_4: 100 ppm, we obtained $E_D = 0.57$ eV, $N_D = 8.9 \times 10^{16}\,cm^{-3}$, $N_A = 1.2 \times 10^{16}\,cm^{-3}$, and the compensation ratio $N_A/N_D = 0.13$. In a film with a P concentration of $2.3 \times 10^{16}\,cm^{-3}$, N_A was $1.5 \times 10^{16}\,cm^{-3}$. With decreasing P concentration below $1 \times 10^{16}\,cm^{-3}$, P-doped films become fully compensated and highly resistant. These results suggest that in the present study, the concentration of compensating defects that exists naturally in the {111} CVD film is of the order of $10^{15}\,cm^{-3}$.

Figure 9.3 shows the Hall mobility of the P-doped films as a function of temperature. The Hall mobility at RT increases from 180 to 660 cm^2/V s with decreasing P concentrations from 3×10^{18} to $7 \times 10^{16}\,cm^{-3}$. The value of 660 cm^2/V s is the highest ever reported for n-type diamond films. At P concentrations below $7 \times 10^{16}\,cm^{-3}$, however, no improvement in the Hall mobility has been achieved. This may be ascribed to the existence of a large amount of compensating acceptors. The Hall mobility for the P-doped film with a P concentration of $7 \times 10^{16}\,cm^{-3}$

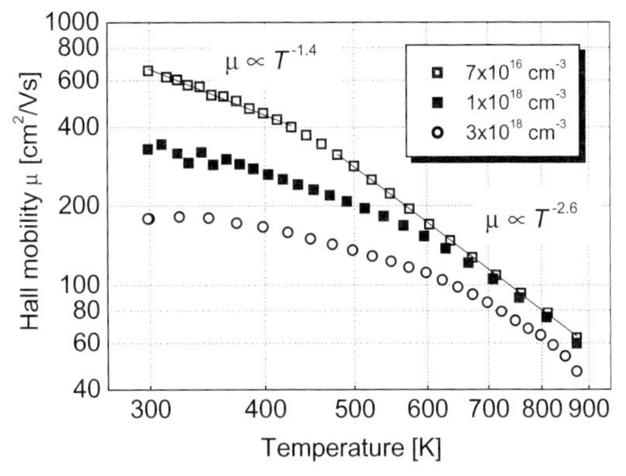

Figure 9.3 Temperature dependence of Hall mobility for P-doped diamond films with P concentrations of 7×10^{16}, $1 \times 10^{18}\,cm^{-3}$ and $3 \times 10^{18}\,cm^{-3}$.

decreases with increasing temperature as $T^{-1.4}$ up to 450 K. This indicates that acoustic phonon scattering dominates. At temperatures above 450 K, the Hall mobility is proportional to $T^{-2.6}$. These two temperature dependencies are representative of the phonon scattering with a small contribution due to impurity (see the next section).

9.2.3 Calculation of the Hall Mobility Temperature Dependence

In order to quantitatively describe the temperature dependence of the Hall mobility, we used the procedure described in Ref. [21] taking into account all electron scattering mechanisms relevant for diamond: intravalley acoustic phonons (ac), intervalley phonons (iph), ionized impurities (ii) and neutral impurities (ni), all within the relaxation time approximation. Six equivalent minima of the conduction band are considered at $2\pi/a$ (0.75, 0, 0) close to the X point of the Brillouin zone associated to the following effective masses for the principal directions: $m_{\|} = 1.81\ m_0$, $m_{\perp} = 0.306\ m_0$ [22]. The value of the deformation potential D_A for the intravalley acoustic phonon scattering is treated as a fitting parameter. Since the diamond is a multivalley semiconductor, the intervalley phonon scattering is included taking into account scattering between opposite valley (g-scattering) and non-opposite valley (f-scattering), in analogy to silicon [23]. The selection rules for phonons in inter-valley scattering processes give the LO mode for g-scattering and LA and TO for f-scattering. Geometrical considerations show that the LO phonon (g-scattering) is at the midpoint (halfway from the boundary) while LA and TO phonons (f-scattering) lie at the edge of the Brillouin zone in the Δ direction. The energies used in the calculations are $\hbar\omega_f = 140$ meV for the f-scattering and $\hbar\omega_g = 165$ meV for the g-scattering [24]. Since these phonons have non-zero energy, the intervalley scattering mode is inelastic, and so the procedure of Farvacque was used [25]. The intervalley coupling constants (D_{if} and D_{ig}) between these phonons and the electrons are treated as adjustable parameters, but in order to limit the number of adjustable parameters they are assumed to be equal. The intravalley scattering mechanism due to optical phonons is forbidden by the selection rules, and was thus neglected. Electron scattering by phonons with a finite energy (intervalley) starts at very high temperature, due to the high energy of phonon in diamond, so that these coupling constants are determined with a lower precision than in the case of acoustic phonons.

For ionized impurity scattering, the relaxation time is inversely proportional to $N_i = n + 2 \times N_A$, N_i being the number of ionized impurity, n the free electron density and N_A the compensating centers (acceptor) density. The neutral impurity (ni) scattering was treated using an effective Rydberg of 570 meV. The relaxation time of the ni mode is inversely proportional to $N_n = N_D - N_A - n$, where N_n is the number of neutral impurities and N_D the phosphorus (donor) density. N_D, N_A and n are obtained from the fit of the Hall electron density to experimental data through the neutrality equation. Then, the total relaxation time was calculated using the Mathiessen rule and the Hall mobility is defined by $\mu_H = r_H e \langle\tau\rangle / m_H^*$. In this equation, the Hall factor r_H is assumed to be unity, $\langle\tau\rangle$ is the average value of τ as defined in Ref. [21] and $m_H^* = (2m_{\|} + m_{\perp})/(2 + m_{\|}/m_{\perp})$ the Hall mobility mass. In order to determine precisely the acoustic deformation potential, calculations were compared with the experimental Hall mobility for several samples over a wide range of temperature (293 K to 873 K).

Figures 9.4a and b show the theoretical contributions of various modes and the total Hall mobility. N_D and N_A are determined from the fit of the electron density temperature dependence using the neutrality equation. The intrinsic diamond parameters determined from the fit of the mobility were the acoustic deformation potential equals to 17.7 eV and the intervalley coupling constants equal to $D_{if} = D_{ig} = 4.2 \times 10^9$ eV/cm [26]. The resulting acoustic deformation potential value is higher than in silicon (9.5 eV [23]) and 4H–SiC (11.4 eV [21]) showing more efficient coupling between electrons and acoustic phonons in diamond. Using another set of effective masses from the literature ($m_{\|} = 1.5m_0$, $m_{\perp} = 0.26m_0$ [27]), the determined acoustical deformation potential would be 21.5 eV. The D_{if} and D_{ig} are effective since in fact, if we want to evaluate the coupling constant of a single phonon, we must consider the number of equivalent valleys to which the electron could be scattered (4 for f-scattering and 1 for g-scattering) and the number of different phonons able to scatter the incident electron with respect to the selection rules (1 phonon LA and 1 phonon TO for f-scattering and 1 phonon LO for g-scattering). In this way, an approximate value of D_i with a physical meaning can be evaluated as 9.3×10^8 eV/cm in good agreement with the 8×10^8 eV/cm determined in Ref [28].

For sample $[P]_{SIMS} = 7 \times 10^{16}$ cm^{-3} (see Figure 9.4a), intravalley acoustic scattering dominates the RT mobility, giving the maximum measured Hall mobility of 660 cm^2/Vs. This is due to the low density of compensating centers ($N_A = 8.8 \times 10^{15}$ cm^{-3}) which induces a low contribution of the ionized impurity mode to the total mobility. Above 700 K intervalley scattering limits the mobility and ionized

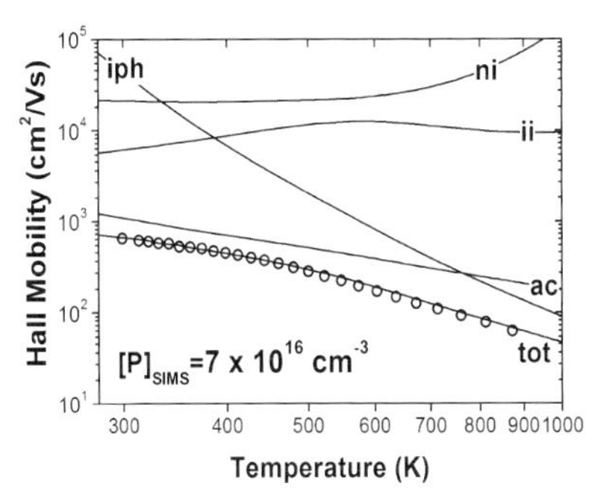

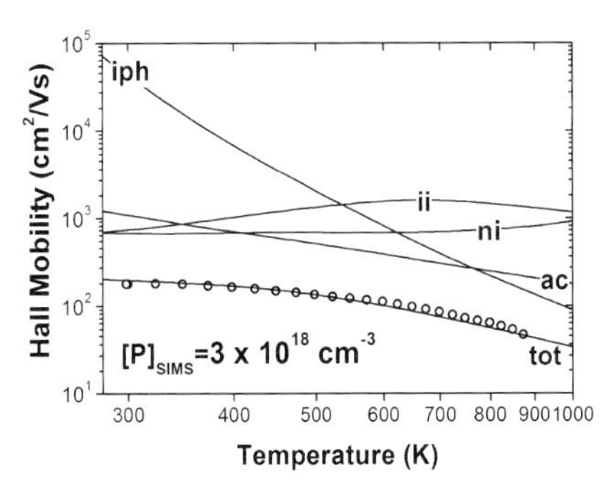

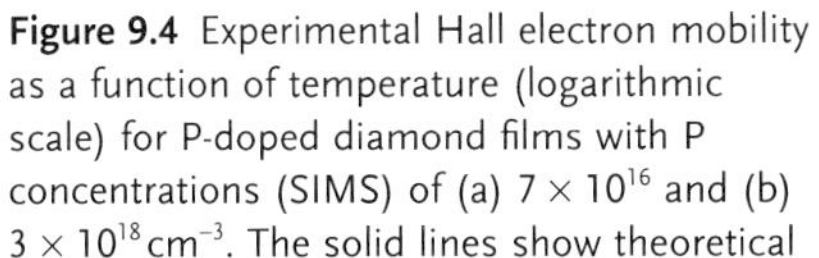

Figure 9.4 Experimental Hall electron mobility as a function of temperature (logarithmic scale) for P-doped diamond films with P concentrations (SIMS) of (a) 7×10^{16} and (b) 3×10^{18} cm^{-3}. The solid lines show theoretical contribution of various modes and total mobility (tot). The different scattering mechanisms are ionized impurities (ii), neutral impurities (ni), acoustic phonons (ac), intervalley phonons (iph).

impurity scattering is negligible. Neutral impurity scattering has no influence over the whole temperature range. In the case of sample $[P]_{SIMS} = 3 \times 10^{18}\,cm^{-3}$ (see Figure 9.4b), the intrinsic scattering modes due to the lattice are the same as in the previous sample. The differences between the mobility curves of the two samples come from the doping level which is about fourty times higher and from the compensating center density ($N_A = 1.12 \times 10^{17}\,cm^{-3}$) which is about twenty times higher. The higher neutral phosphorus concentration increases the effect of the ni scattering on the total mobility. In the same way, the uptake of compensating centers reduces the total mobility through the ii mode. Indeed, at RT the condition $n \ll N_A$ applies and the ii scattering time is inversely proportional to N_A. At room temperature, the ni and ii scattering modes have a comparable influence to the total mobility. The influence of the impurity scattering (ni and ii) can be directly observed in Figure 9.4 for two different doping and compensation levels.

Finally, in order to determine the maximum Hall mobility achievable in n-type diamond at room temperature, we have calculated the mobility for pure and compensated n-type diamond. Three different acceptor densities (compensation) have been considered: $N_A = 0$, $N_A = 5 \times 10^{15}\,cm^{-3}$ and $N_A = 5 \times 10^{16}\,cm^{-3}$. The results are shown Figure 9.5 as a function of the temperature. The phonon scattering mechanism (solid lines ac and iph) is the same for the three compensation levels, since it is intrinsic to diamond. In order to be able to neglect the effect of the neutral impurity scattering, the $N_D - N_A$ concentration has been taken to be constant at $1 \times 10^{15}\,cm^{-3}$ for the three compensation levels. Finally, as shown on Figure 9.5, the ionized impurity scattering is at the origin of the change in the total mobility for the three N_A. At room temperature, the calculated Hall mobility reaches the value

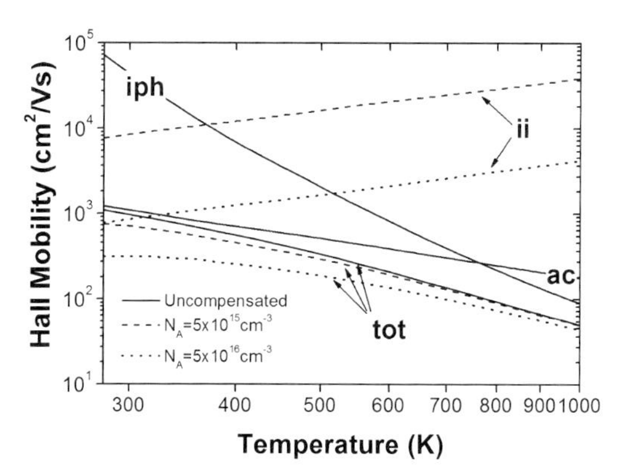

Figure 9.5 Theoretical Hall electron mobility (tot) as a function of temperature (logarithmic scale) for uncompensated and compensated n-type diamond. $N_A = 0$, solid lines, $N_A = 5 \times 10^{15}\,cm^{-3}$, dashed lines and $N_A = 5 \times 10^{16}\,cm^{-3}$, dotted lines. The contributions of the different scattering mechanism are represented: ionized impurities (ii), neutral impurities (ni), acoustic phonons (ac), intervalley phonons (iph).

of $310\,cm^2/Vs$ for $N_A = 5 \times 10^{16}\,cm^{-3}$, $725\,cm^2/Vs$ for $N_A = 5 \times 10^{15}\,cm^{-3}$ and finally $1030\,cm^2/Vs$ for $N_A = 0$. This last value is the maximum low field Hall mobility achievable at room temperature in diamond at thermodynamic equilibrium and is fully governed by phonon scattering. This maximum electron mobility is higher than in the case of 4H–SiC [29].

Higher mobilities were measured using optical techniques to generate free electrons in the conduction band [28, 30–35]. The $T^{-3/2}$ temperature dependence of the electron mobility due to acoustical phonon scattering is also observed for samples in Refs [28, 30, 32, 33] between 120 K and 400 K. However, the different absolute values of these experimental mobilities cannot be described with a classical model using only one deformation potential related to the conduction band of diamond. This discrepancy reveals a poorly understood effect on the carrier mobility when using optical techniques to generate electrons in the conduction band. Such an influence of the measurement conditions on the mobility values remains misunderstood. On the other hand, the mobility in samples at thermodynamic equilibrium can be described qualitatively, with one single deformation potential as is done in this work and Ref. [26]. This seriously limits the relevance of any comparison between mobilities measured at thermodynamic equilibrium and those using optical excitations.

9.3 Electrical Properties of n-Type Diamond–Schottky Diode

Schottky diodes, which are primitive and popular semiconductor devices, provide a lot of useful information on semiconductor materials. n-type diamond Schottky diodes have been made using phosphorus-doped homoepitaxial diamond [36–38]. I–V and C–V properties have shown clear n-type Schottky diode characteristics and the high quality of the phosphorus-doped diamond.

Lateral dot-and-plane (with ring-shaped gap) Schottky diodes were fabricated on P-doped homoepitaxial diamond layers [39]. The schematic cross-sectional view of the sample structure is shown in Figure 9.6. The ohmic contacts were fabricated by evaporating and annealing Au/Pt/Ti (700 °C, 10 minutes in N_2) on selectively grown heavily P-doped sectors [20] before fabricating Schottky contacts. Ni, Pt, Al and Ti were used as Schottky contacts. Each Schottky contact was surrounded by the ohmic contact with a 20 μm gap as shown in Figure 9.7. Prior to any electrode processing, all of the samples were boiled in a mixture of H_2SO_4 and HNO_3 at 230 °C to remove any graphitic layers and surface contaminations. Moreover, the sample surfaces were exposed to oxygen microwave plasma at 500 W for 5 minutes just before evaporating the Schottky contact metal. These procedures could cause oxygen-terminated surfaces.

I–V measurements and C–V measurements were undertaken over a temperature range of 297–773 K to investigate the Schottky diode characteristics, donor characteristics and the barrier heights for Ni, Pt, Al and Ti Schottky contacts on n-type diamond.

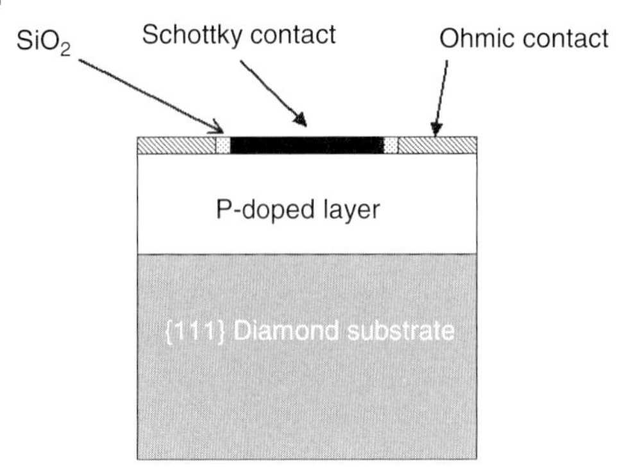

Figure 9.6 Schematic cross-sectional view of the Schottky diode.

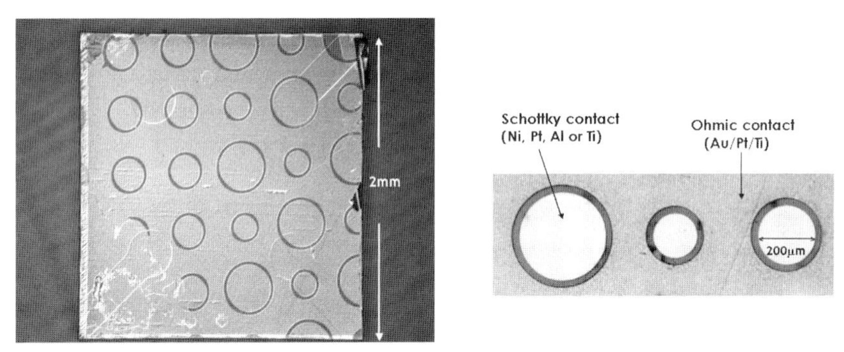

Figure 9.7 Plane view of the electrode pattern of the lateral dot and plane Schottky barrier diode. (a) Whole sample area. (b) Three different areas of Schottky contacts.

9.3.1
Schottky Diode Characteristics

Figure 9.8 shows typical I–V characteristics from 297 K to 773 K of the Ni/n-type diamond Schottky diode. The P concentration was $1.6 \times 10^{16}\,cm^{-3}$. We have obtained excellent Schottky junction properties even at 773 K as shown in this figure. The rectification ratio was ~10^6 at ±10 V for 573 K, and the ideality factor had a minimum value of 1.0 at this temperature. The forward current density at 10 V was $0.2\,A/cm^2$ at 773 K. The reverse current gradually increased with increasing voltage at 573 K as show in Figure 9.9. This relatively large reverse current cannot be explained by the conventional I–V equation of Schottky junctions.

Figure 9.10 shows typical current density (J) -V curves for Ni, Pt, Al and Ti Schottky contacts at 473 K for several P concentrations. We have obtained clear rectifying characteristics for every kind of metal contact. The rectifying characteristics deteriorated with increasing P concentration. It was difficult to obtain good Schottky diodes for high P concentration ($3.4 \times 10^{19}\,cm^{-3}$), and J–V characteristics

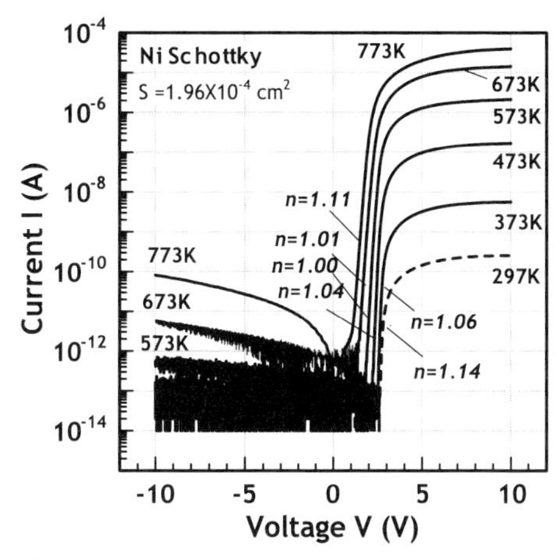

Figure 9.8 Typical I–V curve at 297 K, 373 K, 473 K, 573 K and 773 K of the Ni/n-type diamond Schottky diode.

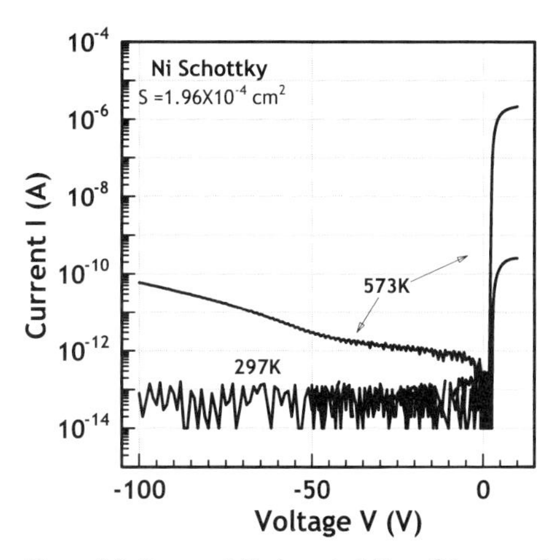

Figure 9.9 Reverse I–V characteristics of the sample shown in Figure 9.8.

came closer to ohmic-like characteristics. The results suggest that the heavily P-doped layer can act as a good contact layer [20].

9.3.2 P-Related Donor Characteristics

Typical results of the capacitance (C) and the conductance divided frequency (G/ω, $w = 2\pi f$) at zero dc bias voltage are shown as a function of frequency (f) for sample temperatures (T = 297–673 K) in Figure 9.11a and b), respectively. The capacitance strongly depends on frequency and temperature. The observed variation in

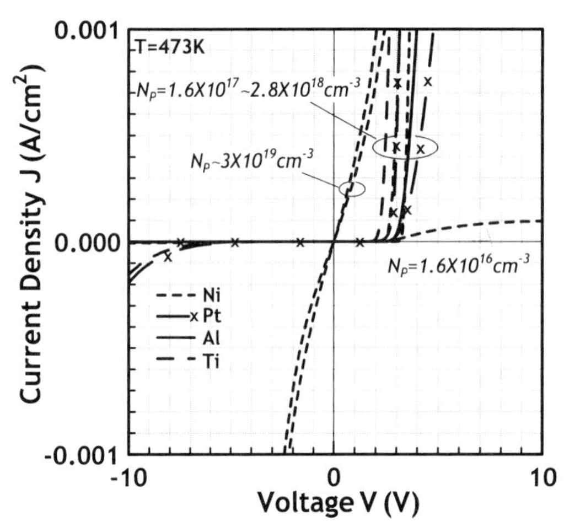

Figure 9.10 Typical J–V curves for Ni, Pt, Al and Ti Schottky contacts on P-doped diamond layers.

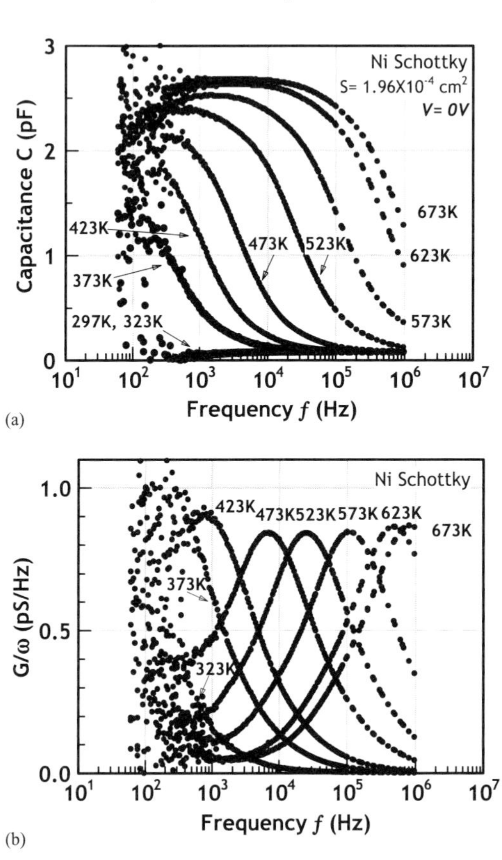

Figure 9.11 A typical result of the capacitance (C) and the conductance divided by frequency (G/ω) at zero bias voltage as a function of frequency (f) for several temperatures (T = 297 K, 323 K, 373 K, 423 K, 523 K, 623 K and 673 K) for P-doped diamond.

capacitance can be due to the high resistivity of the diamond and/or to the well-known dispersion effect, which occurs when a deep level is unable to follow the high frequency voltage modulation and unable to contribute to the net space charge in the depletion region. In most semiconducting materials, the former effect is relatively unimportant since the capacitance values were obtained from the series mode resistance of the inductance–capacitance–resistance (LCR) meter, in which the capacitance is determined independently of any series resistance. We have obtained the activation energy of 0.54 eV for this sample from the slope of the plot of ω_{peak}/T^2 against the reciprocal temperature. Where, ω_{peak} is the peak frequency of the G/ω determined from the results of the G/ω-f measurements. The detailed calculation procedure has been shown in a previous paper (admittance spectroscopy [40]). This value is a little smaller than the P donor activation energy ($\Delta E_D \sim 0.57$ eV) measured by temperature-dependent Hall effect measurements or optical measurements reported in the literature [9, 12, 41, 42]. It is assumed that the applied electric field during C–V measurements gives rise to barrier height lowering. The observed variations in the C–f curves were considered to show the activation and deactivation of P donor. This result indicates the net donor concentration should be determined in the capacitance saturation region from the C–V measurements.

Figure 9.12 shows a typical result of the C–V measurements in such a capacitance saturation region (1 kHz, 373 K) for the same sample as that shown in Figure 9.11. The $1/C^2$–V plot had good linearity as shown in Figure 9.12. The capacitance of the depletion region for the Schottky junction can be written using the following equation [43],

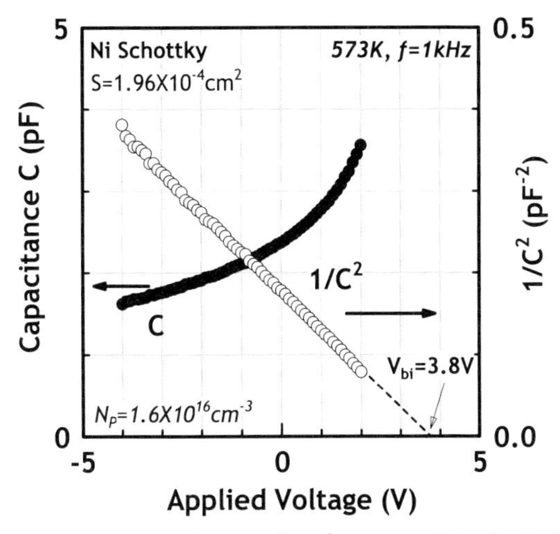

Figure 9.12 Typical C–V and $1/C^2$ curves at 573 K for the Ni/P-doped diamond Schottky diode. The applied frequency was 1 kHz.

$$C = A\sqrt{\frac{q\varepsilon_S(N_D - N_A)}{2(V_{bi} - V)}},\tag{9.2}$$

where $(N_D–N_A)$ is the net donor concentration, V_{bi} is the built-in potential, q is the elementary charge, ε_s is the permittivity of semiconductor ($\varepsilon_s = \varepsilon_{sr}\varepsilon_0$), and A is an area of the Schottky contact. The net donor concentration $(N_D–N_A)$ denotes the net positively charged ionized impurity concentration in the depletion region, and is given by the following equation (from the Equation 9.2)

$$N_D - N_A = \frac{2}{q\varepsilon_S}\left[\frac{-1}{d(1/C^2)/dV}\right]\frac{1}{A^2}.\tag{9.3}$$

It was evaluated to be $1.3 \times 10^{16}\,\text{cm}^{-3}$ from the slope of the $1/C^2$ against voltage plot with relative dielectric constant of diamond, ε_s of 5.7. Figure 9.13 shows the net donor concentration as a function of the P concentration. It was found that P electrical activity, which means the ratio of the net donor concentration to the P concentration, was near 1 for P concentrations of $1.6 \times 10^{16}\,\text{cm}^{-3}$ to $2.7 \times 10^{18}\,\text{cm}^{-3}$. The ratio of the uncompensated P concentration to the total P concentration was considered to be very high. This is a result of the fact that the background acceptor concentration in the P-doped layers is quite low in comparison with the acceptor concentration or P concentration, such as $\sim 10^{15}\,\text{cm}^{-3}$ or less. Figure 9.14 shows the depth profile of P and other residual impurity concentrations for a typical P-doped diamond layer. P concentration is almost constant in the P-doped epitaxial layer, and it is not detected ($<1 \times 10^{15}\,\text{cm}^{-3}$) in the diamond substrate. Hydrogen and boron, which may compensate P donors, are not detected. Although the detection limit of hydrogen concentration ($\sim 10^{18}\,\text{cm}^{-3}$) is comparable to the P doping

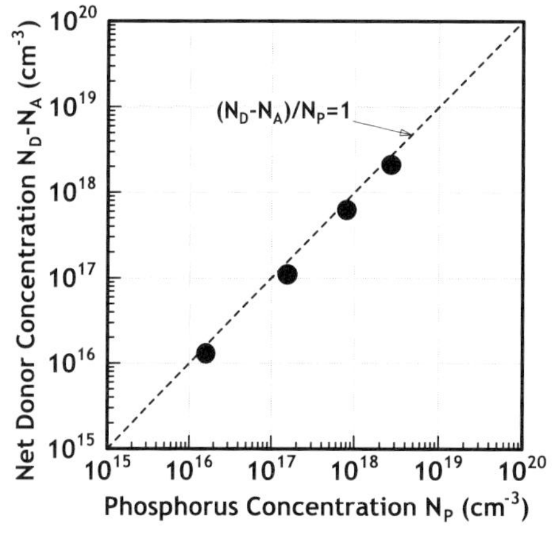

Figure 9.13 Net donor concentration as a function of phosphorus concentration in P-doped diamond.

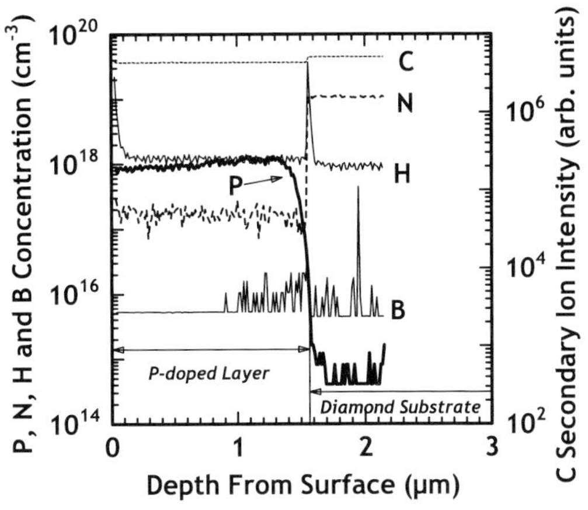

Figure 9.14 Depth profiles of P and residual impurity (H, B, N) concentration determined by SIMS measurements.

level, it is significantly small for compensating P-donors [44, 45]. Moreover, this result of high uncompensated P donor concentration is consistent with the results of the Hall effect measurements [46] and/or the PIXE (particle-induced X-ray emission) measurements [13].

9.3.3 Electrical Properties of Metal/n-Type Diamond Interfaces

9.3.3.1 Schottky Barrier Height Dependence on Metal Work Function

Built-in potential, V_{bi}, can be determined by extrapolation of $1/C^2$ to 0 as shown in the following equation (from Equation 9.2),

$$V - V_{bi} = \frac{q\varepsilon_S(N_D - N_A)}{2}\left(\frac{1}{C^2}\right)\frac{1}{A^2}. \tag{9.4}$$

As shown in Figure 9.12, V_{bi} of 3.8 eV was obtained from the extrapolation of $1/C^2$ to 0 at 573 K for the diode. This value gives the Schottky barrier height of 4.3 eV from the next equation [43],

$$\phi_{Bn} = qV_{bi} + qV_n \tag{9.5}$$

where qV_n is the energy difference between the Fermi level (E_F) and the bottom of the conduction band (E_C). qV_n is given by

$$qV_n = E_C - E_F = \frac{kT}{q}\ln\frac{N_C}{n}, \tag{9.6}$$

where N_C is the density of states in the conduction band, n is the electron concentration at the temperature T, and k is the Boltzmann constant. The qV_n was calculated to be about 0.5 eV at 573 K for the sample. The value of the Schottky barrier height, 4.3 eV, is considerably higher compared to those of other semiconductors. One of the reasons for such a large value is the small value of the electron affinity of diamond. For an ideal Schottky contact between a metal and an n-type semiconductor in the absence of surface states, the barrier height can be equal to the energy difference between the metal work function (ϕ_m) and the electron affinity of the semiconductor (χ),

$$\phi_{Bn} = \phi_m - \chi. \tag{9.7}$$

Assuming that Ni has a work-function of ϕ_{WF} = 5.15 eV [47] and the electron affinity of (111) diamond is assumed to be χ_D = 0.38 eV [48], the Schottky barrier height should be:

$$\phi_{SCH} = \phi_{WF} - \chi_D = 4.77\,eV. \tag{9.8}$$

This value is considered to be consistent with the value using our experimental data, taking into account that the interface between Ni and diamond may not be perfect. On the other hand, in the case that a large density of surface states is present on the diamond surface, or the Fermi level is pinned at the interface, the barrier height is determined from the surface charge neutrality level (pinning level) independent of the metal work function. In this case, the barrier height can also be large if the pinning level is close to the valence band. Therefore, both cases are thought to be consistent with our experimental data for Ni/n-type diamond Schottky junctions. We have investigated the metal work function dependence on the Schottky barrier height to make clear which of the above-mentioned cases is valid for these metal/n-type diamond interfaces. Figure 9.15 shows the Schottky

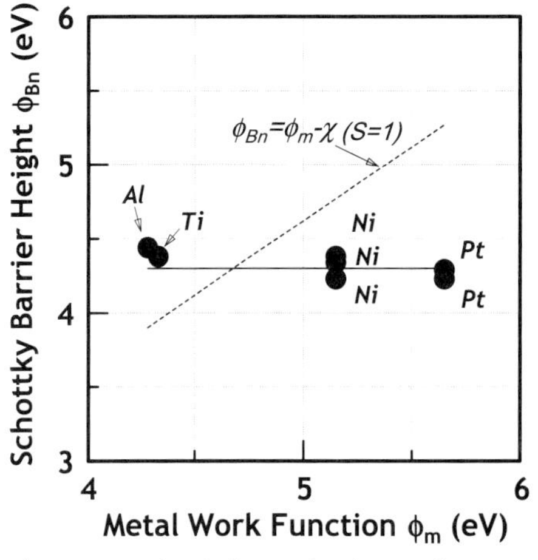

Figure 9.15 Schottky barrier height as a function of metal work function for P-doped diamond.

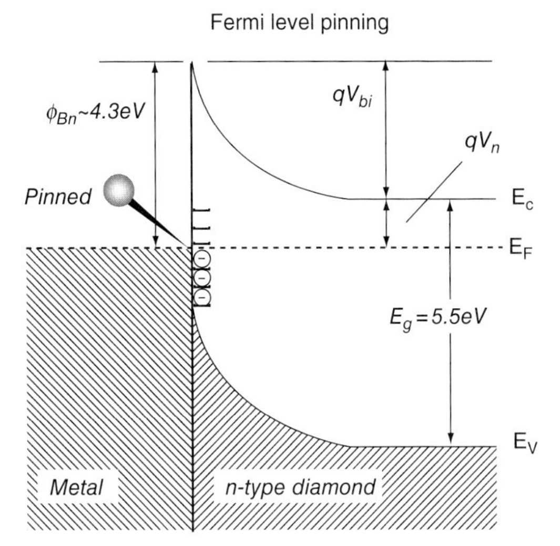

Figure 9.16 Schematic energy-band diagram of metal/n-type diamond under thermal equilibrium condition. Schottky barrier height as a function of metal work function for P-doped diamond.

barrier height as a function of the metal work function. All of the Schottky diodes, shown in this figure, show good linear relationship between $1/C^2$ and V. The barrier height is almost constant at around 4.3 eV, independent of the metal work function. The dotted line shows that the Schottky limit, or S factor, is one. This result that the S factor is almost zero indicates that the Fermi level is strongly pinned at the metal/n-type diamond interfaces below ~4.3 eV from the bottom of the conduction band as shown in Figure 9.16. This strong pinning of the Fermi level at the metal/diamond interfaces may be related to a high density of states at approximately 1.7 eV above the valence band resulting from oxygen termination [49].

9.3.3.2 **Comparison Between n-Type and p-Type Diamond**

Electrical characteristics of Ni/n-type diamond interfaces were compared to those of Ni/p-type diamond interfaces for (111) oriented homoepitaxial diamond. The electrode processing and the surface treatment were the same for n-type Schottky diodes and for p-type Schottky diodes. The built-in potential for the p-type Schottky diode was found to be 1.3 V from the results of C–V measurements as shown in Figure 9.17. The Schottky barrier height was evaluated to be 1.6 eV from Equations 9.4 and 9.5. Thus, the sum of the Ni Schottky barrier heights of p-type and n-type diamond was 5.9 eV. This value was approximately consistent with the bandgap energy of diamond. Considering the surface Fermi level pinning for p-type Schottky junctions on (100) diamond [50], it is possible that the surface Fermi level is pinned at the same energy level in the band gap for p-type and n-type diamond.

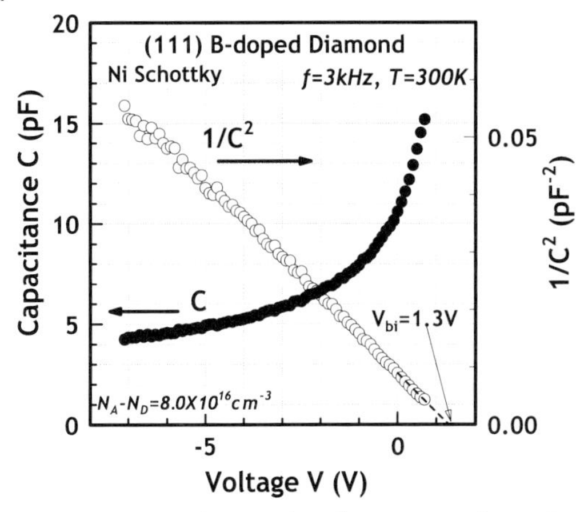

Figure 9.17 Typical C–V and $1/C^2 - V$ curves of Ni/B-doped diamond Schottky diode (300 K, 3 kHz).

9.4 Summary

Lightly phosphorus doped {111} homoepitaxial diamond films have been grown by microwave plasma-assisted CVD. We demonstrated the capability of phosphorus doping control at the concentrations as low as $10^{16}\,cm^{-3}$ with clear signatures of n-type conductivity characterized by Hall measurements. The highest value of the Hall mobility at RT is $660\,cm^2/V\,s$ obtained for a film with a phosphorus concentration of $7 \times 10^{16}\,cm^{-3}$. We conclude that lightly P-doped diamond films with high crystalline perfection can be successfully obtained.

The low field Hall mobility was described as a function of temperature using the relevant scattering modes. The intrinsic parameters of phonons were determined and the scattering mechanisms of the electron Hall mobility established. The maximum Hall mobility achievable in n-type diamond has been determined at room temperature. Evidence is provided for a discrepancy between low field Hall mobility at thermodynamical equilibrium and electron mobilities deduced from optically excited experiments.

I–V and C–V measurements were carried out over a wide range of temperatures and P concentrations to investigate the electrical properties of n-type diamond Schottky barrier diodes and metal/diamond interfaces for P-doped homoepitaxial diamond. We have obtained excellent Schottky diode properties from 297 K to 773 K. The ideality factor is 1.0 at ±10 V for 573 K and the rectification ratio is ~10^6 at these temperatures. The rectifying characteristics deteriorated with increasing P concentration, becoming closer to ohmic-like characteristics when the P concentration is very high ($3.4 \times 10^{19}\,cm^{-3}$). Frequency-dependent capacitance measurements reveal a P-related deep donor state. The donor activation energy is calculated to be 0.54 eV from conductance–frequency measurements. P electrical activity, which means the ratio of the net donor concentration to the P concentrations detected by SIMS, is nearly one even at low P concentration ($1.6 \times 10^{16}\,cm^{-3}$). The Schottky barrier height is ~4.3 eV independent of the metal work function. The result suggests that the Fermi level is strongly pinned at the metal/diamond interfaces ~4.3 eV below the bottom of the conduction band. A comparison with other metals shows no barrier variation, so that Fermi level pinning due to surface defect state is detected. The fact that the sum of Schottky barrier heights of the same metal on p-type and n-type diamond is close to the bandgap energy indicates that oxygen termination may result in a high density of surface states at the same energy level from the valence band maximum on p-type and n-type (111) diamond surface.

References

1 Kajihara, S.A., Antonelli, A. and Bernholc, J. (1993) *Physica B*, **185** (1–4), 144–9.

2 Alexenko, A.E. and Spitsyn, B.V. (1992) *Diamond and Related Materials*, **1**, 705–9.

3 Kamo, M., Yurimoto, H., Ando, T. and Sato, Y. (1991) *Proceeding of the 2nd International Conference on New Diamond Science and Technology* (eds S.R. Messier, J.T. Glass, J.E. Butler and R. Roy), Materials Research Society, Pittsburgh, p. 637.

4 Okano, K., Kiyota, H., Iwasaki, T., Nakamura, Y., Akiba, Y., Kurosu, T., Iida, M. and Nakamura, T. (1990) *Applied Physics A*, **51**, 344–6.

5 Koizumi, S., Ozaki, H., Kamo, M., Sato, Y. and Inuzuka, T. (1997) *Applied Physics Letters*, **71**, 1064–7.

6 Koizumi, S., Kamo, M., Sato, Y., Mita, S., Sawabe, A., Reznik, A., Uzan-Saguy, C. and Kalish, R. (1998) *Diamond and Related Materials*, **7** (2–5), 540–4.

7 Koizumi, S. (1999) *physica status solidi (a)*, **172**, 71–8.

8 Koizumi, S. (2003) *Semiconductors and Semimetals*, **76**, 239–59.

9 Katagiri, M., Isoya, J., Koizumi, S. and Kanda, H. (2004) *Applied Physics Letters*, **85** (26), 6365–7.

10 Nesládek, M., Meykens, K., Haenen, K., Knuyt, G., Stals, L.M., Teraji, T. and Koizumi, S. (1999) *Physical Review B*, **59**, 14852.

11 Gheeraert, E., Koizumi, S., Teraji, T. and Kanda, H. (2000) *Solid State Communications*, **113**, 577.

12 Sternschulte, H., Thonke, K., Sauer, R. and Koizumi, S. (1999) *Physical Review B*, **59**, 12924.

13 Hasegawa, M., Teraji, T. and Koizumi, S. (2001) *Applied Physics Letters*, **79**, 3068.

14 Koizumi, S., Watanabe, K., Hasegawa, M. and Kanda, H. (2001) *Science*, **292**, 1899.

15 BenMoussa, A., Schühle, U., Haenen, K., Nesládek, M. and Koizumi, S. and Hochedez, J.-F. (2004) *physica status solidi (a)*, **201** (11), 2536–41.

16 BenMoussa, A, Schühle, U., Scholze, F., Kroth, U., Haenen, K., Saito, T., Campos, J., Koizumi, S., Laubis, C., Richter, M., Mortet, V., Theissen, A. and Hochedez, J.F. (2006) *Measurement Science and Technology*, **17**, 913–17.

17 Teukam, Z., Chevallier, J., Saguy, C., Kalish, R., Ballutaud, D., Barbe, M., Jomard, F., Tromson-Carli, A., Cytermann, C., Butler, J.E., Bernard, M., Baron, C. and Deneuville, A. (2003) *Nature Materials*, **2** (7), 482–6.

18 Kato, H., Yamasaki, S. and Okushi, H. (2005) *Applied Physics Letters*, **86** (22), 222111.

19 Koizumi, S., Teraji, T. and Kanda, H. (2000) *Diamond and Related Materials*, **9**, 935.

20 Teraji, T., Katagiri, M., Koizumi, S., Ito, T. and Kanda, H. (2003) *Japanese Journal of Applied Physics*, **42**, L882.

21 Pernot, J., Zawadzki, W., Contreras, S., Robert, J.L., Neyret, E. and Di Cioccio, L. (2001) *Journal of Applied Physics*, **90**, 1869.

22 Gheeraert, E., Koizumi, S., Teraji, T. and Kanda, H. (2001) *Diamond and Related Materials*, **10**, 444.

23 Lundstrom, M. (2000) *Fundamentals of Carrier Transport*, Cambridge University Press, Cambridge.

24 Solin, S.A. and Ramdas, A.K. (1970) *Physical Review B*, **1**, 1687.

25 Farvacque, J.L. (2000) *Physical Review B*, **62**, 2536.

26 Pernot, J., Tavares, C., Gheeraert, E., Bustarret, E., Katagiri, M. and Koizumi, S. (2006) *Applied Physics Letters*, **89**, 122111.

27 Willatzen, M., Cardona, M. and Christensen, N.E. (1994) *Physical Review B*, **50**, 18054.

28 Nava, F., Canali, C., Jacoboni, C., Reggiani, L. and Kozlov, S.F. (1980) *Solid State Communications*, **33**, 475.

29 Pernot, J., Contreras, S., Camassel, J., Robert, J.L., Zawadzki, W., Neyret, E. and Di Cioccio, L. (2000) *Applied Physics Letters*, **77**, 4359.

30 Pearlstein, E.A. and Sutton, R.B. (1950) *Physical Review*, **79**, 907.

31 Redfield, A.G. (1954) *Physical Review*, **94**, 526.

32 Konorova, E.A. and Shevchenko, S.A. (1967) *Fiz. Tekh. Poluprovdn. (USSR)*, **1**, 364.

33 Konorova, E.A. and Shevchenko, S.A. (1967) *Soviet Physics–Semiconductors*, **1**, 299.

34 Pan, L.S., Kania, D.R., Pianetta, P., Ager, J.W., III, Lanstrass, M.I. and Han, S. (1993) *Journal of Applied Physics*, **73**, 2888.

35 Isberg, J., Hammersberg, J., Johansson, E., Wikström, T., Twitchen, D.J., Whitehead, A.J., Coe, S.E. and Scarsbrook, G.A. (2002) *Science*, **297**, 1670.

36 Suzuki, M., Yoshida, H., Sakuma, N., Ono, T., Sakai, T. and Koizumi, S. (2004) *Applied Physics Letters*, **84**, 2349.

37 Suzuki, M., Koizumi, S., Katagiri, M., Yoshida, H., Sakuma, N., Ono, T. and Sakai, T. (2004) *Diamond and Related Materials*, **13**, 2037.

38 Suzuki, M., Koizumi, S., Katagiri, M., Ono, T., Sakuma, N., Yoshida, H. and Sakai, T. (2006) *physica status solidi (a)*, **203**, 3128.

39 Suzuki, M., Yoshida, H., Sakuma, N., Ono, T., Sakai, T., Ogura, M., Okushi, H. and Koizumi, S. (2004) *Diamond and Related Materials*, **13**, 198.

40 Koide, Y., Koizumi, S., Kanda, H., Suzuki, M., Yoshida, H., Sakuma, N., Ono, T. and Sakai, T. (2005) *Applied Physics Letters*, **86**, 232105.

41 Nesladek, M. (2005) *Semiconductor Science and Technology*, **20**, 19.

42 Gheeraert, E., Koizumi, S., Teraji, T., Kanda, H. and Nesladek, M. (2000) *Diamond and Related Materials*, **9**, 948.

43 Rhoderik, E.H. and Williams, R.H. (1988) *Metal–Semiconductor Contacts*, 2nd edn, Oxford University Press, New York.

44 Chevallier, J., Jormard, F., Teukam, Z., Koizumi, S., Kanda, H., Sato, Y., Deneuville, A. and Bernard, M. (2002) *Diamond and Related Materials*, **11**, 1566.

45 Ando, J.T., Haneda, H., Akaishi, M., Sato, Y. and Kamo, M. (1996) *Diamond and Related Materials*, **5**, 34.

46 Nesladek, M., Haenen, K. and Vanecek, M. (2003) *Semiconductors and Semimetals 76–Thin Film Diamond I* (eds C.E. Nebel and J. Ristein), Elsevier Academic Press, Amsterdam, p. 325.

47 Michaelson, J.H.B. (1977) *Journal of Applied Physics*, **48**, 4729.

48 Ristein, J., Maiyer, F., Riedel, M., Cui, J.B. and Ley, L. (2000) *physica status solidi (a)*, **181**, 65.

49 Daniesenko, A., Aleksov, A. and Kohn, E. (2001) *Diamond and Related Materials*, **10**, 667.

50 Takeuchi, D., Yamanaka, S., Watanabe, H. and Okushi, H. (2001) *physica status solidi (a)*, **186**, 269.

10
Single Defect Centers in Diamond

Fedor Jelezko and Jörg Wrachtrup
Universität Stuttgart, Physikalisches Institut, Pfaffenwaldring 57, Stuttgart 70550, Germany

10.1
Introduction

The ever increasing demand in computational power and data transmission rates has inspired researchers to investigate new ways to process and communicate information. Among others, physicists explored the usefulness of "non-classical", that is, quantum mechanical systems in the world of information processing. Spectacular achievements like Shor's discovery of the quantum factoring algorithm [1] or the development of quantum secure data communication gave birth to the field of quantum information processing (QIP) [2]. After an initial period where the physical nature of information and the means by which information processing can be carried out by unitary transformation in quantum mechanics, were explored, [3] researchers looked for systems which might be of use as hardware in QIP. From the beginning, it became clear that the restrictions on the hardware of choice were severe, in particular for solid state systems. Hence, in the recent past, scientists working in the development of nanostructured materials and quantum physics have cooperated on different solid state systems to define quantum mechanical two level systems, make them robust against decoherence, and addressable as individual units. While the feasibility of QIP remains to be shown, these endeavors will deepen our understanding of quantum mechanics, and they also mark a new area in material science which now regards diamonds as a potential host material. The usefulness of diamond is based on two properties. First, defects in diamond are often characterized by low electron phonon coupling, mostly due to the low density of phonon states; that is, the high Debye temperature of the material [4]. Secondly, color centers in diamond are usually found to be very stable, even under ambient conditions. This makes them unique among all optically active solid state systems.

The main goal of QIP is the flexible generation of quantum states from individual two level systems (qubits). The state of the individual qubits should be changed coherently and the interaction strength among them should be

Physics and Applications of CVD Diamond. Satoshi Koizumi, Christoph Nebel, and Milos Nesladek

ISBN: 978-3-527-40801-6

controllable. At the same time, those systems which are discussed for data communication must be optically active, which means that they should show a high oscillator strength for an electric dipole transition between their ground and some optically excited state. Individual ions or ion strings have been applied with great success. Currently, up to eight ions in a string have been cooled to their ground state, addressed and manipulated individually [5]. Owing to careful construction of the ion trap, decoherence is reduced to a minimum [6]. Landmark experiments, like teleportation of quantum states among ions [7, 8] and first quantum algorithms, have been shown in these systems [9, 10].

In solid state physics, different types of hardware are discussed for QIP. Because dephasing is fast in most situations in solids, only specific systems allow for controlled generation of a quantum state with preservation of phase coherence for a sufficient time. At present, three systems are under discussion. Superconducting systems are either realized as flux or charge quantized individual units [11]. Their strength lies in their long coherence times and well established control of quantum states. Significant progress has been achieved with quantum dots as individual quantum systems. Initially, the electronic ground, as well as excited states (exciton ground state), have been used in the definition of qubits [12]. Meanwhile, the spin of individual electrons, either in a single quantum dot or coupled GaAs quantum dots, has been subject to controlled experiments [13–15]. Because of the presence of paramagnetic nuclear spins, the electron spin is subject to decoherence or a static inhomogeneous frequency distribution. Hence, a further direction of research is in Si or SiGe quantum dots where practically no paramagnetic nuclear spins play a significant role. The third system under investigation is of phosphorus impurities in silicon [16]. Phosphorus implanted in Si is an electron paramagnetic impurity with a nuclear spin $I = 1/2$. The coherence times are known to be long at low temperature. The electron or nuclear spins form a controllable two level system. The addressing of individual spins is planned via magnetic field gradients. Major obstacles with respect to nanostructuring of the system have been overcome, while the readout of single spins based on spin-to-charge conversion with consecutive detection of charge state has not yet been successful.

10.2
Color Centers in Diamond

There are more then 100 luminescent defects in diamond. A significant fraction has been analyzed in detail, such that their charge and spin state are known under equilibrium conditions [17]. For this chapter, nitrogen related defects are of particular importance. They are most abundant in diamond since nitrogen is a prominent impurity in the material. Nitrogen is a defect which either exists as a single substitutional impurity or in aggregated form. The single substitutional nitrogen has an infrared local mode of vibration at 1344 cm^{-1}. The center is at a C_{3v} symmetry site. It is a deep electron donor, probably 1.7 eV below the conduction band edge. There is an electron paramagnetic resonance (EPR) signal associated with this defect, called P1, which identifies it to be an electron paramagnetic system with $S = 1/2$ ground state [17]. Nitrogen aggregates are, most commonly, pairs of neighboring substitutional atoms, the A aggregates, and groups of four around a vacancy, the B aggregate. All three forms of nitrogen impurities have distinct infrared spectra. Another defect often found in nitrogen rich type Ib diamond samples after irradiation damage is the nitrogen vacancy defect center, see Figure 10.1. This defect gives rise to a strong absorption at 1.945 eV (637 nm) [18]. At low temperature, the absorption is marked by a narrow optical resonance line (zero phonon line) followed by prominent vibronic side bands, see Figure 10.2. Electron spin resonance measurements have indicated that the defect has an electron paramagnetic ground state with electron spin angular momentum $S = 1$ [19]. The zero field splitting parameters were found to be $D = 2.88$GHz and $E = 0$ indicating a

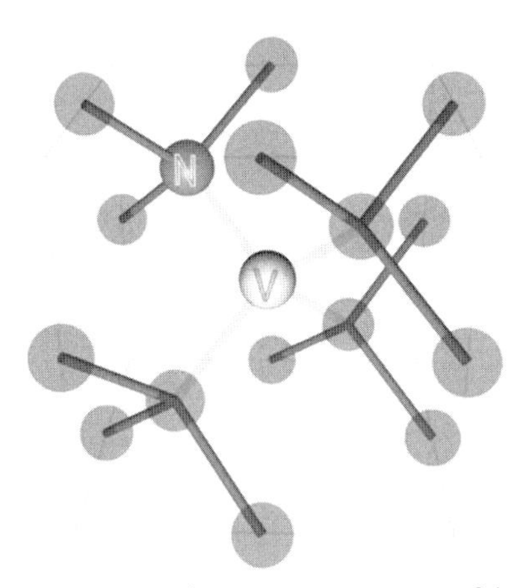

Figure 10.1 Schematic representation of the nitrogen vacancy (NV) center structure.

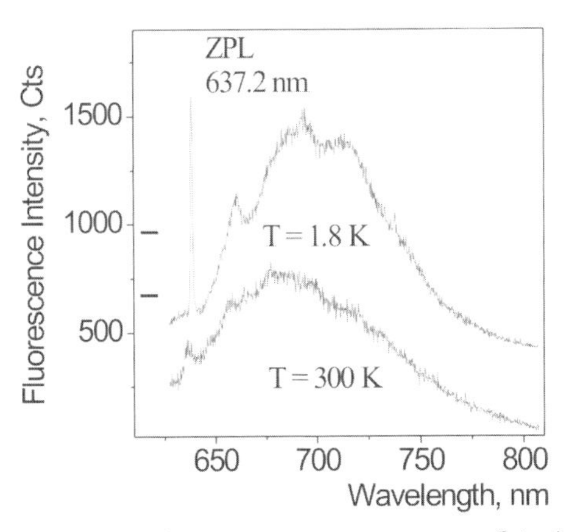

Figure 10.2 Fluorescence emission spectra of single NV centers at room temperature and LHe temperatures. Excitation wavelength was 514 nm.

C_{3v} symmetry of the electron spin wave function. From measurements of the hyperfine coupling, constant to the nitrogen nuclear spin and carbon spins in the first coordination shell, it was concluded that roughly 70% of the unpaired electron spin density is found at the three nearest neighbor carbon atoms, whereas the spin density at the nitrogen is only 2%. Obviously, the electrons spend most of their time at the three carbons next to the vacancy. To explain the triplet ground state, primarily, a six electron model is invoked which requires the defect to be negatively charged; that is, to be NV^- [20]. Hole burning experiments and the high radiative recombination rate (lifetime roughly 11 ns [21], quantum yield (0.7) indicate that the optically excited state is also a spin triplet. The width of the spectral holes burned into the inhomogeneous absorption profile were found to be in the order of 50 MHz [22, 23]. Detailed investigation of the excited state dephasing and hole burning have caused speculations as to whether the excited state is subject to a Jahn–Teller splitting [24, 25]. From group theoretical arguments, it is concluded that the ground state is 3A and the excited state is of 3E symmetry. In the C_{3v} group this state thus comprises two degenerate substrates 3Ex,y with an orthogonal polarization of the optical transition. Photon echo experiments have been interpreted in terms of a Jahn-Teller splitting of 40 cm^{-1} among these two states with fast relaxation between them [24]. However, no further experimental evidence has been found to support this conclusion. Hole burning experiments showed two mechanisms for spectral hole burning: a permanent one and a transient mechanism with a time scale in the order of ms [23]. This is either interpreted as a spin relaxation mechanism in the ground state, or a metastable state in the optical excitation emission cycle. Indeed, it proved difficult to find evidence for this metastable state. Also, number and energetic position relative to the triplet ground and excited state are still subject of debate. Meanwhile, it seems to be clear that at least one singlet state is placed between the two triplet states. As a working hypothesis, it should be assumed, throughout this chapter, that the optical excitation emission cycle is described by three electronic levels.

10.3
Optical Excitation and Spin Polarization

Given the fact that the NV center has an electron spin triplet ground state with an optically allowed transition to a ^3E spin triplet state, one might wonder about the influence of optical excitation on the electron spin properties of the defect. Indeed, in initial experiments, no electron spin resonance (EPR) signal of the defect was detected, except when subject to irradiation in a wavelength range between 450 and 637 nm [19]. Later, it became clear that, in fact, there is an EPR signal, even in the absence of light, yet the signal strength is considerably enhanced upon illumination [26]. EPR lines showed either absorptive or emissive line shape, depending on the spectral position, indicating that only specific spin sublevels are affected by optical excitation [27]. In general, a $S = 1$ electron spin system is described by a spin Hamiltonian of the following form: $H = g_e\beta\vec{B}\hat{S} + \hat{S}D\hat{S}$. Here

g_e is the electronic g-factor ($g_e = 2.0028 \pm 0.0003$); B_o is the external magnetic field and D is the zero field splitting tensor. This tensor comprises the anisotropic dipolar interaction of the two electron spins forming the triplet state averaged over their wave function. The tensor is traceless and thus characterized by two parameters, D and E as already mentioned above. The zero field splitting causes a lifting of the degeneracy of the spin sublevels $m_s = \pm 1.0$ even in the absence of an external magnetic field. Those zero field spin wave functions $T_{x,y,z}$ do not diagonalize the full high field Hamiltonian H, but are related to these functions by

$$T_x = \frac{1}{\sqrt{2}} \beta_1\beta_2 - d_1 d_2\rangle = \frac{1}{\sqrt{2}} |T_{-1} = T_{+1}\rangle$$

$$T_y = \frac{i}{\sqrt{2}} \Big| \beta_1\beta_2 + \alpha_1\alpha_2\rangle = \frac{1}{\sqrt{2}} |T_{-1} + T_{+1}\rangle$$

$$T_z = \frac{1}{\sqrt{2}} |\alpha_1\beta_2 + \beta_1\alpha_2\rangle = |T_O\rangle.$$

The expectation value of S_z for all three wave functions $\langle T_{x,y,z}|S_z|T_{x,y,z}\rangle$ is zero. Hence, there is no magnetization in zero external field. There are different ways to account for the spin polarization process in an excitation scheme involving spin triplets. To first order optical excitation is a spin state conserving process. However, spin orbit (LS) coupling might allow for a spin state change in the course of optical excitation. Cross relaxation processes, on the other hand, might cause a strong spin polarization as it is observed in the optical excitation of various systems, like for example, GaAs. However, optical spectroscopy and hole burning data gave little evidence for non spin conserving excitation processes in the NV center. In two laser hole burning experiments, data have been interpreted by assuming different zero field splitting parameters in ground and excited state ($D^{exc} \approx 2\,GHz, E^{exc} \approx 0{,}8\,GHz$) by an otherwise spin state preserving optical excitation process [28]. Indeed, this is confirmed by later attempts to generate ground state spin coherence via a Raman process [29], which only proves to be possible when ground state spin sublevels are brought close to anticrossing by an external magnetic field. Another spin polarizing mechanism involves a further electronic state in the optical excitation and emission cycle [30, 31]. Though being weak, LS coupling might be strong enough to induce intersystem crossing to states with different spin symmetry, for example, a singlet state. Indeed, the relative position of the ^1A singlet state with respect to the two triplet states has been subject of intense debate. Intersystem crossing is driven by LS induced mixing of singlet character into triplet states. Due to the lack of any emission from the ^1A state or noticeable absorption to other states, no direct evidence for this state is available, at present. However, the kinetics of photo emission from single NV centers strongly suggests the presence of a metastable state in the excitation emission cycle of the state. As described below, the intersystem crossing rates from the excited triplet state to the singlet state are found to be drastically different, whereas the relaxation to the ^3A state might not depend on the spin substate. This provides the required optical excitation dependent relaxation mechanism. Bulk, as well as single center experiments, show that predominantly the $m_s = 0$ (T_z) sublevel in the spin ground state is populated. The polarization in this state is on the order of 80% or higher [27].

10.4 Spin Properties of the NV Center

Because of its paramagnetic spin ground and excited state, the NV center has been the target of numerous investigations regarding its magneto-optical properties. Pioneering work has been carried out in the groups of N.B. Manson [32–39] and Rand [26, 40, 41].

The hyperfine and fine structure splitting of the NV ground state has been used to measure the Autler–Townes splitting induced by a strong pump field in a three level system. Level anticrossing among the $m_s = 0$ and $m_s = -1$ allows for an accurate measurement of the hyperfine coupling constant for the nitrogen nucleus, yielding an axially symmetric hyperfine coupling tensor with $A_{\parallel} = 2.3\,\text{MHz}$ and $A_{\perp} = 2.1\,\text{MHz}$ [42, 43]. The quadrupole coupling constant $P = 5.04\,\text{MHz}$. Because of its convenient access to various transitions in the optical, microwave and radiofrequency domain, the NV center has been used as a model system to study the interaction between matter and radiation in the linear and nonlinear regime. An interesting set of experiments concerns electromagnetically induced transparency in a Λ-type level scheme. The action of a strong pump pulse on one transition in this energy level scheme renders the system transparent for radiation resonant with another transitions. Experiments have been carried out in the microwave frequency domain [44] as well as for optical transitions among the $3A$ ground state and the $3E$ excited state [29]. Here, two electron spin sublevels are brought into near level anticrossing, such that an effective three level system is generated with one excited state spin sublevel and two allowed optical transitions. A 17% increase in transmission is detected for a suitably tuned probe beam.

While much work has been done on vacancy and nitrogen related impurities, comparatively little is known about defects comprising heavy elements. For many years it was difficult to incorporate heavy elements as impurities into the diamond lattice. Only six elements have been identified as bonding to the diamond lattice, namely nitrogen, boron, nickel, silicon, hydrogen and cobalt. Attempts to use ion implantation techniques for incorporation of transition metal ions were unsuccessful. This might be due to the large size of the ions and the small lattice parameters of diamond together with the metastability of the diamond lattice at ambient pressure. Recent developments in crystal growth and thin film technology have made it possible to incorporate various dopants into the diamond lattice during growth. This has enabled studies of nickel defects [45, 46]. Depending on the annealing conditions Ni can form clusters with various vacancies and nitrogen atoms in nearest neighbor sites. Different Ni related centers are listed with NE as a prefix and numbers to identify individual entities. The structure and chemical

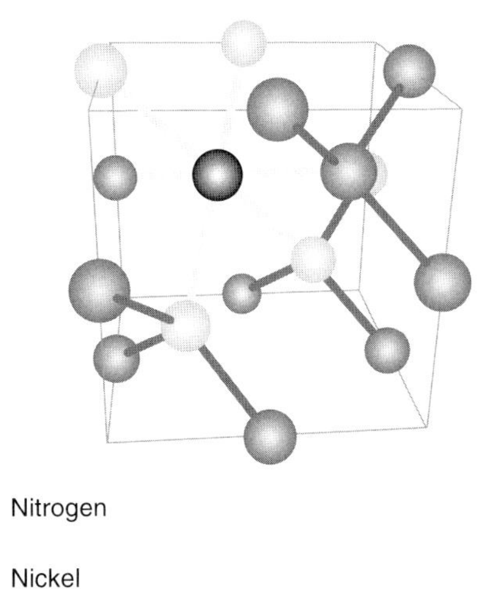

Figure 10.3 Structure of the NE8 center.

composition of defects have mostly been identified by EPR on the basis of the hyperfine coupling to nitrogen nuclei [46]. A particularly rich hyperfine structure has been identified for the NE8 center.

Analysis of the angular dependence of the EPR spectrum for the NE8 center showed that this center has electronic spin $S = 1/2$ and a g-value typical of a d-ion with more than half filled d-shell. The NE8 center has been found not only in HPHT synthetic diamonds, but also in natural diamonds which contain the nickel–nitrogen centers NE1 to NE3 [46]. The structure of the center is shown in Figure 10.3. It comprises four substitutional nitrogen atoms and an interstitial Ni impurity. The EPR signature of the system has been correlated to an optical zero phonon transition at around 794 nm. The relative integral intensity of the zero phonon line and the vibronic side band at room temperature is 0.7 (Debey–Waller factor) [47]. The fluorescence emission statistics of single NE8 emitters shows decay to an as yet unidentified metastable state with a rate of 6 MHz.

10.5 Single Defect Center Experiments

Experiments on single quantum systems in solids have brought about a considerable improvement in the understanding of the dynamics and energetic structure of the respective materials. In addition, a number of quantum optical phenomena, especially when light–matter coupling is concerned, have been investigated. As opposed to atomic systems on which first experiments on single quantum systems are well established, similar experiments with impurity atoms in solids remain challenging. Single quantum systems in solids usually strongly interact with their environment. This has technical as well as physical consequences. First of all, single solid state quantum systems are embedded in an environment which, for example, scatters excitation light. Given a diffraction limited focal volume, usually, the number of matrix atoms exceed those of the quantum systems by 10^6–108. This puts an upper limit on the impurity content of the matrix, or on the efficiency of inelastic scattering processes like, for example, Raman scattering from the matrix. Various systems like single hydrocarbon molecules, proteins, quantum dots and defect centers have been analyzed [48]. Except for some experiments on surface enhanced Raman scattering, the technique usually relies on fluorescence emission. In this technique, an excitation laser in resonance with a strongly allowed optical transition of the system is used to populate the optically excited state (e.g. the ^{3}E state for the NV center), see Figure 10.4a. Depending on the fluorescence emission quantum yield, the system either decays via fluorescence emission or non-radiatively, for example, via intersystem crossing to a metastable state (^{1}A in the case of the NV). The maximum number of photons emitted is given when the optical transition is saturated. In this case, the maximum fluorescence intensity is given as

$$I_{max} = \frac{k_{31}(k_{21}+k_{23})\Phi_F}{2k_{31}+k_{23}}$$

Here k_{31} is the relaxation rate from the metastable to the ground state and k_{21} is the decay rate of the optically excited state, k_{23} is the decay rate to the metastable

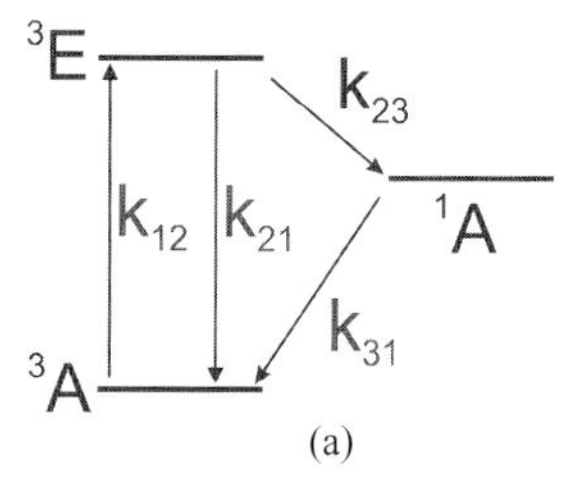

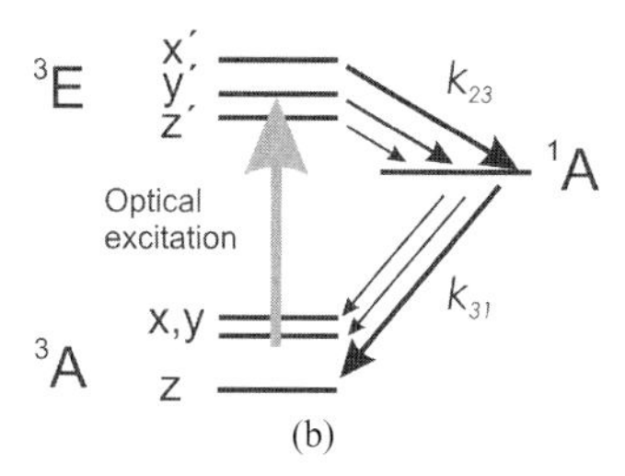

Figure 10.4 (a) Three level scheme describing the optical excitation and emission cycle of single NV centers. ^3A and ^3E are the triplet ground and excited state. ^1A is a metastable singlet state. No information is at hand presently about the number and relative position of singlet levels. The arrows and k_{ij} denote the rates of transition among the various states. (b) More detailed energy level scheme differentiating between triplet sublevels in the ^3A and ^3E state.

state and Φ_F marks the fluorescence quantum yield. For the NV center I_{max} is about 10^7 photon/s. I_{max} critically depends on a number of parameters. First of all, the fluorescence quantum yield limits the maximum emission. A good example to illustrate this is the GR1 center, the neutral vacancy defect in diamond. The overall lifetime of the excited state for this defect is 1 ns at room temperature. However, the radiative lifetime is in the order of 100 ns. Hence Φ_F is in the order of 0.01. Given the usual values for k_{21} and k_{31} this yields an I_{max} which is too low to allow for detecting single GR1 centers with current technology. Figure 10.5 shows the saturation curve of a single NV defect. Indeed, the maximum observable emission rate from the NV center is around 10^5 photons/s which corresponds well to the value estimated above, if we assume a detection efficiency of 0.01. Single NV centers can be observed by standard confocal fluorescence microscopy in type Ib diamond. In confocal microscopy, a laser beam is focused onto a diffraction limited spot in the diamond sample and the fluorescence is collected from that spot. Hence, the focal probe volume is diffraction limited with a volume of roughly $1\,\mu m^3$. In order to be able to detect single centers, it is thus important to control the density of defects. For the NV center, this is done by varying the number of vacancies created in the sample by, for example, choosing an appropriate dose of electron irradiation. Hence, the number of NV centers depends on the number of vacancies created and the number of nitrogen atoms in the sample. Figure 10.6

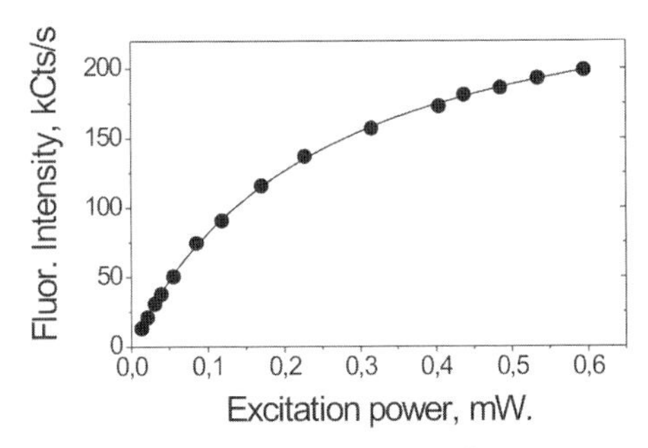

Figure 10.5 Saturation curve of the fluorescence intensity of a single NV center at $T = 300\,K$. Excitation wavelength is 514 nm. The power is measured at the objective entrance.

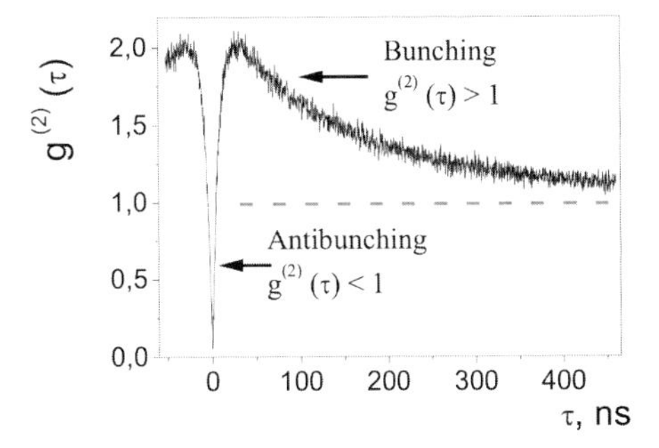

Figure 10.6 Confocal fluorescence image of various diamond samples with different electron irradiation dosages.

shows an image of a diamond sample where the number of defects in the sample is low enough to detect the fluorescence from single color centers [49]. As expected, the image shows diffraction limited spots. From the image alone, it cannot be concluded whether the fluorescence stems from a single quantum system or from aggregates of defects. To determine the number of independent emitters in the focal volume, the emission statistics of the NV center fluorescence can be used [50–52]. The fluorescence photon number statistics of a single quantum mechanical two level system deviates from a classical Poissonian distribution. If one records the fluorescence intensity autocorrelation function

$$g^2(\tau) = \frac{\langle I(t)\cdot I(t+\tau)\rangle}{\langle I/t)\rangle^2}$$

for short time τ one finds $g^2(0) = 0$ if the emission stems from a single defect center (see Figure 10.7). This is due to the fact that the defect has to be excited first, before it can emit a single photon. Hence, a single defect never emits two fluorescence photons simultaneously, in contrast to the case when a number of independent emitters are excited at random. If one adopts the three level scheme from Figure 10.4a, rate equations for temporal changes of populations in the three levels can be set up. The equations are solved by

$$g^2(\tau) = 1 - (k+1)_e^{k_1\tau} + k_e^{k_2\tau}$$

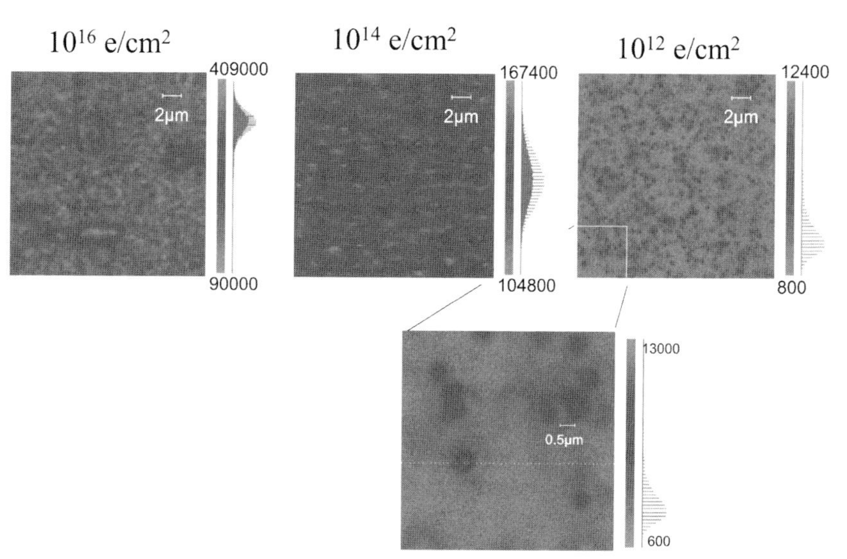

Figure 10.7 Fluorescence intensity autocorrelation function of a single NV defect at room temperature.

with rates

$$k_{1/2} = -\frac{1}{2}P \pm \sqrt{\frac{1}{4}P^2 - Q}.$$

Here

$$P = k_{21} + k_{12} + k_{23} + k_{31} \text{ and } Q = k_{31}(k_{21} + k_{12}) + k_{23}(k_{31} + k_{12})$$

with

$$K = \frac{k_2 + k_{31} - k_{12}\dfrac{k_{23}}{k_{31}}}{k_1 - k_2}.$$

This function reproduces the dip in the correlation function $g^2(\tau)$ for $\tau \to 0$ shown in Figure 10.7, which indicates that the light detected originates from a single NV. The slope of the curve around $\tau = 0$ is determined by the pumping power of the laser k_{12} and the decay rate k_{21}. For larger times τ a decay of the correlation function becomes visible. This decay marks the ISC process from the excited triplet ^3E to the metastable singlet state ^1A. Besides the spin quantum jumps detected at low temperature, the photon statistics measurements are the best indication for detection of single centers. It should be noted that the radiative decay time depends on the refractive index of the surrounding medium as $1/n_{\text{medium}}$. Because n_{medium} of diamond is 2.4 the decay time should increase significantly if the refractive index of the surrounding is reduced. This is indeed observed for NV centers in diamond nanocrystals [51]. It should be noted that, owing to their stability, single defect centers in diamond are prime candidates for single photon sources under ambient conditions. Such sources are important for linear optics, quantum computing, and quantum cryptography. Indeed quantum key distribution has been successful with fluorescence emission from single defect centers [53].

An important merit of single photon sources is their signal to background ratio, given, for example, by the amplitude of the correlation function at $\tau = 0$. This ratio should be as high as possible to ensure that a single bit of information is encoded in a single photon only. The NV center has a broad emission range which does not allow efficient filtering of background signals. This is in sharp contrast to the NE^8 defect, which shows a very narrow spectrum, only 1.2 nm wide. As a consequence, the NE^8 emission can be filtered out efficiently [47]. The correlation function resembles the one from the NV center. Indeed, the photophysical parameters of the NV and NE8 are similar, yet, under comparable experimental conditions, the NE8 shows an order of magnitude improvement in signal-to-background ratio because of the narrower emission range.

Besides application in single photon generation, photon statistical measurements also allow us to derive conclusions on photoionization and photochromism of single defects. Most notably, it is speculated that the NV center exists in two charge forms, the negatively charged NV with zero phonon absorption at 637 nm, and the neutral from NV^0 with absorption around 575 nm [20, 54]. Although evidence existed that both absorption lines stem from the same defect, no direct charge interconversion has been shown in bulk experiments. The best example for a spectroscopically resolved charge transfer in diamond is the vacancy, which exists in two stable charge states. In order to observe the charge transfer from NV to NV^0 , photon statistical measurements similar to the ones described have been carried out, except for a splitting of photons depending on the emission wavelength [55]. This two channel set up allows us to detect the emission of NV^0 in one and NV in another detector arm. Figure 10.8 shows the experimental result. For delay time $\tau = 0$, g^2 (τ) shows a dip, indicating the sub-Poissonian statistics of the light emitted. It should be noted that, for recording Figure 10.8, $g^2(\tau)$ has been detected in a cross correlation type of measurement between NV^0 and NV photons. The data acquisition was started by detection of a NV and stopped by the detection of a NV^0 photon. The coincidence rate $(\tau = 0)$ in Figure 10.8 is zero. Hence, it must be concluded that there is a continuous interconversion between the two spectral positions. Detailed time resolved experiments show that switching from NV^0 to NV is photoinduced, whereas the reverse process NV to NV^0 occurs under dark conditions with a time constant between 0.3 and 3.6 μs.

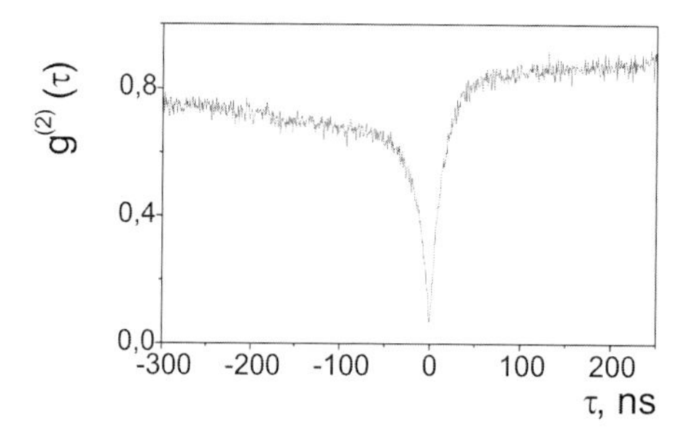

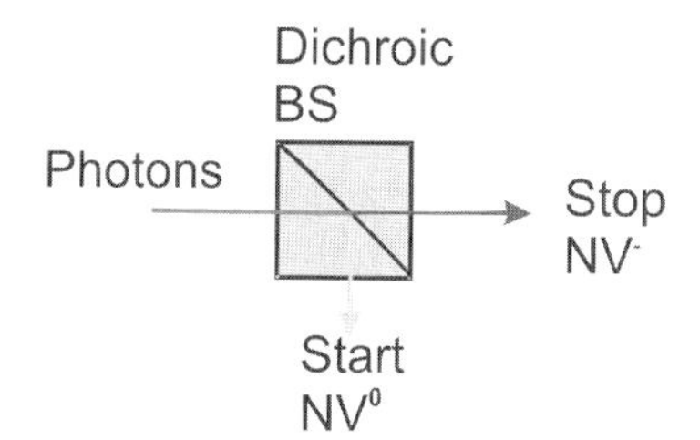

Figure 10.8 Fluorescence cross correlation function between the NV^0 and NV emission of a single defect.

10.6 Spin Physics of Single Defects

The controlled generation of quantum states from individual quantum objects is a current research topic that has received considerable interest during the last decade. In part, this is motivated by possible applications in quantum information processing. On the other hand, the simulation of quantum systems itself, for example, to investigate the physics of quantum phase transitions, is of interest. Basically, the control of a wave function of a collective quantum state requires control over the quantum state of interacting qubits:

$$\psi = \sum_{i}^{N} a_i |\alpha_i\rangle \; .$$

The evolution of ψ is subject to unitary transformations: $\psi' = U\psi$. In general, it is necessary to be able to manipulate coherently each individual qubit and control the strength of interaction among them. This puts certain restrictions on the system parameters. To allow for nontrivial unitary operations U, a certain phase coherence time together with interaction strength and speed of operation is required. While the interaction strength and control speed of individual qubits are limited by technical means, the dephasing times in solids are usually short. Spins are certainly among the most promising systems owing to long coherence times together with availability of fast control of individual qubits and relatively strong spin–spin coupling. Although such robust control of spin states plus adjustment of spin–spin interactions are common practice in electron and nuclear magnetic resonance, the measurement of single spin states is a fierce experimental challenge. Only a few solid state systems currently allow for single spin state detection. Most notably, single spin state measurements have been successful in III–V quantum dots and in P center defects in silicon single electron transistor (SET) structures. A system where single spin control and state measurement are well developed is the NV center using optical technique. It is remarkable that, for spins associated with defects in diamond, phase memory times can be long even under ambient conditions. As an example, the electron spin lattice time is reported to be 1.8 ms at $T = 300\,\text{K}$. The long dephasing times are attributed to the low phonon density of states in diamond even at room temperature. In the NV center the spin state is detected via fluorescence. As discussed above, the fluorescence intensity I_{max} depends on the spin state via the ISC rate k_{23}. Upon changing the spin state, this rate is changed from some kHz by more than three orders of magnitude towards some MHZ. Given the other parameters, this results in a change of roughly 30% of I_{max}. Taking into account the photon shot noise and an average Imax of 10^5 photocounts per second, this change in fluorescence intensity can be detected with some ms averaging time. Figure 10.9 shows an example of an optically detected magnetic resonance (ODMR) spectrum of a single NV defect. The spin Hamiltonian describing the spectrum is

$$H = g_e \beta_1 \vec{B}_O \vec{S} + \vec{S} A \vec{I} - g_N \beta_N \vec{I} \vec{B}$$

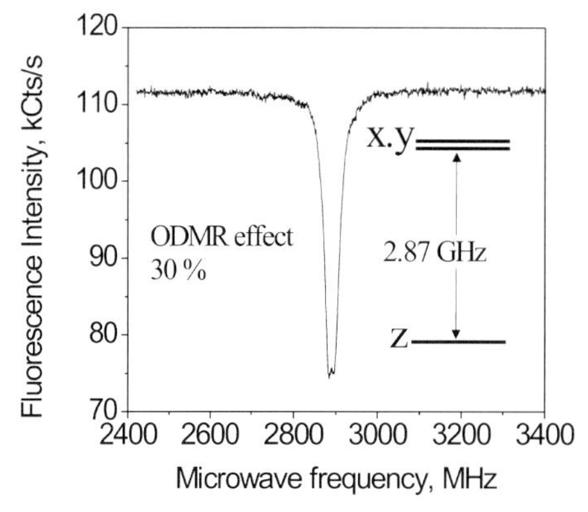

Figure 10.9 Optically detected magnetic resonance (ODMR) spectrum of a single defect. The spectrum has been recorded at room temperature with 514 nm irradiation without an external B_0.

The spectra have been taken without an external $\vec{B}_O$ field. Hence only the fine structure (***SDS***) and hyperfine term ***SAI*** need to be considered. As mentioned above, due to the C_{3v} symmetry two of the three spin sublevels are degenerated ($E = 0$). Hence, only a single ODMR line is seen in the spectrum. Upon application of a $\vec{B}_O$ field the two degenerated levels split and two lines become visible. The hyperfine coupling to the ^{14}N nucleus is not resolved in these spectra because of the large optical pumping rate used [56]. Since the hyperfine and quadrupole coupling constants are 2 and 5 MHz only, the corresponding splitting are easily masked by the homogeneous transition line width. For the NV center, this line width depends on the optical excitation intensity, since at least one of the levels is optically excited to the ^3E state. Since the optical Rabi frequency easily achieves some MHz, the line width broadens correspondingly. As a consequence, for low excitation power, the hyperfine and quadruple structure gets resolved, as shown in Figure 10.10a. The hyperfine structure in Figure 10.10 has been analyzed in detail [56] and corresponds to the known value of the hyperfine and quadruple coupling of the ^{14}N nucleus of the NV center. It should be noted that these spectra provide an opportunity to verify the mechanism by which the defect has been generated. There are two mechanisms by which NV centers can be created in diamonds. First, vacancies are generated and the intrinsic nitrogen present in the material is used to create NV centers. Alternatively, the nitrogen atoms are implanted in nitrogen-free diamond and the vacancies which are generated during the implantation form NV defects. To ensure that a defect center originates from an implanted nitrogen, ^{15}N isotope, which has a natural abundance of only 0.1% and is a $I = 1/2$ nucleus, can be used.

A corresponding ODMR spectrum is shown in Figure 10.10a and is clearly different from the ^{14}N case. A completely different set of hyperfine coupling parameters is measured when a ^{13}C nucleus is found in the shell of first nearest neighbors

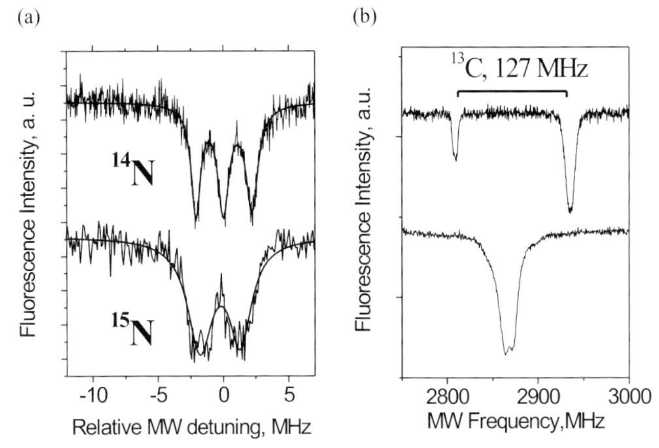

Figure 10.10 ODMR spectra with resolved ^{14}N and ^{15}N hyperfine structure (a) and ^{13}C hyperfine structure (b). The spectra have been acquired at room temperature without the application of an external magnetic field.

around the vacancy. Because of the much higher spin density of the electron at these carbon positions, the measured coupling parameter is around 130 MHz (see Figure 10.10b).

10.7 Coherence and Single Spin States

The generation of a coherent state superposition is achieved by a short microwave pulse in resonance with the transition in Figure 10.9 or one of the transitions in Figure 10.10. In order to generate a state superposition with arbitrary expansion coefficients of the two eigenstates, for example, $|\alpha\rangle$ and $|\beta\rangle$ one uses microwave pulses of variable length, such that $\psi(t) = \sin \Omega^{MW} t|ms = 0\rangle + \cos \Omega_{MW} t\ |ms{\pm}1\rangle$ (here Ω_{MW} is the microwave Rabi frequency). Depending on the magnitude of $\cos 2\,\Omega_{MW} t$ the fluorescence will change. Hence, when plotting the fluorescence intensity as a function of pulse length, a periodical variation of the fluorescence is seen (see Figure 10.11). The frequency of these oscillations (Rabi nutations) depends linearly on the MW field amplitude, as can be seen in Figure 10.11 [57]. Rabi frequencies of up to 140 MHz have been achieved with miniaturized coupling loops or wire structures. For the nutation curve in Figure 10.11, a decay of the amplitude is expected. The corresponding decay constant is related to the dephasing time T_2 but not equivalent to T_2. Rather T_2 has to be measured in the absence of any microwave field. This is achieved by the application of a Hahn echo sequence. In this pulse sequence, all inhomogeneous distributions of resonance frequencies are refocused, while fluctuations of transition frequency or random phase jumps cause an echo decay upon increasing the time between pulses. Figure 10.12 shows

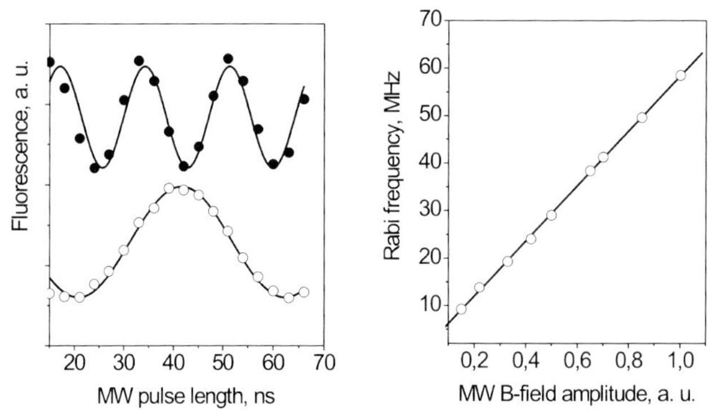

Figure 10.11 Optically detected Rabi nutations of a single NV electron spin. The points represent experimental data and the line is a fit with a $\cos^2 \Omega t$ function, where Ω is the spin Rabi frequency. Figure on right: Dependence of Rabi frequency on MW filed amplitude. The solid line represents a linear fit.

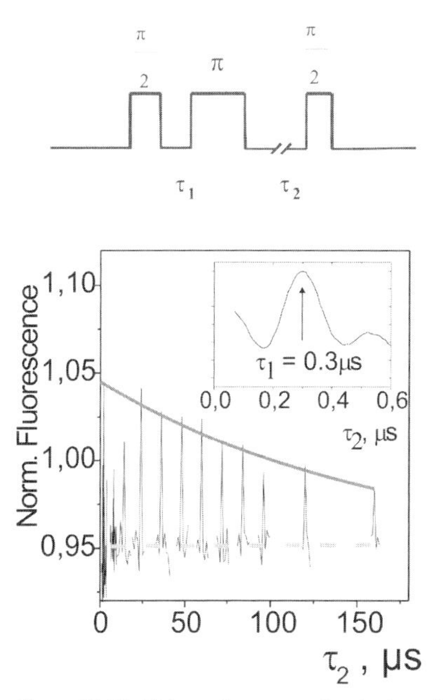

Figure 10.12 Hahn echo trace of a single spin. The upper part of the figure shows the microwave pulse sequence used. The lower part depicts experimental data. The inset shows the Hahn echo itself. The main figure demonstrates the Hahn echo decay as a function of $\tau_1 = \tau_2$. Printed are the individual echoes at different delay times together with a fitted decay curve of the amplitude.

an example of a Hahn echo train with variable delay. An echo decay is visible which can be fitted with a monoexponential decay time of 350 μs. The chief causes of dephasing in diamond are electron paramagnetic impurities in the lattice [58]. These impurities show dipolar coupling to the NV center and, hence, may undergo energy conserving spin flip–flop processes with the NV spin. These processes result in a loss of phase memory of the NV spin. It has been demonstrated that the NV center T_2 time depends on the concentration of impurities in the lattice and the dephasing time was shown to decrease up to some hundred ns for nitrogen-rich diamond. In defects which do show a hyperfine coupling to a 13C nucleus, in addition to electron spin, nuclear spin nutations can be detected, also [59]. Because nuclear spin wave functions do not couple to the optical transition outside of level anticrossing, all changes in nuclear spin wave function must be mapped into the electron spin states to be detectable. A single electron plus nuclear spin system is described by a four level system. To first order only electron spin, resonance transitions with $\Delta m_s = 1$ and $\Delta m_i = 0$ are allowed, indicated by the two arrows in Figure 10.13. In order to drive nuclear magnetic resonance transitions, radio frequency has to be irradiated at transition energy between level 1 and 2 (or level 3 and 4). For the 1–2 transition, this corresponds to the hyperfine splitting observed in Figure 10.10b.

^{13}C nuclear relaxation times in diamond vary between 1.4 and 36 hours. The T_2 time can be estimated from the width of ^{13}C NMR spectra to be in the order of ms for those nuclei that are not detuned from the dipolar nuclear spin bath. Hence, nuclear spin states should allow for coherent state preparation. To observe nuclear spin transients, a microwave radio frequency double resonance experiment has been carried out. The experiment comprises π pulses separated by time τ. During this time interval, a radio frequency pulse of variable length in resonance with, for

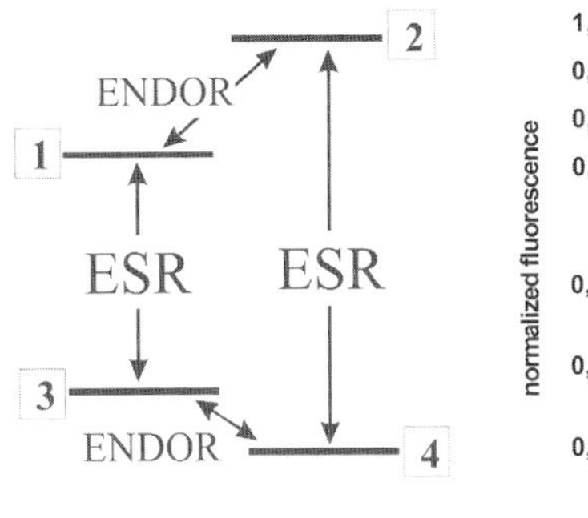

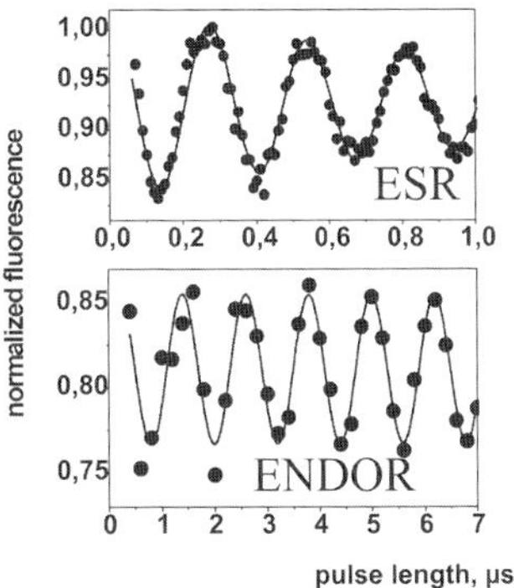

Figure 10.13 Energy level scheme and Rabi nutations of a single NV electron spin coupled to a single ^{13}C spin. The left part of the figure shows the relevant spin levels for the coupled electron–nuclear spin system. The allowed electron spin resonance (ESR) and nuclear magnetic resonance (NMR) transitions are shown in the figure. The right part of the figure shows Rabi nutations of an electron spin (upper trace) and nuclear spin (lower trace).

example, the 2–1 transition is applied. The strength of the EPR signal is measured on the 3–1 transition. Figure 10.13 shows an example of a nuclear transient measured in this way. The amplitude of the oscillations corresponds to the amplitude of the ODMR signal itself, that is, 30% of the fluorescence intensity. The approach corresponds to the well known electron nuclear double resonance experiments. With two spins at hand, it is possible to carry out basic quantum computation experiments like, for example, the conditional not gate (CNOT). It can be shown that two gate operations are sufficient to perform all operations necessary for full quantum computation. These two gates are the single quantum bit NOT gate, which corresponds to an inversion of the bit value and the CNOT gate which is the inversion of one bit conditioned on the value of a second bit. In this nomenclature, a single qubit corresponds to a single spin with either of the two eigenvalues $|0>$ or $<1|$ (spin-up or spin-down). A CNOT gate would flip, for example, the electron spin depending on the state of the nuclear spin. Such a scenario can be easily realized in a situation shown in Figure 10.13, that is, in a four level system with coherent control over (at least) two transitions. For example, the nuclear spin will be inverted only by RF irradiation in resonance with the 1–2 transition when the electron is in spin-up configuration, that is, in state 1. A simplified version of the CNOT gate is the CNOT transformation. The two operations are identical to each other except for a phase factor which can be achieved by a rotation around the z-axis. The CROT itself is only a π pulse. The action of the pulse only corresponds to an ideal gate in the limit of infinitely narrow spectral lines, that is, long T_2 and rectangular microwave pulses. Under realistic conditions, this is not the case. Performing more complex quantum information operations requires a certain precision of operations. Hence, it is useful to control the quality of gates. The result of such a gate tomography experiment is shown in Figure 10.14. Here a pulse has been applied to transition 1–2 for various initial conditions, that is, the system being in state 1, 3, 4 before start of the π pulse. For an ideal situation, one would expect the system to end up in state 2 when it started in 1 with the only non zero matrix element of the 4×4 density matrix describing the system to be $\rho 22$ after the pulse. However because of the finite dephasing time T_2 ($T_2 = 2\,\mu s$ in the present case) this is not the case here. Instead, the inversion of the spin is not complete and coherences (no diagonal elements) are created. The behavior can be reproduced when a full density matrix treatment of the pulse acting on the four level system is carried out, as shown on the right side of Figure 10.14.

10.8
Controlled Generation of Defect Centers

The nitrogen vacancy center in diamond is traditionally observed in radiation damaged, nitrogen rich diamond. The center is formed from the vacancy after annealing with temperatures larger than 600 °C The vacancy gets mobile and forms a stable NV complex under these conditions. Well controlled generation of NV color centers has been achieved by this approach [60]. Electron (400 keV) and

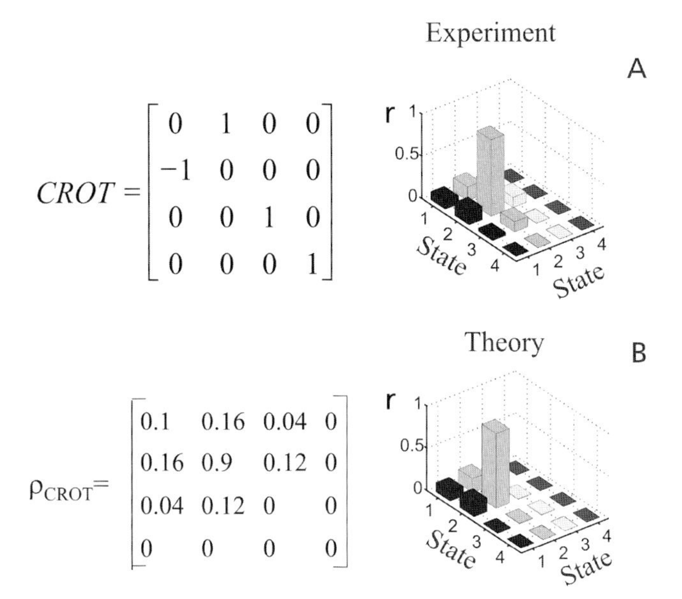

Figure 10.14 Schematic representation of the experimental and theoretical outcome of a CROT gate on a coupled singlet electron spin nuclear spin system. The upper left part shows the matrix representing the unitary transformation CROT. The other figure represents the density matrix of the two spin systems after the CROT gate, in a graphical representation. Part B represents a numerical calculation of the experimental results.

Ga (30 keV) ion beams were used to generate localized areas of NV centers in Ib diamond. For 30 keV Ga ions, the nominal penetration depth of ions inside the material is 15 nm. Patterns of NV centers have been generated with some ten thousand Ga ions used per irradiated dot. From the experiments, the diffusion constant of vacancies in diamond has been determined to be $D = 1.1\,\mathrm{nm}^2/\mathrm{s}$. The activation energy for vacancy diffusion is calculated to be 2.4 eV. Electrons with 400 keV penetrate some μm inside the diamond sample. At high irradiation doses, an increased generation of NV^0 centers was observed. For a localized generation of NV defects, these approaches do have the disadvantage of the relatively high diffusion constant of the vacancy, plus the large "natural" abundance of nitrogen. For generating NV defects with large spin dephasing times, for example, it would be preferable to implant defects into nitrogen free samples. This is possible by implanting nitrogen directly into type IIa diamonds. In a first attempt, 2 MeV nitrogen atoms have been implanted into type IIa diamond substrates [61] (Figure 10.15a). STRIM calculations suggest that the ions should be found 1 μm below the surface. The lateral scattering in the end position of the nitrogen (straggling) should be 0.5 μm. With a displacement energy of 55 eV for carbon and a density of $3.5\,\mathrm{g/cm}^3$ about 200 vacancies should be produced during a single nitrogen implantation. Indeed, the optical spectra of the implantation areas do show mostly fluorescence emission from neutral vacancies prior to annealing. After annealing,

(a)

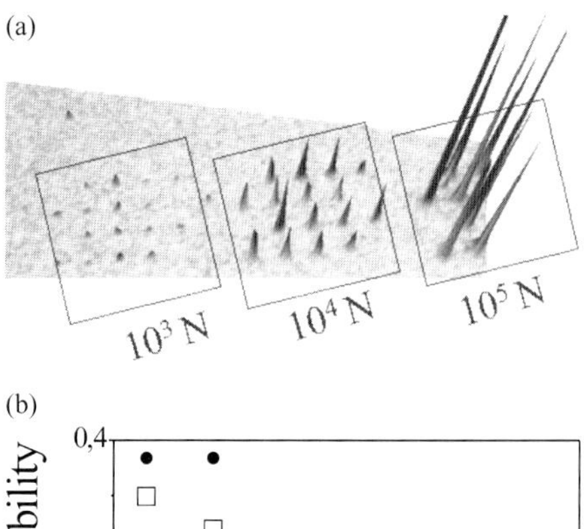

(b)

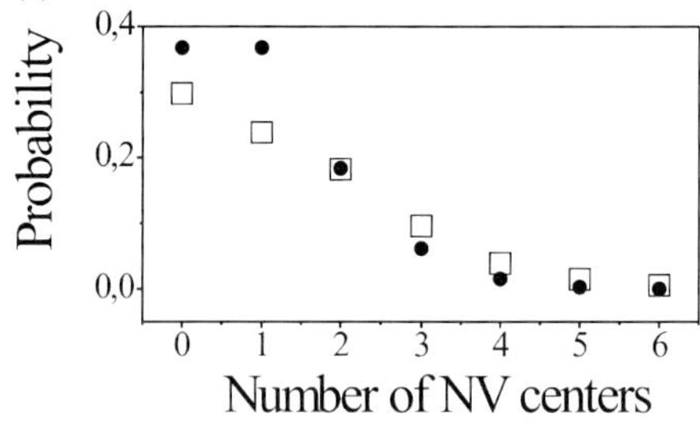

Figure 10.15 Fluorescence image of a type IIa diamond irradiated with a N^+ ion micro beam at different nitrogen dosages (a). (b) Probability to generate the given number of defects for a deposition of two nitrogen ions per spot. Open squares are measured data and filled circles are calculated values.

fluorescence from mostly NV centers is found. In these studies, the number of nitrogens implanted per spot has been decreased gradually. Single center emission was observed when, on average, two ions are implanted in a single spot (see Figure 10.15b). It is, however, difficult to ensure that the generated NV is made from the nitrogen implanted, and not from an abundant one. Even with a nitrogen concentration below 0.1 ppm, there would be 30 native nitrogen atoms in a spherical volume of 150 nm surrounding the end range where the Bragg peak in the stopping power creates the maximum concentration of vacancies. Hence, there is a considerable chance in Figure 10.15 that the NV fluorescence does not stem from an implanted nitrogen. To be able to separate implanted from impurity nitrogen, ^{15}N has been used as an implantation ion. ^{15}N has a nuclear spin angular momentum $I = 1/2$ with a characteristically different ODMR spectrum from ^{14}N (see Figure 10.10a) [62]. Experimentally 14 keV $^{15}N_2^+$ ions have been implanted. From a comparison of the number ^{15}NV defect centers to the number of implanted ^{15}N the efficiency of NV defect generation has been calculated. Under the conditions employed in the experiment, this efficiency was about 2.5%. It should be noted however, that a 14 keV implantation of N_2 results in a penetration depth of only a few nm. This close proximity to the surface might lead to a loss of vacancies, due to diffusion to the surface. In any case, the experiments demonstrate that single defects can be generated close to a diamond surface.

10.9 Fluorescent Defects in Diamond Nanocrystals

Due to its chemical inertness and biocompatibility, diamond is a material of choice in biosensing. Recent progresses in cell biology on the other hand have resulted in an increased demand for biolabels. Here, important factors are biochemical variability, biocompatibility and photostability. These last two requisites are ideally met in diamond. Diamond nanocrystals have to be used in order to allow for labeling in cell biological applications. Diamond nanocrystals can be made either by milling bulk diamond, or via detonation growth. Both forms are available commercially. While stable NV fluorescence has been demonstrated in milled diamond nanocrystals, no such observations have been published for explosion grown diamond, so far. Fluorescence from NV centers has been detected in diamond nanocrystals with sizes down to 20 nm. In a recent report, the biocompatibility of diamond nanocrystals was investigated [63]. Upon addition of diamond nanocrystals to the buffer medium of a cell culture, no deterioration in cell survival was detected. Most notably, the endocytosis of diamond nanocrystals in cells has been seen. However, it should be noted that an increased cell death rate was measured in other studies [64, 65].

10.10 Conclusions

The application of color centers in diamond will rely on progresses in quality control of diamond material, and in the precise generation of defects. In the first area, it will be important to fabricate diamond with controlled impurity content. This refers to paramagnetic impurities like nitrogen, but also to ^{13}C. In addition, it will be important to fabricate two- and three dimensional diamond structures with photoactive defect centers inside. One aim would be to build optical resonators out of diamond and to detect the coupling between the cavity mode and color centers implanted in the diamond. This would be similar to what has been shown for quantum dots. This not only requires structuring and material growth, but also controlled defect implantation. Currently, the precision of nitrogen implantation is limited by the quality of the ion beam optics and the ion source. A substantial improvement might involve implantation through metal nano apertures, as has been shown for the implantation of phosphorus in silicon. Here, lateral position accuracies in the order of 10 nm might be achieved using currently available technology. This might be sufficient for photonics applications, for example, or for the positioning of defects inside cavities or wave guides. However, for coupling of defects via their magnetic dipole interaction, a further improvement is necessary. Here, accuracies down to 2 nm might be necessary. Also, for charge transport based read out in structures like single electron transistors, a positioning accuracy on this order of magnitude might be essential. Extraction of nitrogen ions from traps is a promising approach. A recent proposal suggests that nitrogen ions are sympathetically cooled down to some mK in a radio frequency trap. Subsequently, they should be extracted from the trap and imaged onto a diamond sample. It is proposed that this would result in a sub-nm resolution. Achieving such accuracy in single defect center positioning would open the door to new device fabrication technology, which might be of fundamental importance to future quantum technology.

Acknowledgments

The work has been supported by the European Commission (via integrated project "Quantum applications"), ARO, DFG (via SFB/TR 21 and graduate college "Magnetische Resonanz"), and the Landestiftung BW (via the program "Atomoptik").

References

1 Shor, P.W. (1996) *Proceedings of the 37th Symposium on the Foundation of Computer Science*, IEEE Press, Los Alamitos, CA, pp. 56–65.

2 Nielsen, M.A. and Chung, I.L. (2000) *Quantum Computation and Quantum Information*, Cambridge University Press, Cambridge, UK.

3 Deutsch, D. (1985) Quantum-Theory, the Church–Turing Principle and the Universal Quantum Computer. *Proceedings of the Royal Society of London A: Mathematical, Physical and Engineering Sciences*, **400** (1818), 97–117.

4 Davies, G., Lawson, S.C. *et al.* (1992) Vacancy-related centres in diamond. *Physical Review B*, **46** (20), 13157–70.

5 Haffner, H., Hansel, W. *et al.* (2005) Scalable multiparticle entanglement of trapped ions. *Nature*, **438** (7068), 643–6.

6 Haffner, H., Schmidt-Kaler, F. *et al.* (2005) Robust entanglement. *Applied Physics B: Lasers and Optics*, **81** (2–3), 151–3.

7 Riebe, M., Haffner, H. *et al.* (2004) Deterministic quantum teleportation with atoms. *Nature*, **429** (6993), 734–7.

8 Barrett, M.D., Chiaverini, J. *et al.* (2004) Deterministic quantum teleportation of atomic qubits. *Nature*, **429** (6993), 737–9.

9 Gulde, S., Riebe, M. *et al.* (2003) Implementation of the Deutsch–Jozsa algorithm on an ion-trap quantum computer. *Nature*, **421** (6918), 48–50.

10 Wineland, D.J., Barrett, M. *et al.* (2003) Quantum information processing with trapped ions. *Philosophical Transactions of the Royal Society of London A: Mathematical, Physical and Engineering Sciences*, **361** (1808), 1349–61.

11 You, J.Q. and Nori, F. (2005) Superconducting circuits and quantum information. *Physics Today*, **58** (11), 42–7.

12 Li, X.Q., Wu, Y.W. *et al.* (2003) An all-optical quantum gate in a semiconductor quantum dot. *Science*, **301** (5634), 809–11.

13 Petta, J.R., Johnson, A.C. *et al.* (2005) Coherent manipulation of coupled electron spins in semiconductor quantum dots. *Science*, **309** (5744), 2180–4.

14 Koppens, F.H.L., Folk, J.A. *et al.* (2005) Control and detection of singlet–triplet mixing in a random nuclear field. *Science*, **309** (5739), 1346–50.

15 Hanson, R., van Beveren, L.H.W. *et al.* (2005) Single-shot readout of electron spin states in a quantum dot using spindependent tunnel rates. *Physical Review Letters*, **94** (19), 196802.

16 Kane, B.E. (2000) Silicon-based quantum computation. *Fortschritte der Physik – Progress of Physics*, **48** (9–11), 1023–41.

17 Davies, G. (1994) *Properties and Growth of Diamond*, EMIS Data Review Series,

INSPEC, The Institution of Electrical Engineers, London.
18 Davies, G. and Hamer, M.F. (1976) Optical Studies of 1.945 eV vibronic band in diamond. *Proceedings of the Royal Society of London A: Mathematical, Physical and Engineering Sciences*, **348** (1653), 285–98.
19 Loubser, J. and Vanwyk, J.A. (1978) Electron–spin resonance in study of diamond. *Reports on Progress in Physics*, **41** (8), 1201–48.
20 Mita, Y. (1996) Change of absorption spectra in type-Ib diamond with heavy neutron irradiation. *Physical Review B*, **53** (17), 11360–4.
21 Collins, A.T., Thomaz, M.F. *et al.* (1983) Luminescence decay time of the 1.945 eV centre in type Ib diamond. *Journal of Physics C: Solid State Physics*, **16** (11), 2177–81.
22 Harley, R.T., Henderson, M.J. *et al.* (1984) Persistent spectral hole burning of color-centres in diamond. *Journal of Physics C: Solid State Physics*, **17** (8), L233–6.
23 Redman, D., Brown, S. *et al.* (1992) Origin of persistent hole burning of N–V centres in diamond. *Journal of the Optical Society of America B: Optical Physics*, **9** (5), 768–74.
24 Lenef, A., Brown, S.W. *et al.* (1996) Electronic structure of the N–V centre in diamond: experiments. *Physical Review B*, **53** (20), 13427–40.
25 Lenef, A. and Rand, S.C. (1996) Electronic structure of the N–V centre in diamond: theory. *Physical Review B*, **53** (20), 13441–55.
26 Redman, D.A., Brown, S. *et al.* (1991) Spin dynamics and electronic states of N–V centres in diamond by EPR and 4-wave-mixing spectroscopy. *Physical Review Letters*, **67** (24), 3420–3.
27 Harrison, J., Sellars, M.J. *et al.* (2004) Optical spin polarisation of the N–V centre in diamond. *Journal of Luminescence*, **107** (1–4), 245–8.
28 Reddy, N.R.S., Manson, N.B. *et al.* (1987) Laser spectral hole burning in a color centre in diamond. *Journal of Luminescence*, **38** (1–6), 46–7.
29 Hemmer, P.R., Turukhin, A.V. *et al.* (2001) Raman-excited spin coherences in nitrogen-vacancy color centres in diamond. *Optics Letters*, **26** (6), 361–3.
30 Nizovtsev, A.P., Kilin, S.Y. *et al.* (2003) Spin-selective low temperature spectroscopy on single molecules with a triplet-triplet optical transition: application to the NV defect centre in diamond. *Optics and Spectroscopy*, **94** (6), 848–58.
31 Nizovtsev, A.P., Kilin, S.Y. *et al.* (2003) NV centres in diamond: spin-selective photokinetics, optical ground-state spin alignment and hole burning. *Physica B: Condensed Matter*, **340**, 106–10.
32 Zhang, L.S., Feng, X.M. *et al.* (2004) Coherent transient in dressed-state and transient spectra of Autler–Townes doublet. *Physical Review A*, **70** (6), 063404.
33 Windsor, A.S.M., Wei, C.J. *et al.* (1998) Experimental studies of a strongly driven Rabi transition. *Physical Review Letters*, **80** (14), 3045–8.
34 Holmstrom, S.A., Wei, C.J. *et al.* (1997) Spin echo at the Rabi frequency in solids. *Physical Review Letters*, **78** (2), 302–5.
35 Manson, N.B., Wei, C.J. *et al.* (1996) Response of a two-level system driven by two strong fields. *Physical Review Letters*, **76** (21), 3943–6.
36 Wei, C.J., Manson, N.B. *et al.* (1995) Dressed state nutation and dynamic Stark switching. *Physical Review Letters*, **74** (7), 1083–6.
37 Glasbeek, M. and Vanoort, E. (1991) Coherent transients of the N–V centre in diamond. *Radiation Effects and Defects in Solids*, **119**, 301–6.
38 Vanoort, E. and Glasbeek, M. (1990) Electric-field-induced modulation of spin echoes of N–V centres in diamond. *Chemical Physics Letters*, **168** (6), 529–32.
39 Vanoort, E., Manson, N.B. *et al.* (1988) Optically detected spin coherence of the diamond N–V centre in its triplet ground-state. *Journal of Physics C: Solid State Physics*, **21** (23), 4385–91.
40 Rand, S.C., Lenef, A. *et al.* (1994) Zeeman coherence and quantum beats in ultrafast photon-echoes of N–V centres in diamond. *Journal of Luminescence*, **60** (1), 739–41.
41 Redman, D., Shu, Q. *et al.* (1992) 2-Beam coupling by nitrogen-vacancy centres in diamond. *Optics Letters*, **17** (3), 175–7.
42 He, X.F., Manson, N.B. *et al.* (1993) Paramagnetic-resonance of photoexcited

N–V defects in diamond. 1. level anticrossing in the (3)a ground-state. *Physical Review B*, **47** (14), 8809–15.

43 He, X.F., Manson, N.B. *et al.* (1993) Paramagnetic-resonance of photoexcited N–V defects in diamond. 2. Hyperfine interaction with the N-14 nucleus. *Physical Review B*, **47** (14), 8816–22.

44 Wei, C.J. and Manson, N.B. (1999) Observation of electromagnetically induced transparency within an electron spin resonance transition. *Journal of Optics B: Quantum and Semiclassical Optics*, **1** (4), 464–8.

45 Isoya, J., Kanda, H. *et al.* (1990) Fourier-transform and continuous-wave EPR studies of nickel in synthetic diamond–site and spin multiplicity. *Physical Review A*, **41** (7), 3905–13.

46 Nadolinny, V.A., Yelisseyev, A.P. *et al.* (1999) A study of C-13 hyperfine structure in the EPR of nickel-nitrogencontaining centres in diamond and correlation with their optical properties. *Journal of Physics: Condensed Matter*, **11** (38), 7357–76.

47 Gaebel, T., Popa, I. *et al.* (2004) Stable single-photon source in the near infrared. *New Journal of Physics*, **6**, 98.

48 Orrit, M. (2002) Single-molecule spectroscopy: The road ahead. *Journal of Chemical Physics*, **117** (24), 10938–46.

49 Gruber, A., Drabenstedt, A. *et al.* (1997) Scanning confocal optical microscopy and magnetic resonance on single defect centres. *Science*, **276** (5321), 2012–14.

50 Kurtsiefer, C., Mayer, S. *et al.* (2000) Stable solid-state source of single photons. *Physical Review Letters*, **85** (2), 290–3.

51 Beveratos, A., Brouri, R. *et al.* (2001) Nonclassical radiation from diamond nanocrystals. *Physical Review A*, **64** (6), 061802.

52 Brouri, R., Beveratos, A. *et al.* (2000) Photon antibunching in the fluorescence of individual color centres in diamond. *Optics Letters*, **25** (17), 1294–6.

53 Alleaume, R., Treussart, F. *et al.* (2004) Experimental open-air quantum key distribution with a single-photon source. *New Journal of Physics*, **6**, 92.

54 Mainwood, A. and Stoneham, A.M. (1997) *The Vacancy (V – 0, V+, V–) in Diamond: The Challenge of the Excited States and the GR2–GR8 Lines*. Proceedings of the 13th International Conference on Defects in Insulating Materials – ICDIM 96, Vol. 239-2.

55 Gaebel, T., Wittmann, C., Popa, I., Jelezko, F., Rabeau, J., Greentree, A., Prawer, S., Trajkov, E., Hemmer, P.R. and Wrachtrup, J. (2005) Photochromism in single nitrogen-vacancy defects in diamond. *Applied Physics B: Lasers and Optics*, DOI: 10.1007/s00340-005-2056-2.

56 Popa, I., Gaebel, T. *et al.* (2004) Energy levels and decoherence properties of single electron and nuclear spins in a defect centre in diamond. *Physical Review B*, **70** (20), 201203.

57 Jelezko, F., Gaebel, T. *et al.* (2004) Observation of coherent oscillation of a single nuclear spin and realization of a two-qubit conditional quantum gate. *Physical Review Letters*, **93** (13), 130501.

58 Kennedy, T.A., Colton, J.S. *et al.* (2003) Long coherence times at 300 K for nitrogen-vacancy centre spins in diamond grown by chemical vapor deposition. *Applied Physics Letters*, **83** (20), 4190–2.

59 Jelezko, F., Gaebel, T. *et al.* (2004) Observation of coherent oscillations in a single electron spin. *Physical Review Letters*, **92** (7), 076401.

60 Martin, J., Wannemacher, R., Teichert, J., Bischoff, L. and Kohler, B. (1996) Generation and detection of fluorescent color centres in diamond with submicron resolution. *Applied Physics Letters*, **20**, 3096–8.

61 Meijer, J., Burchard, B., Domhan, M., Wittmann, C., Gaebel, T., Popa, I., Jelezko, F. and Wrachtrup, J. (2006) Generation of Single Colour Centres by Focussed Nitrogen Implantation, condmat/0505063.

62 Rabeau, J.R.R.P., Tamanyan, G., Jamieson, D.N., Prawer, S., Jelezko, F., Gaebel, T., Popa, I., Domhan, M. and Wrachtrup, J. (2006) Implantation of Labelled Single Nitrogen Vacancy Defect Centre in Diamond Using 15N, condmat/0511722.

63 Yu, S.-J., Kang, M.-W., Chang, H.-C., Chen, K.-M. and Yu, Y.-C. (2005) Bright fluorescent nanodiamonds: no photobleaching and low cytotoxicity. *Journal of the American Chemical Society*, **127**, 17604–5.

64 Puzyr, A.P., Neshumayev, D.A. *et al.* (2004) Destruction of human blood cells in interaction with detonation nanodiamonds in experiments in vitro. *Diamond and Related Materials*, **13** (11/12), 2020–3.

65 Bondar, V.S. and Puzyr, A.P. (2004) Nanodiamonds for biological investigations. *Physics of the Solid State*, **46** (4), 716–19.

11
Bose–Einstein Statistical Properties of Dense Exciton Gases in Diamond

Hideyo Okushi,[1,3] *Hideyuki Watanabe,*[2,3] *Satoshi Yamasaki*[1,3] *and Shoukichi Kanno*[1,3]
[1]Institute of Nanotechnology, AIST (National Institute of Advanced Industrial Science and Technology), Tsukuba, Ibaraki 305-8568, Japan
[2]Diamond Research Center, AIST, Tsukuba, Ibaraki 305-8568, Japan
[3]CREST (Core Research for Evolutional Science and Technology) /JST (Japan Science and Technology Corporation), c/o AIST, Tsukuba, Ibaraki 305-8568, Japan

11.1
Introduction

Excitons in diamond have a large binding energy (80 meV) with a small Bohr radius (1.57 nm) because of the relatively low dielectric constant [1, 2]. Therefore, a dense exciton gas in which Bose–EinsteinBose–Einstein (BE) statistical properties appear can be generated even at high temperature. Indeed, recent cathodoluminescence (CL) spectra of free excitonic emission from high quality homoepitaxial chemical vapor deposition (CVD) diamond films indicate that BE statistical properties appear in case of dense exciton gases at thermodynamic quasi-equilibrium [3–5]. From these results, it seems that diamond is a perfect material for the generation and detection of Bose–Einstein condensation (BEC), [6] which is one of the most characteristic features of the BE statistics [7–9].

In this chapter, recent studies on excitonic emission from dense exciton gases in CVD diamond films are described and the possibility of exciton BEC in diamond is discussed: Section 11.2 gives a description about the emission properties of free exciton from dense exciton gases in CVD diamond obtained by CL experiments, followed by a discussion of the general characteristics of excitons in diamond. Section 11.3 firstly gives a detailed description of lineshape analysis based on BE statistics of indirect semiconductors and taking into account collisions of excitons. Second, results of lineshape analysis of observed spectra and their validity are described, and discussed in terms of the BE statistical characteristics. In Section 11.4, the possibility of exciton BEC in diamond is discussed, and we demonstrate that the chemical potential of an exciton gas approaches zero ($\mu = 0$) at around 38 K gas temperature. Please note that $\mu = 0$ is a striking indication of BEC [7]. Finally, a summary is given in Section 11.5.

Physics and Applications of CVD Diamond. Satoshi Koizumi, Christoph Nebel, and Milos Nesladek

ISBN: 978-3-527-40801-6

11.2
Lineshape of Excitonic Emission from Diamond

11.2.1
Excitonic Emission Spectra from High Quality CVD Diamond Films

Figure 11.1 shows a typical near band edge CL spectrum from an undoped CVD diamond film at the sample holder temperature of $T_{ob} = 5.7$ K excited by an electron beam current of $I_{beam} = 30$ μA at an acceleration voltage of $E_{acc} = 15$ kV with a beam diameter of 100 μm [3, 4, 10]. Four emission peaks are observed. The spectrum consists of a prominent line and three relatively weaker lines. These are intrinsic features of diamond arising from the recombination of free excitons with phonons. The emission lines, labeled as FE^{TA}, FE^{TO} and FE^{LO}, are specified by the emission of (i) transverse acoustic (TA) phonon, (ii) transverse optical (TO) phonon, and (iii) longitudinal optical (LO) phonon, which have energies of 87 ± 2 meV, 141 ± 1 meV and 163 ± 1 meV, respectively [1]. The line labeled $FE^{TO+O\Gamma}$ represents the recombination of free excitons with a second excited transverse optical phonon.

The interpretation of the features in these spectra was first discussed by Dean *et al.* [1] based on indirect exciton recombination processes that occur in diamond, as shown in Figure 11.1b, where the dispersion curves for the optical and acoustic phonons are superimposed on the band structure of diamond. The conduction band in diamond has six equivalent minima along the [100] axis of the Brillouin zone, located at a wave vector $k = 0.76\ (2\pi/a)$ (symmetry Δ_1), where a is the lattice constant (3.567 Å), while the valence band has its maximum at $k = 0$ (symmetry Γ_{25}). The valence band at $k = 0$ is generally split by spin–orbit interaction. The

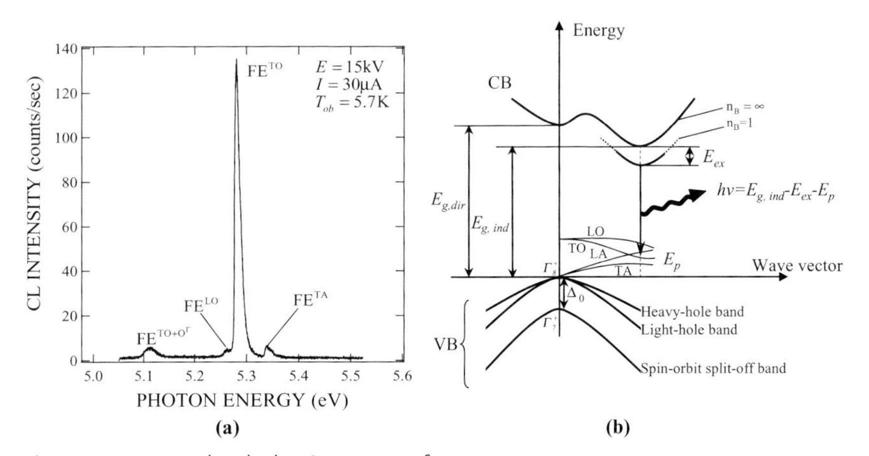

Figure 11.1 (a) Near band edge CL spectrum from an undoped homoepitaxial CVD diamond film at $T_{ob} = 5.7$ K under $I_{beam} = 30$ μA at $E_{acc} = 15$ kV. (b) Schematic diagram of exciton recombination *vs.* band gap for diamond. Δ_0 is the spin–orbit split energy and n_B is the principal quantum number.

spin–orbit interaction in diamond results in splitting into the Γ_8^+ and Γ_7^+ levels with splitting energy Δ_0 as shown in Figure 11.1b, where $\Delta_0 = 6 \pm 1$ meV from cyclotron resonance measurements [11] and $\Delta_0 = 13$ meV from *ab initio* band structure calculations (N. Orita, private communication). Further, it is known that the Γ_8^+ level consists of two bands: heavy and light hole bands, as shown in the figure.

As shown in Figure 11.1a, no emission other than these intrinsic emissions can be detected, indicating that this CVD diamond sample is of high quality. Recently, Collins *et al.* found that both FE^{TO} and FE^{TA} features consist of at least four components at 15 K for HPHT synthetic IIa diamond [12, 13]. Regarding the subcomponents or finestructures in the FE^{TO} line, Sauer *et al.* also reported fine structures in the FE^{TO} line in their CL spectrum of HPHT synthetic IIa diamond [14, 15]. At present, the origin of this fine structure is not clear. However, such fine structures were not observed in homoepitaxial CVD diamond films under the same experimental conditions, as reported by Collins *et al.* We assume that the appearance of fine structures in the lineshape of excitonic emission depends on the quality of diamond samples.

11.2.2
Temperature Dependence of Emission Spectra from a Dense Exciton Gas

Figure 11.2a shows T_{ob} dependence of the band edge emission CL spectra of FE^{TO} + FE^{LO} lines from an undoped homoepitaxial CVD diamond film under the excitation conditions of $I_{beam} = 15$ μA at $E_{acc} = 15$ kV with a beam diameter of 100 μm.

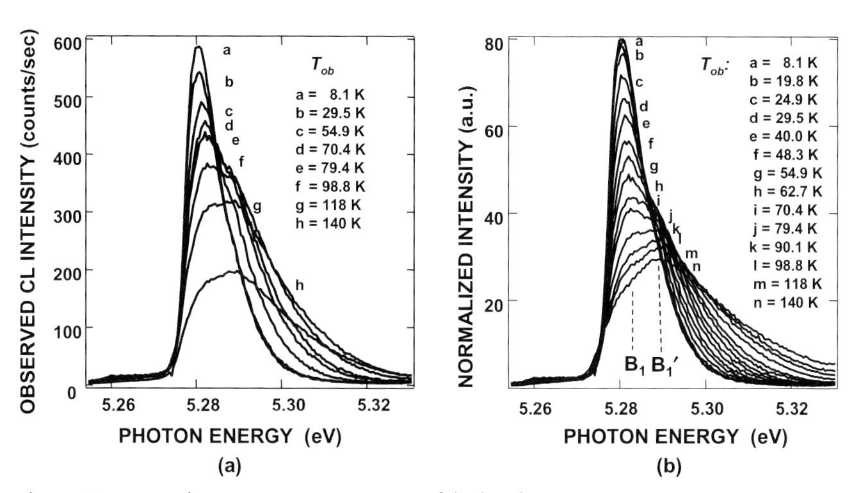

Figure 11.2 Dependence on temperature (T_{ob}) of the band edge emission CL spectra due to FE^{TO} + FE^{LO} lines observed in the high quality homoepitaxial CVD diamond film under $I_{beam} = 15$ μA at $E_{acc} = 15$ kV. (a) Observed emission spectra $I(\omega)$. (b) Normalized emission spectra $\tilde{I}(\omega)$.

These excitation parameters were chosen in order to realize a dense exciton gas in which the BE statistical properties may occur.

I_{beam} = 15 μA was selected by the following considerations: a dense exciton gas whose density is above 10^{18} cm^{-3} can be realized by I_{beam} = 15 μA at E_{acc} = 15 kV with a beam diameter of 100 μm if the exciton lifetime τ_{ex} is given by the radiative exciton lifetime τ_{rad}, which is reported to be 2–3 × 10^{-6} s [16]. Generally, τ_{ex} in diamond is reported to be ≈10^{-8} s, which is determined by transient photoluminescence experiments [13, 16, 17]. However, in the case of CL measurements which are excited by a continuous high beam current, it is expected that $\tau_{ex} \approx \tau_{rad}$ as described in detail in Section 11.3.5. The exciton density n_{ex} above 10^{18} cm^{-3} is in the same order of magnitude as the critical density n_c where BEC in diamond can be expected, namely at exciton gas temperatures T_{ex} around 30 K [3, 4].

E_{acc} = 15 kV was selected so that the penetration depth of the electron beam becomes about 1.8 μm, taking into account the sample thickness of 2.8 μm. A beam spot diameter of 100 μm was selected to minimize the effect of diffusion of excitons. In the case of a much smaller diameter of the spot than the exciton diffusion length, the area of excitonic emission on the sample surface becomes much larger than the beam irradiation area, so that it is difficult to correctly estimate the excitonic emission area of the sample surface.

Figure 11.2b shows the normalized emission spectra of the observed spectra shown in Figure 11.2a. If the detected spectra are denoted as $I(\omega)$, the normalized emission spectra are defined as (see Equation 11.14 in Section 11.3.1 and Ref. [5])

$$\tilde{I}(\omega) = I(\omega) \Big/ \int_{-\infty}^{\infty} d\omega' I(\omega') \tag{11.1}$$

The shape of $I(\omega)$ is not affected by experimental conditions, and it can be characterized by the BE distribution function (see Equation 11.4 in Section 11.3.1) which is a function of α, T_{ex} and ε_p, where $\alpha = -\mu/k_B T_{ex}$ and the chemical potential μ is measured with respect to the total exciton energy at $\varepsilon_p = 0$.

As shown in Figure 11.2b, the lineshape of $\tilde{I}(\omega)$ rapidly increases with decreasing T_{ob} and the spectra width becomes sharper. At T_{ob} = 8.1 K, the peak position of the FE^{TO} line is located at 5.285 eV (λ = 234.6 nm) and its full width at half maximum (FWHM) is less than 10 meV. At high temperatures, the FWHM of the spectra becomes larger than 20 meV, indicating that the FE^{TO} line consists of two subcomponents B_1 and B_1' as is indicated in Figure 11.2b.

The FWHM of 10 meV at T_{ob} = 8.1 K is larger than the FWHM of 3 meV observed in high pressure high temperature (HPHT) synthetic diamond at T_{ob} = 15 K [12, 13]. The large FWHM in the CVD diamond sample originates from thermal broadening, due to the increased temperature in the emitting region, and the exciton gas temperature (T_{ex}). In addition, it also originates from collision related broadening, due to the exciton–exciton scattering mechanism expected for $n_{ex} \geq 10^{18}$ cm^{-3}. This will be also described in detail, in Section 11.3.3.

As observed at higher temperatures, above T_{ob} = 60 K, the spectrum consists of two subcomponents. As shown in Figure 11.2b, the FE^{TO} line is composed of two subcomponents labeled as B_1 and B_1'. The emission intensity ratio B_1 and B_1' decreases with increasing T_{ob}. This result was originally reported by Dean *et al.* in 1965, when the FE^{TO} line with two subcomponents of B_1 and B_1' was observed on natural *p*-type semiconducting diamond (type IIb), by CL measurement that used a 20-μA beam of 60-keV electrons to excite excitonic emission at a temperature T_{ob} of about 100 K [1]. Dean *et al.* proposed that both of the two subcomponents can be explained by theoretical calculations using Maxwell–Boltzmann (MB) statistics, resulting in good agreement between experimental and theoretical data [1].

11.2.3
Origin of the Excitonic Emission Line Splitting

Concerning the origin of the two subcomponents, Dean *et al.* pointed out that this originates from valence band splitting due to spin–orbit interaction [1] based on the experimental results for cyclotron resonance measurement [11]. The origin of splitting, however, involves a model based on electron–hole (e–h) exchange interaction (spin singlet–triplet splitting) that has recently been proposed by Sauer *et al.* [14]. They analyzed the high resolution CL spectra of free and bound excitons from the HPHT synthetic IIa diamond and found that the emission of both free and bound excitons has a common splitting structure at about 11 meV. From the results, they concluded the two subcomponents originate from the splitting of exciton states by the e–h exchange interaction [14].

The origin of splitting of the exciton state is very important for understanding the properties of the exciton gases, in particular, of the dense exciton gases which show the BE statistical properties. If the splitting of the exciton spin states occurs, the exciton spin multiplicity g_{ex} of the exciton states is altered and g_{ex} is directly related to n_{ex} in the BE statistics, as will be shown in Equation 11.16 in Section 11.3.1. For example, the critical density n_c of the excitons undergoing BEC at T_{ex} is given by [7],

$$n_c = 2.612 g_{ex} (m k_B T_{ex} / 2\pi\hbar^2)^{3/2}, \tag{11.2}$$

where g_{ex} is the exciton spin multiplicity of the ground states.

As a principle, the splitting of the exciton states by the e–h exchange interaction should be taken into account in estimating g_{ex}. One exciton state composed of one e–h pair is degenerated fourfold in the spin states and separated into two kinds of spin states by the e–h exchange interaction. Since the level of multiplicity of the spin sates is given by $g_{ex} = 2S + 1$, where S is the magnitude of the exciton spin, the e–h exchange interaction produces the singlet ($S = 0$; para-exciton) and the triplet ($S = 1$; ortho-exciton) states, which are separated by the e–h exchange energy. It is known that the singlet state is usually lower than the triplet state in semiconductors [18]. The exchange energy in diamond has been theoretical calculated by Cardona *et al.* to be about 3.9 meV [19].

Since the spin singlet and triplet states related to the same exciton mass, the e–h exchange interaction model, namely the spin singlet–triplet splitting model is reasonably accepted as the origin of splitting of B_1 and B_1'. In this case, the exciton ground state (B_1 in FE^{TO} and C_1 in FE^{LO}) and the first excited state (B_1' in FE^{TO} and C_1' in FE^{LO}) are respectively associated with the spin singlet state and the spin triplet state. Furthermore, it is understood that $g_{ex} = 6$ for B_1 or C_1 and $g_{ex} = 18$ for B_1' or C_1', since the exciton states of the diamond are composed of six electrons in the sixfold degenerated conduction band minima, and one hole in the heavy hole band at the Γ_8^+ level.

On the other hand, if we take the model based on the spin–orbit interaction as the origin of splitting of B_1 and B_1', the exciton state related to B_1 is composed of the electrons in the sixfold degenerated conduction band minima and the holes in the Γ_8^+ level, while B_1' is of holes the Γ_7^+ level. The mass of excitons related to B_1 is different from those to B_1' because the effective masses of holes at the band edge at the Γ_8^+ and Γ_7^+ levels are generally different from each other [20]. In this case, the e–h exchange interaction should be negligible, and the exciton states remain degenerated, in order to explain the experimental results. Therefore, in the case of the spin–orbit interaction model, g_{ex} for both B_1 and B_1' lines are given by $g_{ex} = 24$ if the light holes in the Γ_8^+ and Γ_7^+ levels do not contribute to the emission lines, and $g_{ex} = 48$ if they contribute.

Thus, there is an essential difference in g_{ex} between these two models for the origin of splitting for B_1 and B_1'. This difference is important in judging whether n_{ex} can reach the BE statistical regime or not, because n_{ex} in the BE statistics is strongly related to g_{ex} (as already mentioned above). As will be described in the following sections, the spin singlet–triplet splitting model is applied to construct a theoretical excitonic emission spectrum. The validity of this model has been confirmed by the results of curve fitting between calculated and observed emission lines [4].

11.3
Lineshape Analysis Based on Bose–Einstein Statistics

11.3.1
Theoretical Excitonic Emission Spectra from Diamond

Theoretically excitonic emission spectra from diamond are considered based on taking into account collision of excitons in an indirect semiconductor [3–5]. The light emission spectrum $I(\omega)$, in the case of recombination of excitons which have momentum and energy $(p, \varepsilon_p + \varepsilon_x)$, whereby a photon $(q, c|q|)$ (c is the speed of light and the refractive index is put bunity for simplicity) and a phonon $(p - q)$ are emitted, is generally given by

$$I(\omega) = D(\omega)\frac{V}{(2\pi)^3}\int d^3pS(\omega + \omega_p^{(ph)}, p)|T_p|^2 n_p \tag{11.3}$$

where ε_x is the minimum exciton energy, c is the velocity of light, $D(\omega)$ is the density of photon energy per unit volume and can be replaced by constant $D(\omega)$

for variable ω, V is the emission volume, $S(\omega, p)$ is the spectral function that is assumed to be Lorentzian, and $|T_p|^2$ is the transition matrix element of emission. Density $n_{ex}(\varepsilon_p)$ of excitons having kinetic energy ε_p which is given by the Bose–Einstein distribution function

$$n_{ex}(\varepsilon_p) = \frac{1}{\exp(\varepsilon_p/k_B T_{ex} + \alpha) - 1} \tag{11.4}$$

where $\alpha = -\mu/k_B T_{ex}$ and the chemical potential μ is measured with respect to the total exciton energy at $\varepsilon_p = 0$. $S(\omega, p)$ in Equation 11.3 is given as

$$S(\omega, p) = \frac{2\gamma}{(\omega - \varepsilon_p - \varepsilon_x)^2 + \gamma^2} \tag{11.5}$$

where γ is the inverse of the exciton collision time. Equation 11.3 can be solved numerically by expanding, ε_p, $\omega_p^{(ph)}$, and $|T_p|^2$ are at the momentum p_0 where the exciton energy is the lowest, as follows:

$$\varepsilon_p = \frac{(p - p_0)^2}{2m} + D_0 \frac{a}{2m} (e \cdot (p - p_0))^3 \tag{11.6}$$

$$\omega_p^{(ph)} = \omega_{p0}^{(ph)} + v \cdot (p - p_0) \tag{11.7}$$

$$|T_p|^2 = |T_{p0}|^2 (1 + C_{10} a e \cdot (p - p_0) + C_{20} a^2 (p - p_0)^2) \tag{11.8}$$

where e is the unit vector of p_0 direction, $|v|$ is the velocity of phonons at momentum p_0, D_0, C_{10} and C_{20} are numerical constants whose order is 1, and a is the lattice constant. Please note that C_0, D_0 and v have only to be considered in the case of an indirect semiconductor. These equations are expressed by a unit of $\hbar = 1$.

Using Equations 11.2–11.7, the complex function of $I(\omega)$ given by Equation 11.3 can be transformed into the summation of elementary functions

$$I(\omega) = D(\omega_0) |T_p|^2 \frac{2\pi g V}{\lambda^3} i(\omega) \tag{11.9}$$

where

$$\lambda = \sqrt{2\pi/m k_B T_{ex}} \tag{11.10}$$

$$i(\omega) = i_0(\omega) + C_1 i_1(\omega) + C_2 i_2(\omega) \tag{11.11}$$

$$C_1 = C_{10} m a v \tag{11.12}$$

$$C_2 = C_{20} 2 m a^2 k_B T_{ex} \tag{11.13}$$

Finally, the normalized spectra $\tilde{I}(\omega)$ is given in the following form [5]:

$$\tilde{I}(\omega) = \frac{i(\omega)}{k_B T_{ex}(g_{3/2}(z) + 3/2C_2 g_{5/2}(z))} \tag{11.14}$$

with $z = e^{-\alpha}$, where, for $k = 3/2$ and $k = 5/2$,

$$g_k(z) = \int_0^\infty \frac{x^{2k-1}}{z^{-1}e^{x^2} - 1} dx \Big/ \int_0^\infty x^{2k-1} e^{-x^2} dx = \sum_{n=1}^\infty \frac{z^n}{n^k} \tag{11.15}$$

It should be noted that $\tilde{I}(\omega)$ is independent of both the average transition matrix element of emission $|T_{P0}|^2$ and the emission volume V, and depends only three variables: α, T_{ex} and ε_p [4]. If α and T_{ex} are known, n_{ex} can be calculated by using the following equation:

$$\begin{aligned} n_{ex} &= \frac{g_{ex}}{4\pi^2}\left(\frac{2m}{\hbar^2}\right)\int_0^\infty \frac{E^{1/2}}{e^{E/k_B T_{ex} + \alpha} - 1} dE \\ &= g_{ex}(mK_B T_{ex}/2\pi\hbar^2)^{3/2} \sum_{K=1}^\infty \frac{(e^{-\alpha})^K}{k^{3/2}} = \frac{n_c}{2.612} \sum_{K=1}^\infty \frac{(e^{-\alpha})^K}{k^{3/2}} \end{aligned} \tag{11.16}$$

where n_c is the critical density of the BEC phase boundary at T_{ex} given by Equation 11.2.

If we set

$$n_{ex0}(\alpha, T_{ex}) = (mk_B T_{ex}/2\pi\hbar^2)^{3/2} \sum_{K=1}^\infty \frac{(e^{-\alpha})^K}{k^{3/2}} \tag{11.17}$$

the exciton density for the ground state (n_{exG}) and the excited state (n_{exE}) are given as follows:

$$n_{exG} = g_{ex} n_{ex0}(\alpha, T_{ex}) = 6n_{ex0}(\alpha, T_{ex}) \tag{11.18}$$

$$n_{exE} = g_{ex} n_{ex0}(\alpha + \Delta E_s/k_B T_{ex}, T_{ex}) = 18n_{ex0}(\alpha + \Delta E_s/k_B T_{ex}, T_{ex}) \tag{11.19}$$

if the spin singlet–triplet splitting model are assumed [4].

11.3.2
Calculation of Theoretical Exciton Emission Spectra

Based on the theoretically derived Equation 11.14, theoretical emission spectra from dense exciton gases in diamond were calculated, which can be used to fit to experimentally observed spectra, such as the results shown in Figure 11.2. The following basic assumptions are used to calculate the spectra ($\tilde{I}(\omega)$):

1. The observed emission spectra are composed of two emission lines of FE^{TO} and FE^{TO}. The total spectrum is given by the sum of both. The fine structures

for free exciton emission, as observed by previous investigations [12–15], are not taken into account.

2. The FE^{TO} and FE^{LO} lines consist of two subcomponents B_1 and B_1' for FE^{TO} and C_1 and C_1' for FE^{LO}, respectively. For the two subcomponents, the spin singlet–triplet splitting model is applied, so that the exciton ground state (B_1 in FE^{TO} and C_1 in FE^{LO}) and the excited state (B_1' in FE^{TO} and C_1' in FE^{LO}) have the same masses, and are associated with the spin singlet state ($S = 0$ para-exciton) and the spin triplet state ($S = 1$ ortho-exciton).
3. The spectra represent at thermodynamic quasi-equilibrium (chemical equilibrium), so that the effective chemical potentials for the ground state of B_1 (or C_1), and the excited state of B_1' (or C_1'), are given by α and $\alpha + \Delta E_s/k_B T_{ex}$, respectively, where $\alpha = -\mu/k_B T_e$, μ is the chemical potential of exciton gas, ΔE_s is the splitting energy, and T_{ex} is the exciton gas temperature.
4. $\tilde{I}(\omega)$ can be determined by three variables, namely: α, T_{ex} and ε_p, where ε_p is the energy of the ground state of free exciton with TO phonons.

Therefore, based on these assumptions, the following function to calculate $\tilde{I}(\omega)$ is used:

$$\begin{aligned}\tilde{I}(\omega) &= [I_{B_1}(\alpha, T_{ex}, \varepsilon_p) + R_{TO} I_{B_1'}(\alpha + \Delta E_s/k_B T_{ex}, T_{ex}, \varepsilon_p + \Delta E_s)] \\ &+ R_{TL}[I_{C_1}(\alpha, T_{ex}, \varepsilon_p - \Delta E_{TL}) + R_{LO} I_{C_1'}(\alpha + \Delta E_s/k_B T_{ex}, T_{ex}, \varepsilon_p - \Delta E_{TL} + \Delta E_s)]\end{aligned} \tag{11.20}$$

where I_{B1}, I_{B1}', I_{C1} and I_{C1}' are the respective emission lines of B_1, B_1', C_1, and C_1' and R_{TO}, R_{LO} and R_{TL} are the respective emission intensity ratios between B_1 and B_1', C_1 and C_1', and FE^{TO} and FE^{LO}. ΔE_{TL} is the energy difference between the emission peaks FE^{TO} and FE^{LO}.

Besides the basic assumptions and the parameters appearing in Equation 11.20, the following parameters are used for calculating the excitonic emission spectra in diamond: (i) the energy related to the phonon velocity, v, (ii) the inverse of the collision time, γ, (iii) the coefficients of the momentum term for the transition matrix element of emission, C_1 and C_2, (iv) the coefficient of the third order energy term, D_0.

The excitonic emission of an indirect semiconductor arises at low energy corresponding to the phonon velocity [5]. According to the reported phonon-dispersion relationship in diamond, the energies corresponding to the velocity of phonons at $\boldsymbol{k}_c = 0.76(2\pi/a)$, which is defined by Equation 11.6, for TO and LO phonons, E_{vTO} and E_{vLO} can be calculated to be 0.138 meV and 0.095 meV, respectively.

The parameter of inverse collision time (γ) is an important factor as it significantly affects the lineshape [5]. In the case of the dense exciton gases larger than $n_{ex} = 10^{17-18}\ cm^{-3}$ at thermodynamic quasi-equilibrium, it is reasonable to assume that exciton–exciton scattering dominates over exciton–phonon scattering or exciton–exciton Auger recombination [7–9] γ, for exciton–exciton scattering is given by

$$\gamma = \frac{\hbar}{\tau_{coll}} = \sigma n_{ex} v_{av} \tag{11.21}$$

where τ_{coll} is the collision time, σ is the exciton–exciton scattering cross section, and v_{av} is the average thermal velocity of excitons. If the excitons coexist in both the ground and excited states, different types of collisions should be taken into account for excitons in the ground and excited states, respectively; these are (i) collisions between ground-state excitons, (ii) collisions between ground and excited-state excitons and (iii) collisions between excited-state excitons. In this case, γ is more complex than Equation 11.21, since the scattering cross sections for the ground and excited states are different from each other [4]. According to Ref. [4], the inverse of the collision times for ground (γ_{G}) and exited states (γ_{E}) are given as follows, by assuming that the cross section of collision between ground state excitons is given as $\sigma^0 = \pi(a_{\mathrm{ex}})^2$, where $a_{\mathrm{ex}} = 1.57$ nm is the Bohr radius of excitons in diamond:

$$\gamma_G = 0.49 \times 10^{-19}[9n_{ex0}(\alpha, T_{ex}) + 18n_{ex0}(\alpha + \Delta E_s/k_B T_{ex}, T_{ex})]\sqrt{140/T_{ex}} \cdot k_B T_{ex} \tag{11.22}$$

$$\gamma_E = 0.49 \times 10^{-19}[6n_{ex0}(\alpha, T_{ex}) + 21n_{ex0}(\alpha + \Delta E_s/k_B T_{ex}, T_{ex})]\sqrt{140/T_{ex}} \cdot k_B T_{ex} \tag{11.23}$$

The coefficients for the momentum term of the transition matrix element of emission, C_1 given by Equation 11.12 and C_2 given by Equation 11.13, and the coefficient of the third order energy term D_0 given by Equation 11.5 are also used as calculation parameters of the spectra. Please note that these parameters, except α, T_{ex} and E_{p}, were firstly determined by the curve fitting on the observed spectra over a wide temperature range, such as the results shown in Figure 11.2. Detailed procedures for the determination of these parameters are described in Ref. [4]. The results are shown in Table 11.1. Then, we can calculate the theoretical $\tilde{I}(\omega)$ by using only three variable parameters of α, T_{ex} and ε_{p} and the parameters listed in Table 11.1 [4].

Table 11.1 The parameters determined by the curve fitting.

Parameter		Parameter	
$E_{v\mathrm{TO}}$	0.138 meV	R_{TL}	0.015
$E_{v\mathrm{LO}}$	0.095 meV	γ	0.25
ΔE_{s}	9.02 meV	C_1	0
R_{TO}	0.5	C_2	$-3k_{\mathrm{B}}T_{\mathrm{ex}}$
R_{LO}	0.5	D_0	0
ΔE_{TL}	20.4 meV		

11.3.3
α-Dependence of Theoretical Emission Spectra

Figure 11.3 shows α-dependence of theoretical $\tilde{I}(\omega)$ for the FE^{TO} + FE^{LO} lines at T_{ex} = 38 K and ε_p = 5.285 eV, where the vertical axis of right side figure is scaled by a linear unit, and that of left side by a logarithmic unit, in order to understand the lineshape of $\tilde{I}(\omega)$ in detail. At T_{ex} = 38 K, α = 0, 0.1, 0.5, 1 and 2 corresponds to $n_{ex} = 6.8 \times 10^{18}$ cm^{-3}, 4.0×10^{18} cm^{-3}, 2.0×10^{18} cm^{-3}, 1.0×10^{18} cm^{-3}, and 3.5×10^{17} cm^{-3}, respectively. As shown in the figure, the FWHM of spectrum increases as α approaches 0. It seems that this result has an opposite trend, in comparison with the expected trend where the lineshape becomes shaper as α approach to 0. Please note that the spectrum of α = 0 in Figure 11.3 represents the emission from normal excitons, and not from condense excitons which would show a δ-function like line at ε_p.

The result of Figure 11.3 indicates that τ_{coll} becomes shorter as an increase of n_{ex} and then, the FWHM increases by the uncertainty principle, that is, *a collisional broadening* occurs [4]. In actuality, the estimated collision time at α = 0–0.5 and T_{ex} = 38 K by using Equations 11.22 and 11.23 and $\sigma^0 = \pi(a_{ex})^2$ becomes less than 10^{-13} s, which gives a larger FWHM than 12 meV by the uncertainty principle. On the other hand, in the case of large α values larger than $\alpha \geq 2$ ($n_{ex} \leq 3.6 \times 10^{17}$ cm^{-3}), the effective collisional broadening does not appear in the spectra, and their lineshapes can be described by the MB statistics.

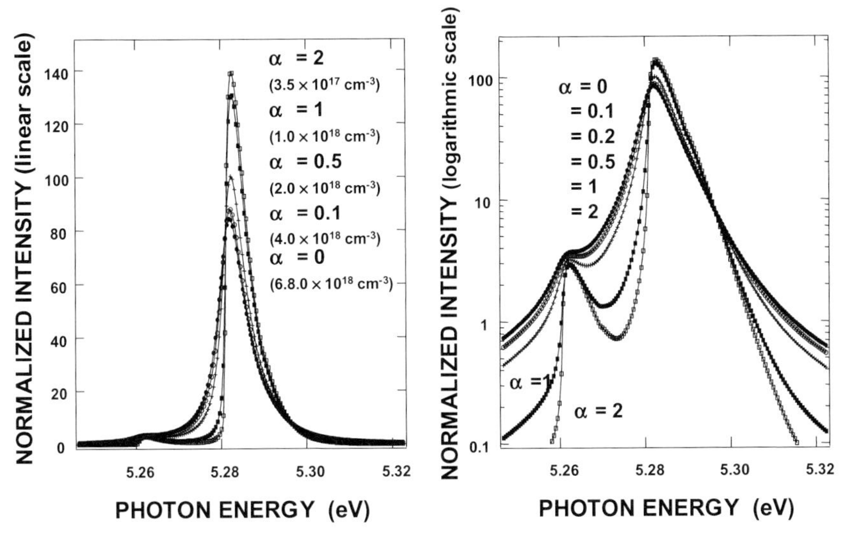

Figure 11.3 α-dependence of theoretical normalized emission spectra of excitons of FE^{TO} + FE^{LO} lines in diamond at T_{ex} = 38 K and ε_p = 5.275 eV, where the vertical axis of the right side figure is scaled by a linear and that of the left side by a logarithmic.

As mentioned above, at $\alpha = 0$ the experimental observed spectrum would have a δ-function like line due to the condense excitons at the ground state (ε_p). In order to fit the observed spectrum for $\alpha = 0$, the δ-function like line should be add to the theoretical spectrum of $\alpha = 0$ shown in Figure 11.3. However, theoretically, its contribution to the total spectrum is small, and it can be considered to be less than 5% (S. Kanno, private communication).

11.3.4
Curve Fitting Results on Temperature-Dependence of Spectra

Next, we performed the curve fitting of the observed $\tilde{I}(\omega)$ of Figure 11.2, using the theoretical $\tilde{I}(\omega)$ described in Section 11.3.3. The fixed parameters listed in Table 11.1 and three variable parameters (α, T_{ex} and ε_p) were used to fit all the experimental emission spectra in Figure 11.2. Figure 11.4 presents typical results. As shown in the figure, the fitted curves agree well with the observed spectra with a small value of χ^2 for over the wide temperature range from $T_{ob} = 8.1$ K to 140 K. Here, χ^2 is used as fitting as fitting quality parameter given by:

$$\chi^2 = \sum (\tilde{I}(\omega)_{ob} - \tilde{I}(\omega)_{cal})^2 \tag{11.24}$$

where $\tilde{I}(\omega)_{ob}$ and $\tilde{I}(\omega)_{cal}$ are the observed and the calculated $\tilde{I}(\omega)$, respectively.

The close match between $\tilde{I}(\omega)_{ob}$ and $\tilde{I}(\omega)_{cal}$ indicates the validity of the basic models used in the lineshape analysis. In particular, it is clarified that (i) $\tilde{I}(\omega)$

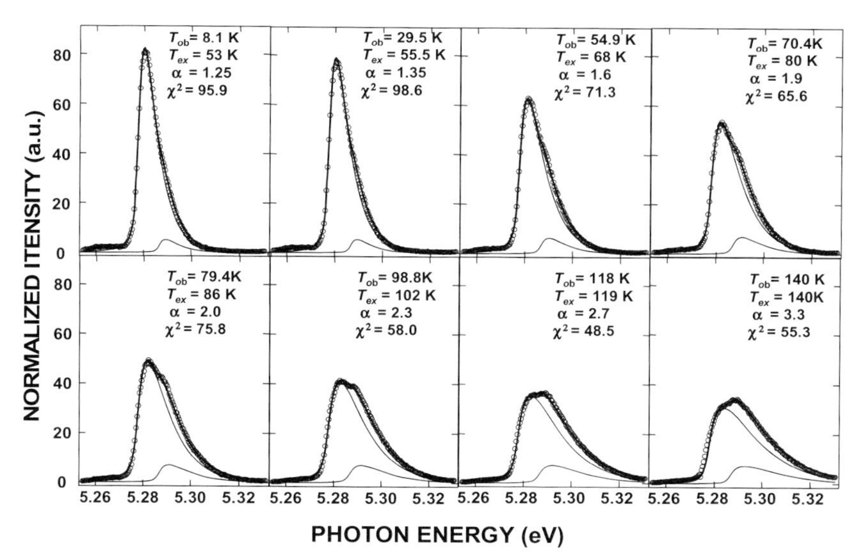

Figure 11.4 Typical results for the curve fitting by using the fitting parameters listed in Table 11.1 and only three variable parameters (α, T_{ex} and ε_p). The χ^2 term is a fitting barometer defined by Equation 11.24.

represents at thermodynamic quasi-equilibrium (chemical equilibrium) and (ii) the origin of the two components of the FE^{TO} (B_1 and B_1') and EF^{LO} (C_1 and C_1') lines are associated with the ground and excited states of the same mass, so that (iii) the spin singlet–triplet splitting model are reasonable one.

11.3.4.1 **Validity of the Parameters Used in the Curve Fitting**

Here, the validity of the parameters listed in Table 11.1 which are used in the curve fitting procedures is discussed. As already mentioned in Section 11.3.1, the values for the kinetic energies corresponding to the phonon velocity, v, defined by Equation 11.7 for the TO and LO phonons (E_{vTO} and E_{vLO}), are smaller than our spectral resolution and, thus, cannot be determined by curve fitting. Therefore, we use the values 0.138 meV and 0.095 meV for TO and LO phonons.

The splitting energy, ΔE_s, between B_1 and B_1' or C_1 and C_1' is obtained as 9.02 meV. This value is larger than the approximately 7 meV reported by Dean *et al.* [1] and the theoretical calculation value of 3.9 meV for the exchange energy reported by Cardona *et al.*, but it is smaller than the approximately 11 meV reported by Sauer *et al.* [14] Taking into account experimental uncertainty, the result is in reasonable agreement with these data.

The emission intensity ratio between B_1 and B_1' or C_1 and C_1', R_{TO} or R_{LO} was obtained as $R_{TO} = R_{LO} = 0.5$. A comparison with published data is, therefore, not possible. The excitonic emission efficiency in the excited states is just one half of that in the ground states. This result has an important physical meaning, particularly when considering the superlinear phenomena of the excitonic emission observed in the high quality CVD diamond films [21, 22].

The values for the energy difference, ΔE_{TL}, and the emission intensity, R_{TL}, between the EF^{TO} and EF^{LO} lines were obtained as 20.4 meV and 0.015, respectively. These values agree with those estimated directly from the experimental data.

In respect of the coefficients for the momentum term of the transition matrix element of emission, C_1, given by Equation 11.12 and C_2 given by Equation 11.13, and the coefficient of the third order energy term D_0 given by Equation 11.6, we obtained $C_1 = 0$, $C_2 = -3k_BT_{ex}$ and $D_0 = 0$. The result $C_1 = D_0 = 0$ means that the contribution to the emission due to the terms related to C_1 and D_0 can be neglected. However, the term related to C_2 cannot be neglected; although the implication of $C_2 = -3k_BT_{ex}$ as obtained by our calculations is within the expected theoretically range [5].

11.3.4.2 **Exciton Gas Temperature (T_{ex}) vs Observed Temperature (T_{ob})**

Figure 11.5 shows the relationship between T_{ob} and T_{ex} that was obtained by curve fitting in the case of $I_{beam} = 15$ μA at $E_{acc} = 15$ kV with 100 μm of beam diameter. The figure shows that T_{ob} deviates from T_{ex} below 100 K. T_{ex} is obviously higher than T_{ob}. This results from heating by a thermalization of excited electron and holes and non-radiative recombination. At $T_{ob} > 100$ K, the heating becomes negligible.

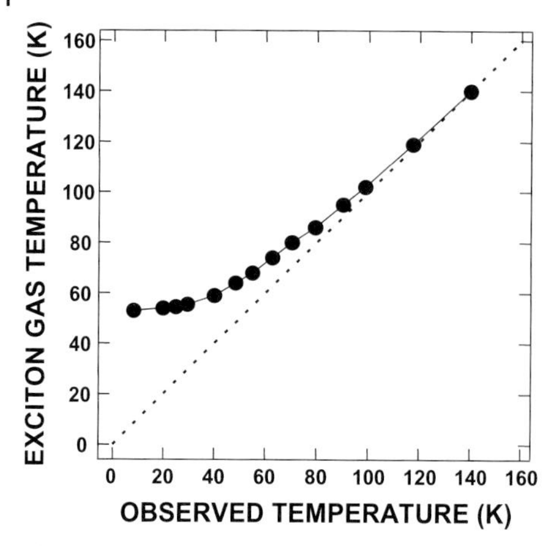

Figure 11.5 Relationship between T_{ob} and T_{ex} obtained by curve fitting for the CL excitonic emission spectra in diamond obtained by I_{beam} = 15 μA at E_{acc} = 15 kV.

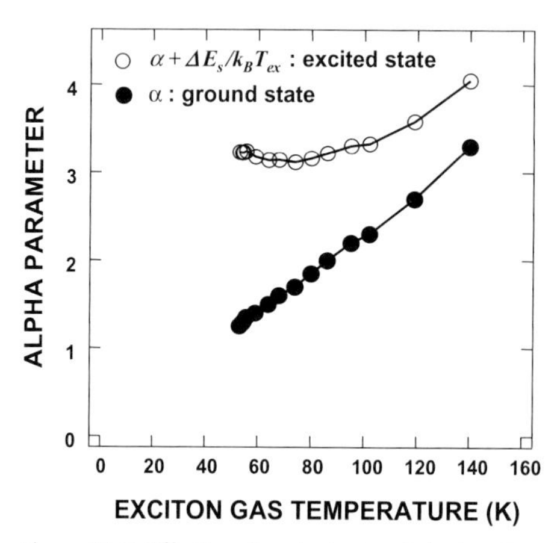

Figure 11.6 Effective chemical potential related to the ground state, α, and the excited state, $\alpha + \Delta E_s/k_B T_{ex}$, as a function of T_{ex}.

The increase of T_{ex} to 45 K at T_{ob} = 8.1 K arises from experimental conditions as the electron beam continuously supplies an energy of 225 mW. In comparison with other materials, it seems that this heating effect is very low. The reason for this is that diamond is a superior thermal conductor of 100 W/cm K at 100 K, which is much higher than that of other materials.

11.3.4.3 Temperature Dependence of α (μ) and n_{ex} by α and T_{ex}

Figure 11.6 shows the effective chemical potential related to the ground state α and the excited state $\alpha + \Delta E_s/k_B T_{ex}$ as a function of T_{ex}. It is apparent that α decreases nearly linearly as T_{ex} decreases to a value of less than 2 for $T_{ex} < 90$ K, while $\alpha + \Delta E_s/k_B T_{ex}$ is always larger than 3. Since α is determined by the relationship between the density of exciton states, which can be taken at a certain finite temperature, and the exciton density, the result of the decrease in α with decreasing T_{ex} (as shown in Figure 11.6) is easy to understand. As mentioned in Section 11.3.3, the lineshape can be explained by classical MB statistics in the case of $\alpha \geq 2$, but not by the BE statistics in case of $\alpha < 2$ [7]. Therefore, the lineshape of the ground state (B_1 and C_1) exciton emission at $T_{ex} < 90$ K must be described by the quantum BE statistics, while the lineshape of the excited state (B_1' and C_1') recombination is still described by the MB statistics.

Dean *et al.* [1] claim that both components can be explained by MB statistics and the total line shape of FE^{TO} could be satisfactorily fitted by the sum of the two MB components. Considering their CL experiments used a 20–μA beam of $E_{acc} = 60$ kV at about 100 K and the level of sample quality, it seems that α related to the ground state (B_1) is larger than 2 and that both lineshapes of B_1 and B_1' can be fitted by the MB statistics.

Based on data shown in Figure 11.6, the exciton density in the ground, n_{exG}, and in the excited states n_{exE}, can be calculated as a function of T_{ex} using Equations 11.25 and 11.26. The result is shown in Figure 11.7, in which n_{exG}, n_{exE} and the total n_{ex} ($n_{exG} + n_{exE}$) are plotted. The temperature dependence of n_c of the BEC boundary in diamond is also shown. It is found that n_{exG} is almost constant for $T_{ex} < 90$ K and at 1×10^{18} cm^{-3} under conditions of 15 kV of E_{acc} and 15–μA of I_{beam} with 100 μm of spot diameter.

On the other hand, n_{exE} increases with increasing of T_{ex} in the low T_{ex} range less than 100 K where $n_{exE} < n_{exG} \simeq 1 \times 10^{18}$ cm^{-3}. Then it decreases with further increasing T_{ex} showing a maximum at around $T_{ex} = 100$ K. The total n_{ex} shows also a maximum of 2×10^{18} cm^{-3} at around $T_{ex} = 100$ K, but the variation is relatively weak.

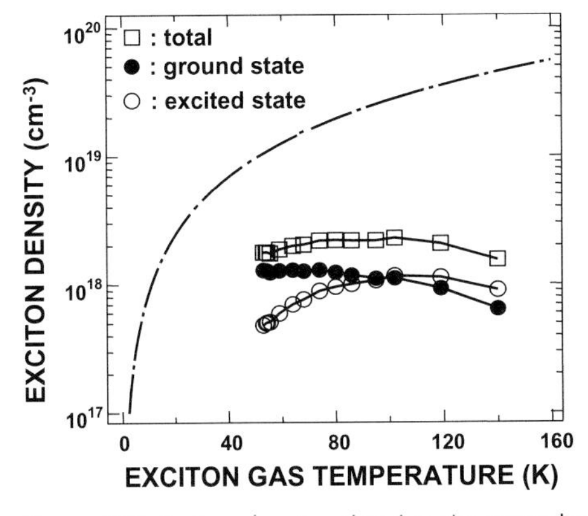

Figure 11.7 Exciton density related to the ground state, n_{exG}, and the excited state, n_{exE}, and the total exciton density as a function of T_{ex}. The dashed line denotes the critical density of the BEC boundary.

The results for the total n_{ex} follow the same trend as that for the integrated emission intensity of FE emission spectra shown in Figure 11.2a. The reason why the total n_{ex} and the integrated emission intensity show a maximum at around 100 K is attributed to τ_{ex} that takes a maximum value around 100 K under the present experimental conditions.

11.3.4.4 Exciton–Exciton Scattering and its Characteristics

As shown in Figure 11.7, a high density of excitons of around 2×10^{18} cm^{-3} was generated at thermodynamic quasi-equilibrium by continuous electron beam excitation. This result is obtained by applying Equations 11.22 and 11.23 to calculate α and T_{ex} values by curve fitting, in which the BE statistical characteristics ($\alpha < 2$) appeared. Therefore, the validity of this result depends on the assumption of whether or not the BE characteristics in a dense exciton gas in diamond can be observed at thermodynamic quasi-equilibrium.

As already mentioned, a collisional broadening in the lineshape analysis has been considered, in which the exciton–exciton scattering is assumed as dominant scattering mechanism. As shown in Equation 11.21, the inverse of the collision time of exciton–exciton scattering γ is directly related to n_{ex}. The determination of n_{ex} as a function of α and T_{ex} by curve fitting takes into account the relationship between γ and $n_{ex0}(\alpha, T_{ex})$ self-consistently through Equations 11.22 and 11.23.

Figure 11.8 shows τ_{coll} ($\hbar/\gamma_G$ and $\hbar/\gamma_E$) of the exciton gas in the ground and excited states as a function of T_{ex}. The values were obtained by calculating Equations 11.22 and 11.23, in which the scattering cross section of $\sigma^0 = \pi(a_{ex})^2$. Please note that the conventional scattering cross section is expected to be $\sigma^0 = \pi(2a_{ex})^2$ for an ideal hard sphere model. Though the result of $\sigma^0 = \pi(a_{ex})^2$ is different from the ideal one, it can be still discussed on the hard sphere scattering model [8].

As shown in the figure, τ_{coll} of both ground and excited exciton is 6×10^{-13}–1×10^{-12} s for T_{ex} ranging between of 52.5–140 K. τ_{coll} has a minimum at around 100 K, which corresponds to n_{ex} having a maximum value at around 100 K, as is shown

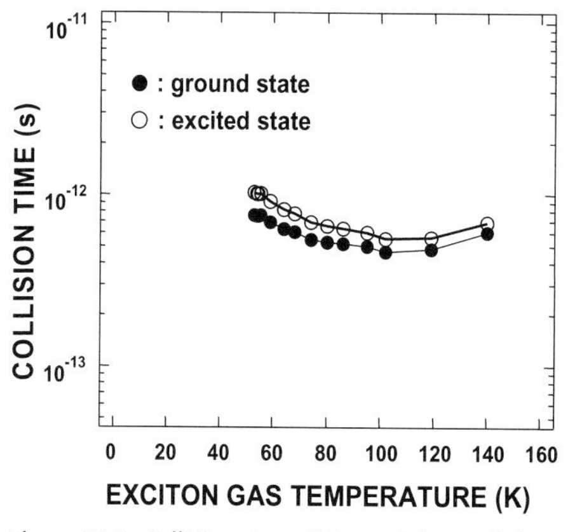

Figure 11.8 Collision time ($\hbar/\gamma_G$ and $\hbar/\gamma_E$) of the exciton gas in the ground and excited states as a function of T_{ex}.

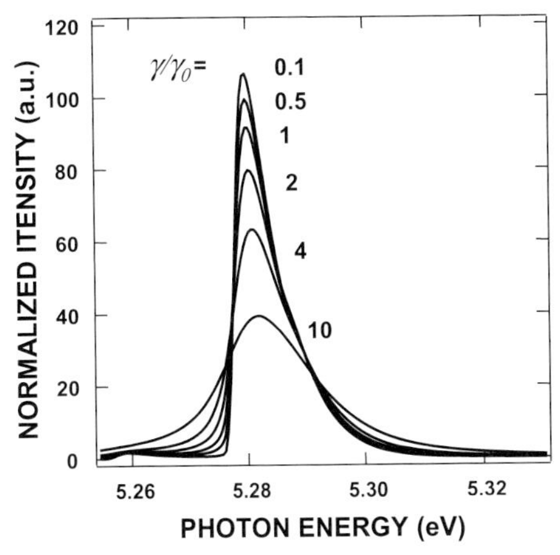

Figure 11.9 Dependence on inverse collision time γ/γ_0 of the calculated the FE emission lines at $\alpha = 1.5$, $T_{ex} = 50$ K and $\varepsilon_p = 5.2775$ eV, where the parameters listed in Table 11.1 is used in the calculation.

in Figure 11.9. Considering the Bohr radius of exciton in diamond (1.57 nm) and n_{ex} of 2×10^{18} cm^{-3}, τ_{coll} of 6×10^{-13}–1×10^{-12} s is consistent with the hard sphere scattering model.

The relationship between τ_{coll} ($\hbar/\gamma$) and γ is governed by the uncertainty principle so that τ_{coll} of 10^{-13}–10^{-12} s gives a large FWHM of 1.2–12 meV for emission lines. Figure 11.9 shows the spectral dependence on the inverse collision time γ/γ_0 at $\alpha = 1.5$, $T_{ex} = 50$ K and $\varepsilon_p = 5.285$ eV, where the parameters listed in Table 11.1 are used in the calculation. γ_0 in the figure corresponds to the value obtained from curve fitting shown in Figure 11.4. It is clear that the lineshape broadening is strongly dependent on γ/γ_0. For $\gamma/\gamma_0 = 1$, the FWHM is about 9 meV, broadening to about 18 meV for $\gamma/\gamma_0 = 10$.

Since the γ value is directly proportional to n_{ex} through Equations 11.22 and 11.23, the result of Figure 11.9 means that n_{ex} can be determined from γ with sufficient accuracy. As already mentioned, n_{ex} was determined as a function of α and T_{ex} by curve fitting, taking into account the relationship between γ and $n_{ex0}(\alpha, T_{ex})$ self-consistently. From this argument, it can be claimed that the dense exciton gas around 2×10^{18} cm^{-3} was actually achieved at thermodynamic quasi-equilibrium.

11.3.5 Validity of the Curve Fitting Results

The result of the dense exciton gases larger than $n_{ex} \approx 2 \times 10^{18}$ cm^{-3} for experimental conditions of $I_{beam} = 15$ μA at $E_{acc} = 15$ kV beam a with 100 μm of spot diameter is compatible with the following empirical relationship for n_{ex} in a semiconductor excited by an electron beam [23],

$$n_{ex} = \frac{I_{beam}}{e} \frac{E_{ie}}{E_i} \frac{\tau_{ex}}{V_{ex}} \tag{11.25}$$

where E_{ie} is the energy of the electron beam, E_i the ionization energy to generate excitons, and V_{ex} is the effective volume of exciton emission. For diamond, E_i has been determined experimentally as being 16 [23]. If we assume that $\tau_{ex} = \tau_{rad}$, where τ_{rad} is the radiative exciton lifetime, under the continuously high excitation of beam, $I_{beam} \simeq 15\ \mu A$ correspond to the generation rate of electron–hole (e–h) pairs (G) is around $5.8 \times 10^{16}\ s^{-1}$ and that I_{beam} uniformly flows through a beam irradiation area with 100-μm diameter, and the excitons are uniformly present in the whole region of the sample thickness, we can put $V_{ex} \simeq 7.0 \times 10^{-9}\ cm^{-3}$ for the sample thickness is 2.8 μm and $\tau_{ex} = \tau_{rad} \simeq 2.3 \times 10^{-6}$ s, which is reported value in the HPHT synthetic IIa diamond [16]. Then, we estimate $n_{ex} \simeq 2 \times 10^{18}\ cm^{-3}$ by Equation 11.25. This value agrees with the curve fitting result within the margin of the estimation error.

However, some questions arise for the assumption that $\tau_{ex} = \tau_{rad}$ in the continuous high excitations, so that the realization of $n_{ex} \geq 10^{18}\ cm^{-3}$. As mentioned in Section 11.2.2, τ_{ex} is generally considered to be $\sim 10^{-8}$ s which were measured by PL decay experiments for the HPHT synthetic IIa diamond [13, 16, 17]. If $\tau_{ex} \simeq 10^{-8}$ s, n_{ex} is estimated to be $8.2 \times 0^{16}\ cm^{-3}$ by assuming of $V_{ex} = 7.0 \times 10^{-9}\ cm^{-3}$, which is inconsistent with the curve fitting results.

For these questions, we can answer with the following arguments. Firstly, it is pointed out that τ_{ex} is generally given by the following relation,

$$\frac{1}{\tau_{ex}} = \frac{1}{\tau_{rad}} + \frac{1}{\tau_{deep}} \tag{11.26}$$

where τ_{deep} is the exciton recombination time at deep levels due to impurities and/or defects and it strongly depends on the density of deep levels. The reported value of $\tau_{ex} \approx 10^{-8}$ s means that $\tau_{deep} << \tau_{rad}$ and $\tau_{ex} \approx \tau_{deep}$. This result may hold in the case of low G such as $G << 10^{17}\ s^{-1}$ or in the case of transient states, such as in the PL decay experiments.

However, the relation of $\tau_{ex} \approx \tau_{deep}$ does not hold in the case of continuous high excitations like $G \approx 10^{17}\ s^{-1}$. We have observed a superlinear relationship between the intensity of the free exciton emission and I_{beam} in CL spectra at 300 K, where the superlinear phenomenon occurs at $G \approx 10^{17}\ s^{-1}$ [21, 22] This took place without any new emission line or peak shift due to many-particle interaction such as the exchange and correlation effects that can be expected for an electron–hole liquid (EHL) [24–26]. We have also observed such the superlinear phenomena of excitonic emission in electroluminescence (EL) spectra from diamond light emitting diodes (LEDs) at high temperature over 300 K, where the superlinear phenomena also occur at $G \simeq 10^{17}\ s^{-1}$ [27]. Since the excitonic emission intensity is proportional to n_{ex} and n_{ex} is proportional to τ_{ex}, these superlinear phenomena strongly indicate an increase of τ_{ex}, with the result that τ_{deep} decreases and the relationship of $\tau_{ex} \approx$

τ_{rad} is attained at high current level from Equation 11.26. Actually, according to detailed analyses on the superlinear phenomena in both CL and EL spectra, it has been confirmed that the relationship of $\tau_{ex} \approx \tau_{rad}$ holds at large $G \simeq 10^{17} s^{-1}$ for high temperature over 300 K (T. Makino *et al.*, unpublished results). Please note that the relationship of $\tau_{ex} \approx \tau_{rad}$ holds at lower G such as $G \approx 10^{15-16}$ s^{-1} for low temperature. The results of Figure 11.2 were obtained in the low temperature range less than 150 K so that the assumption of $\tau_{ex} \approx \tau_{rad}$ is reasonable for the case of Figure 11.2.

Secondly, it is found by spatial distribution of excitonic emission from the CVD diamond sample irradiated by the excitation with $G \simeq 10^{15-17}$ s^{-1} that the dense exciton gases larger than 10^{18} cm^{-3} show an abnormal diffusion process and form a spatial condense region of the exciton gas, due to attractive interaction among the dense exciton gases, suggesting that this spatial condense of the exciton gas is close to a liquid phase (H. Okushi *et al.*, unpublished results). The result means that the excitons are not present uniformly in the whole beam irradiation region of the sample and V_{ex} with n_{ex} over 10^{18} cm^{-3} becomes smaller than those of the uniformly distribution cases.

In conclusion, the realization of ultra dense exciton gas larger than 10^{18} cm^{-3} in the case of Figure 11.2 is enough possible by taking into account of $G \approx 10^{17}$ s^{-1}, $\tau_{ex} = 2.3 \times 10^{-6}$ s, smaller value of V_{ex} than 7.0×10^{-9} cm^{-3} and the low temperature experiments.

11.4
Exciton BEC in Diamond

11.4.1
Bose–Einstein Condensation of Excitons

Excitons in semiconductors obey the BE statistics. They form a nearly ideal (weakly interacting) gas which will condense according to the BE statistics governed by parameters which will be discussed in detail in the following [7, 8]. At low n_{ex}, μ or α in the BE statistics is large and negative, and approaches zero as n_{ex} is increased. At a given temperature T_{ex}, n_{ex} cannot exceed a critical density n_c as the population of exciton states is finite even at $\mu = 0$ which gives rise to saturation [7]. As already mentioned, n_c for this saturation at $\mu = 0$ is given by Equation 11.2. According to Einstein, once the population of exciton states saturates, further increase of excitons must be accommodated by the ground state which gives rise to BEC [5]. Thus, detection of $\mu = 0$ or $\alpha = -\mu/k_B T_{ex} = 0$ is a striking indication of BEC.

In most previous studies on exciton BEC, for example in Cu_2O, time resolved photoluminescence (PL) experiments have been carried out by using pulsed laser excitations. Based on such experiments, several groups have reported distributions which are expected for $\mu = 0$ or closes to 0, detected for a few tens of nanoseconds in the excitonic emission spectra of Cu_2O [28–31]. However, it was pointed out that the observed distribution in such transient states could be explained, even without assuming the BE statistic [32, 33]. In order to overcome this problem, one must study exciton BEC at thermodynamic quasi-equilibrium, and must strictly and precisely determine μ of the exciton gas [3, 4].

11.4.2
Possibility of Exciton BEC in Diamond

Figure 11.10 shows a simple phase diagram for electron–hole pairs in diamond at thermodynamic quasi-equilibrium, in which Equation 11.24 has been plotted by using $g_{exG} = 6$, $g_{exE} = 18$ and $m = 0.80\ m_0$, where m_0 is the electron resting mass and m was estimated from the experimental result of the exciton binding energy. The critical densities of excitons $n_F = 3/(4\pi a_{ex}{}^3) = 6.2 \times 10^{19}\ \text{cm}^{-3}$, and $n_M{\sim}1/a_{ex}{}^3 = 2.6 \times 10^{20}\ \text{cm}^{-3}$ are also shown in Figure 11.10. If the density of the e–h pairs exceeds n_F, the excitonic gas cannot exist any longer and becomes an e–h plasma (Fermi gas). The transformation from an insulating excitonic ensemble to metallic plasma is generally referred to as the exciton Mott transition.

According to the study on the phase diagram of diamond, the Mott density, at which the screening length of the Coulomb potential equals to a_{ex}, is given by the Thomas–Fermi screening approximation $n_{Mott}{}^{TF} = 1/(4a_{ex})^3 = 4 \times 10^{18}\ \text{cm}^{-3}$ [34]. Nagai *et al.* have suggested that the detection of BEC is very unlikely, taking into account this Mott density and their time resolved PL data which have been detected far from thermal equilibrium [34]. This may be true for transient excitations, but not in the case where the e-h pairs with a density in the BEC regime are gradually increased by a continuously excitation at thermodynamic quasi-equilibrium.

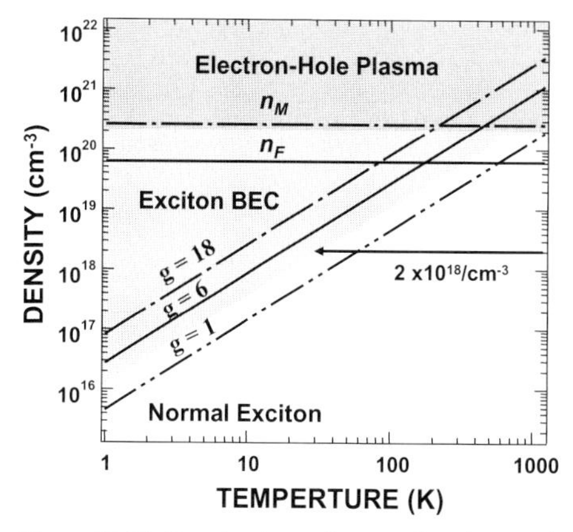

Figure 11.10 Simple phase diagram for electron–hole pairs in diamond at thermodynamic quasi-equilibrium, in which Equation 11.23 has been plotted in the cases of $g_{ex} = 1$, 6, and 18 using $m = 0.80\ m_0$, where m_0 is the electron resting mass and m can be estimated from the exciton binding energy (81 meV). The critical densities, n_M and n_F, are also plotted in the figure.

As mentioned in Section 11.3, the dense exciton gases with n_{ex} of around 2×10^{18} cm^{-3} in T_{ex} 53–140 K are realized at thermodynamic quasi-equilibrium by the CL measurements of continuous high excitation. As is indicated by the phase diagram shown in Figure 11.10, n_{ex} of 2×10^{18} cm^{-3} corresponds n_c of BEC at $T_{ex} \approx 20$ K. Therefore, it can be expected that exciton BEC is achieved by using the CVD diamond films. Furthermore, diamond has many advantages in realizing the exciton BEC. Diamond is the best material in terms of its superior physical and optical properties, for example. In particular, it has a high thermal conductivity of 100 W/cm K at 100 K, which is much higher than that of other materials. The diamonds used in this work were of high quality, without a high density of deep defects that act as non-radiative recombination centers. In this respect, as already mentioned in Section 11.1, diamond seems to be the perfect material for the generation and detection of BEC properties.

However, although the dense exciton gases of $n_{ex} = 2 \times 10^{18}$ cm^{-3} are realized in Figure 11.2, T_{ex} could not be decreased to less than 53 K due to the generation of heat. In order to remove this difficulty and to approach the BEC boundary, it is necessary to search suitable experimental conditions for the exciton BEC in diamond.

11.4.3
Realization of $\mu = 0$ ($\alpha = 0$)

Approaching the BEC boundary, the excitonic emission spectra from the same sample used in Figure 11.2 at $T_{ob} = 20$ K by varying I_{beam} from 0.5 to 10 μA at $E_{acc} = 15$ kV [4] were applied. Figure 11.11 shows the observed spectra, where (a) and (b) represents observed $I(\omega)$ and $\tilde{I}(\omega)$ as a function of photon energy, respectively [4]. As shown in the figure, the spectra become much sharper (FWHM < 10 meV) than the spectra in Figure 11.2 (FWHM≃20 meV), indicating that T_{ex} is low in the case of Figure 11.11, compared to that in the case of Figure 11.2. These results come from the following reason: in the case of Figure 11.11, the heating by the beam current was decreased in comparison to the case of Figure 11.2, so that T_{ex} as well as the sample temperature was decreased; at low T_{ex}, the excitonic emission is governed only by ground state transition, while at $T_{ex} > 53$ K for the case of Figure 11.2, the emission involves both the ground and excited states transitions so that the FWHM of spectra increase compared to the low T_{ex} case.

From these spectra, α and T_{ex} were calculated in the same manner as mentioned in Section 11.3.4. The calculated α is plotted in Figure 11.12a as a function of T_{ex} for different I_{beam}. α and μ become nearly zero for $I_{beam} = 0.5$ μA to 1.0 μA as shown Figures 11b and 12a. n_{ex} estimated from Equation 11.25 is plotted in Figure 11.12b. The result shows n_{ex} reaches n_c of the BEC boundary. The result of $\alpha \approx 0$, $\mu \approx 0$ and $n_{ex} \approx n_c$ was reproducible, and achieved by additional experiments on other high quality CVD diamond films.

Figure 11.13 shows observed $\tilde{I}(\omega)$ for CVD diamond sample different from that used in both Figure 11.2 and Figure 11.11. The spectrum was observed at $T_{ob} = 30$ K under the continuous excitation of $I_{beam} = 0.5$ μA at $E_{acc} = 13$ kV with 100 μm

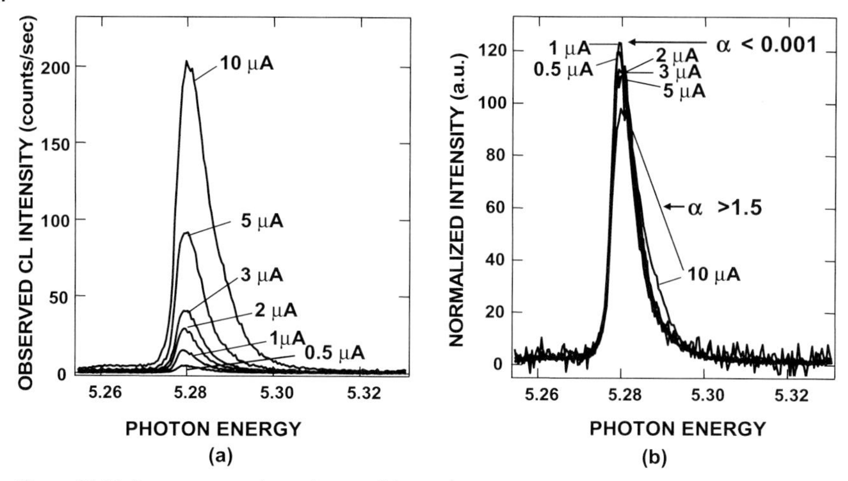

Figure 11.11 Beam current dependence of the FE lines in the same sample of Figure 11.2 (a) at T_{ob} = 20 K. Observed emission intensity. (b) Normalized emission intensity.

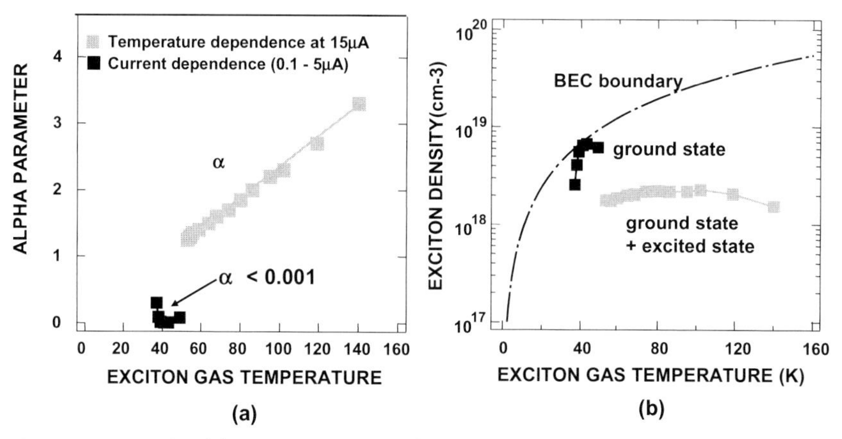

Figure 11.12 (a) The alpha parameter, α, as a function of T_{ex} for the ground state obtained from the beam current dependence (Figure 11.12) of the FE emission lines as well as the temperature dependence of the FE lines (Figure 11.2). (b) Exciton density related to the ground states, n_{exG}, as a function of T_{ex} estimated from the result of (a).

of beam diameter. In the figure, the theoretical $\tilde{I}(\omega)$ for $\alpha = 0$, 0.1 and 0.5 at T_{ex} = 38 K are also plotted in order to demonstrate the accuracy of the determination of α or n_{ex}, where the vertical axis of right side figure is scaled by a linear unit and that of left side by a logarithmic unit.

The observed $\tilde{I}(\omega)$ agrees well with the theoretical $\tilde{I}(\omega)$ for $\alpha = 0$ at T_{ob} = 38 K, even in case of low intensity region, as can be seen in the figure plotted by the

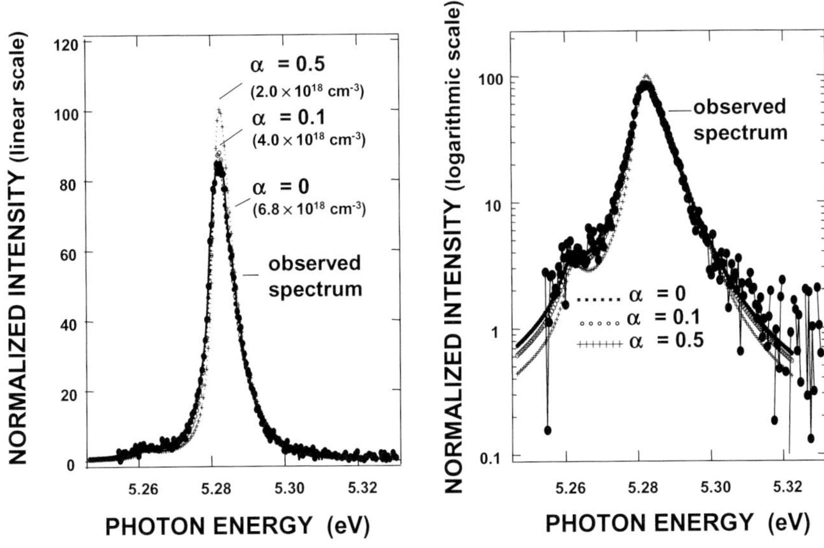

Figure 11.13 A typical example of observed free exciton emission spectrum for CVD diamond and the theoretical emission spectra for $\alpha = 0$, 0.1 and 0.5, where the vertical axis of the right side figure is scaled by a linear and that of the left side by a logarithmic.

logarithmic scale, while the observed $\tilde{I}(\omega)$ can be distinguished from the theoretical spectrum of $\alpha = 0.1$ or $\alpha = 0.5$, as can be seen in the figure plotted by the linear scale. From this, it is certain that $\alpha \approx 0$ and, consequently, $\mu \approx 0$ are achieved in the present CVD diamond sample. Please note that the estimated n_{ex} for $\alpha = 0$ and $\alpha = 0.1$ at $T_{ex} = 38$ K are 6.8×10^{18} cm^{-3} and 4.0×10^{18} cm^{-3}, respectively. In conventional cases, this difference might be within the margin of experimental and theoretical estimation errors.

As is mentioned in Sections 11.3.1 and 11.3.2, the collisional broadening was considered in the lineshape analysis, in which the exciton–exciton scattering is assumed as dominant scattering mechanism. In the case of Figure 11.13, τ_{coll} is estimated to be 1.7×10^{-13} s, by setting the scattering cross section of exciton is $\pi(a_{ex})^2$ and $n_{ex} \approx n_c$. This very short τ_{coll} gives rise to a spectral broadening of about 6 meV (FWHM) by the uncertainty principle, which agrees well with the observed FWHM of $\tilde{I}(\omega)$ as shown in the figure.

On the other hand, as shown in both Figures 11.12 and 11.13, the achievement of $\alpha \approx 0$ was performed by the relatively low beam current of I_{beam}~0.5 μA and not by larger I_{beam} than ~2 μA. In view of the generation rate of excitons (G), I_{beam} = 0.5 μA corresponds to $G \sim 10^{15-16}$ s^{-1}. As already mentioned in Section 11.3.5, the dense exciton gases larger than 10^{18} cm^{-3} show an abnormal diffusion process, and form a spatial condense region of the exciton gas due to attractive interaction among high n_{ex} (H. Okushi *et al.*, unpublished results). Due to this interaction, it is possible to realize n_c of BEC, even in the case of $G \sim 10^{15-16}$ s^{-1}, if the effective emission volume V_{ex} is 10^{-10} cm^{-3}. In actuality, direct observation of the spatial distribution of the exciton emission intensity indicates that a strong emission region is spatially localized in the whole beam irradiation region of the sample at $\alpha \approx 0$ (H. Okushi *et al.*, unpublished results).

It is considered that the reason why α~0 is not achieved when $I_{beam} > \sim 2$ μA even at low T_{ob} is due to the generation of heat by high current level so that T_{ex} increases as well as the sample temperature and V_{ex} becomes larger by diffusion of excitons at high T_{ex}, and then α increases from 0 and n_{ex} decreases from n_c. Thus, in order to realize n_c of exciton BEC, it is important to suppress the generation of heat by I_{beam}. Detailed discussion of this content will be reported in elsewhere.

From arguments mentioned above, the experimental results, as well as the theoretical calculations, indicate that $\mu = 0$ ($\alpha = 0$) that is: the exciton BEC, is achieved in the present CVD diamond samples at thermodynamic quasi-equilibrium (T_{ex} is around 38 K). However, direct evidence of condensate exciton gas has not yet been obtained. It is clear that the results mentioned here need further evaluation to be confirmed. In particular, investigations seeking direct evidence of exciton BEC in diamond should be conducted in the near future.

11.5
Summary

In this chapter, we have provided a systematic study on the dense exciton gases in diamond. The CL spectra of excitonic emission excited by continuous high excitations were used to characterize properties of the dense exciton gases at thermodynamic quasi-equilibrium. Theoretical excitonic emission spectra from diamond were calculated based on Bose–Einstein (BE) statistics, and taking into account collision of excitons in an indirect semiconductor. The theoretical spectra agree well with the observed spectra over a wide temperature range for high quality homoepitaxial chemical vapor deposition diamond films, indicating that the dense exciton gases used in the experiment have densities of larger than 10^{18} cm^{-3}. From the results obtained by the lineshape analysis for the observed spectra, the emission properties from the dense exciton gases have been described in terms of the BE statistical properties. Furthermore, the possibility of exciton BEC in diamond has been discussed, and the realization that the chemical potential of the exciton gas is zero ($\mu = 0$) at around 38 K of the gas temperature has been demonstrated, where $\mu = 0$ is one of the striking indications of BEC.

Acknowledgments

The authors express their gratitude to Drs K. Kajimura and K. Shinohara (CREST), and to K. Tanaka and K. Arai (AIST) for their encouragements and fruitful discussions, as well as thanks to Drs S. Abe, S. Kawabata, C.E. Nebel, T. Miyazaki and N. Orita (AIST), and H. Kume (NIES) for their valuable discussions.

References

1 Dean, P.J., Lightowlers, E.C. and Wight, D.R. (1965) *Physics Review*, **140**, A352.
2 Kawarada, H., Matsuyama, H., Yokota, Y., Sogi, T., Yamanaka, A. and Hiraki, A. (1993) *Physical Review B*, **47**, 3633.
3 Okushi, H., Watanabe, H. and Kanno, S. (2005) *physica status solidi (a)*, **202**, 2051.
4 Okushi, H., Watanabe, H., Yamasaki, S. and Kanno, S. (2005) *physica status solidi (a)*, **203**, 3226.
5 Kanno, S. and Okushi, H. (2007) *physica status solidi (b)*, **244**, 1381.
6 Einstein, A. (1924) *Sitzungsberichte der Preussischen Akademie der Wissenschaften zu Berlin*, **22**, 261.
7 Wolfe, J.P., Kin, J.L. and Snoke, D.W. (2002) *Bose–Einstein Condensation* (eds A. Griffin, D.W. Snoke and S. Strnigari), Cambridge University Press, Cambridge and references therein.
8 Snoke, D.W., Braun, D. and Cardona, M. (1991) *Physical Review B*, **44**, 2991.
9 Myssyrowicz, A., Hulin, D. and Benoit a la Guillaume, C. (1981) *Journal of Luminescence*, **24/25**, 629.
10 Watanabe, H., Hayashi, K., Takeuchi, D., Yamanaka, S., Okushi, H. and Kajimura, K. (1998) *Applied Physics Letters*, **73**, 981.
11 Rauch, C.J. (1962) Proceedings of the International Conference on Semiconductor Physics, Institute of Physics and the Physical Society, London, 1963, p. 276.
12 Collins, A.T., Lightowlers, E.C., Higgs, V., Allers, L. and Sharp, S.J. (1994) *Advances in New Diamond Science and Technology (Proceedings of the Fourth International Conference on New Diamond Science and Technology)* (eds S. Saito, N. Fujimori, O. Fukunaga, M. Kamo, K. Kobashi and M. Yoshikawa), MYU, Tokyo, p. 307.
13 Sharp, S.J. and Collins, A.T. (1996) *Material Research Society Symposium Proceeding*, **416**, 125.
14 Sauer, R., Stemshulute, H., Wahl, S. and Thonke, K. (2000) *Physical Review Letters*, **84**, 4172.
15 Teofilov, N., Sauer, R., Thonke, K., Anthony, T.R. and Kanda, H. (2001) Proceedings of the 25th International Conference on the Physics of Semiconductors, PTS I and II 87: 95.
16 Fujii, A., Takiyama, K., Maki, R. and Fujita, T. (2001) *Journal of Luminescence*, **94/95**, 355.
17 Takiyama, K., Abd-Elrahman, M.I., Fujita, T. and Oda, T. (1996) *Solid State Communications*, **99**, 793.
18 Snoke, D.W. and Wolfe, J.P. (1990) *Physical Review B*, **41**, 11171.
19 Cardona, M., Ruf, T. and Serrano, J. (2001) *Physical Review Letters*, **86**, 3923.
20 Willatzen, M., Cardona, M. and Christensen, N.E. (1994) *Physical Review B*, **50**, 18054.
21 Watanabe, H. and Okushi, H. (2000) *Japanese Journal of Applied Physics*, **39**, L835.
22 Watanabe, H., Sekiguchi, T. and Okushi, H. (2001) *Solid State Phenomena*, **78**, 165.
23 Klein, C.C. (1968) *Journal of Applied Physics*, **39**, 2029.
24 Thonke, K., Schliesing, R., Teofilov, N., Zacharias, H., Sauer, R., Zaitsev, A.M., Kanda, H. and Anthony, T.R. (2000) *Diamond and Related Materials*, **9**, 428.
25 Shimano, R., Nagai, M., Horiuchi, K. and Kuwata-Gonokami, M. (2002) *Physical Review Letters*, **88**, 574041.
26 Sauer, R., Teofilov, N. and Thonke, K. (2004) *Diamond and Related Materials*, **13**, 691.
27 Makino, T., Tokuda, N., Kato, H., Ogura, M., Watanabe, H., Ri, S.G., Yamasaki, S. and Okushi, H. (2006) *Japanese Journal of Applied Physics*, **45**, L1042.
28 Snoke, D.W., Wolfe, J.P. and Mysyrowicz, A. (1980) *Physical Review B*, **41**, 11171.
29 Snoke, D.W., Wolfe, J.P. and Mysyrowicz, A. (1987) *Physical Review Letters*, **59**, 827.
30 Goto, T., Shen, M.Y., Koyama, S. and Yokouchi, T. (1997) *Physical Review B*, **55**, 7609.
31 Naka, N., Kono, S., Hasuno, M. and Nagasawa, N. (1996) *Progress in Crystal Growth and Characterization of Materials*, **33**, 89.
32 O'Hara, K.E. and Wolfe, J.P. (2000) *Physical Review B*, **62**, 12909.
33 O'Hara, K.E., Suilleabhain, L.O. and Wolfe, J.P. (1999) *Physical Review B*, **60**, 10565.
34 Nagai, M., Shimano, R., Horiuchi, K. and Kuwata, M. (2003) *physica status solidi (b)*, **238**, 509.

12
High Mobility Diamonds and Particle Detectors

Heinz Pernegger
CERN, Physics Department, Bat. 301-1-104, CH-1211 Geneva, Switzerland

12.1
Introduction

CVD diamond possesses some remarkable properties which make it an attractive material for the fabrication of solid state particle detectors. Increasingly, solid state particle detectors are required to have fast signals, operate at high rates, allow for very high segmentation, achieve good spatial resolution and, very often, operate reliably in high radiation environments for several years. While silicon, the de facto standard for solid state detectors, is very well established in the market, and for scientific applications, diamond detectors compete in applications where their required properties are clearly superior, or where silicon detectors cannot be used.

Table 12.1 summarizes some basic properties of diamond and silicon that are relevant for particle detectors [1–3]. Diamond detectors, with their high band gap, can be operated with very low leakage currents in high radiation environments. Low leakage currents, hence low power dissipation in the detector, and its high thermal conductivity make diamond an interesting material for application where active detector cooling is impractical or impossible. Its lower dielectric constant results in a lower capacitive load on the detector's front end (FE) electronics, resulting in better noise performance. Its high saturation velocity and mobility combined with a high breakdown voltage enables us to design systems with fast signal response. Although diamond appears ideal in many respects, one its "good" properties, the high band gap, also limits the primary ionization charge. Additionally, charge trapping in diamond further limits the obtained signal. While the energy to form an electron hole pair is an intrinsic parameter, significant improvements have been made over recent years to grow CVD diamond films of high electrical quality with detection properties suitable for a large array of detector applications. During the nineteen-seventies, the attractive properties of diamonds for particle detectors were demonstrated on highly selected monocrystalline natural diamonds with exceptional electrical properties [4, 5]. However, only the availability of micro-

Physics and Applications of CVD Diamond. Satoshi Koizumi, Christoph Nebel, and Milos Nesladek

ISBN: 978-3-527-40801-6

Table 12.1 Properties of diamond and silicon [1–3].

Property	Diamond	Si
Band Gap (eV)	5.5	1.12
Breakdown field (V/cm)	107	3×10^5
Resistivity (-cm)	$>10^{11}$	2.3×10^5
Intrinsic Carrier Density (cm^{-3})	$<10^3$	1.5×10^{10}
Mass Density (g cm^{-3})	3.52	2.33
Atomic Charge	6	14
Dielectric Constant	5.7	11.9
Displacement Energy (eV/atom)	43	13–20
Energy to create e–h pair (eV)	13	3.6
Radiation Length (cm)	12.2	9.4
Avg. Signal Created / μm (e)	36	89
Avg. Signal Created/0.1% Rad. Length X_0 (e)	4400	8400

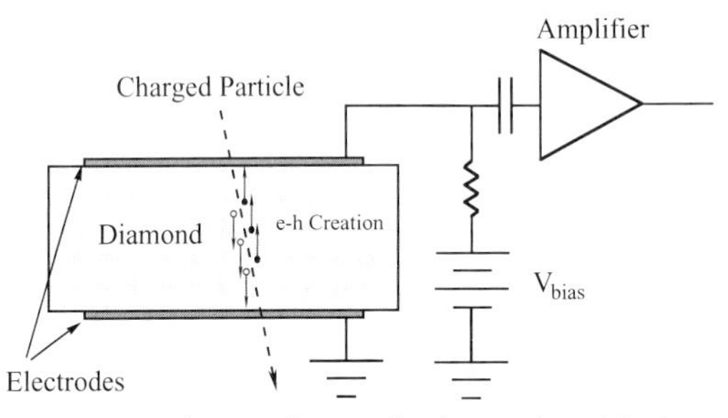

Figure 12.1 Schematic layout of a diamond particle detector.

wave plasma assisted chemical vapor deposition growth processes made diamonds a viable option for particle detectors, for obvious reasons: sample availability, large detector size and reduced variability.

12.2
Polycrystalline CVD Diamond Films as Particle Detectors

Typical detector designs for diamond particle detectors are based on a bulk of intrinsic CVD diamond, usually a few hundreds of micrometers thick, with electrodes on opposite sides of the diamond bulk as shown in Figure 12.1. Detector surfaces typically used for device characterization measure from approximately 0.5–1 cm^2, up to ~10 cm^2, depending on application. Prior to deposition of contacts, the diamond surfaces are polished to smooth the surface on the growth side and reduce low grade material from the substrate side. Metal contacts, often chromium or titanium, to form suitable carbides, are evaporated or sputtered on both diamond surfaces and annealed. A covering layer of gold, for example, is applied to allow connection to the electronics by wire bonding. The dimensions of electrodes can be varied through lithography, in a range of ten micrometers to centimeters. The electrodes can also be removed from the diamond substrate and a different electrode pattern reapplied. In this way, one and the same diamond can be tested consecutively, for example, as a device with large pads, fine strips or micropixels – a considerable advantage for detector development.

For detector operation, a bias voltage is applied between the electrodes to generate a drift field. A traversing charged particle will ionize the atoms in the crystal lattice and leave a trail of primary ionization charge, 36 electron–hole pairs per micrometer [6, 7] denoted as q_0, along its path. The drift of electrons and holes in the applied electric field induces a current pulse at the electrodes. The induced initial current i_0 can be calculated by the Shockley–Ramo theorem [8, 9] for a uniform constant field between the two electrodes as

$$i_0 = Q_0 \frac{v}{d} \tag{12.1}$$

where $Q_o = q_0 d$ denotes the total generated ionization charge, v the drift velocity, and d the gap between the electrodes, which is equal to the thickness of the detector. The connected FE electronics then measure either the current amplitude or, in case of charge sensitive amplifiers, the current integral Q_m. However, the ionization charge is reduced by charge trapping during its drift. One advantage for the characterization of CVD diamond detectors is the mean distance electrons and holes drift apart before being trapped, called the charge collection distance (CCD),

$$CCD = d \frac{Q_m}{Q_0} \tag{12.2}$$

which is also often defined through the product of electron and hole mobility μ and lifetime τ as $CCD = (\mu_e \tau_e + \mu_h \tau_h) E$ under the assumption that the detector thickness is larger than CCD, and the field is uniform. As diamond detectors are usually operated at high field strength, the charge collection distance is quoted here as charge collection distance measured at 1 V/μm.

One of the main efforts in the development of diamonds as particle detectors is towards the continuous improvement of charge collection distance. For application as tracking detectors, for example, an initial charge collection distance beyond 200 μm is required, in order for diamond to be a viable alternative to silicon detectors for future accelerators. To encourage the development of such diamonds, CERN has launched the CERN RD42 collaboration [10] for the development of diamond tracking detectors for experiments at the Large Hadron Collider (LHC). In cooperation with Element Six Ltd,* RD42 is currently developing high quality polycrystalline CVD (pCVD) diamond wafers large enough for full scale detectors. Figure 12.2 (top) shows a recent five inch pCVD wafer ready for tests with contact

*Element Six Ltd, King's Ride Park, Ascot, Berkshire SL5 8BP, UK.

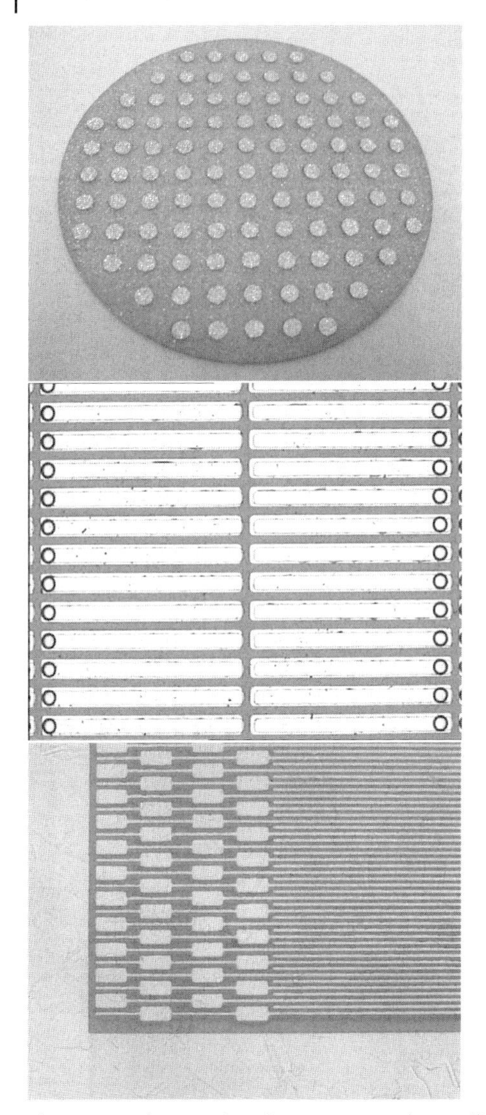

Figure 12.2 Photographs of an as-grown polycrystalline wafer with test contacts (top, courtesy of Element Six Ltd), a pCVD diamond pixel detector (middle, courtesy of SiLab, Bonn) and a 25 μm pitch strip detector (bottom).

pads spaced at 1-cm intervals. Figure 12.2 (middle) shows a section of a pCVD diamond with a pixel electrode pattern with pads spaced by 50 μm × 400 μm, Figure 12.2 (bottom) shows a strip detector with a strip pitch of 25 μm.

Beyond the presence of charge traps, which reduce the measured signal, the polycrystalline nature of the diamond also causes a non-uniformity of the signal response (beyond intrinsic energy loss straggling), resulting in a large spread of signal amplitudes. The non-uniformity of CCD related to the grain structure has been investigated in ion beam induced conductivity measurements, for example [11]. For thick polycrystalline detectors, RD42 has studied the grain size related variation in signal response with large event samples, which were recorded in high energy pion beam tests. At the same time, the CCD increases with diamond thickness almost linearly, being lowest on the substrate side where the crystal size is small, and highest on the growth side where the crystals are large. Therefore, diamonds for particle detectors are regularly thinned down by removing material from the substrate side (e.g. from 800 μm to 500 μm thickness). This has been shown to improve the overall CCD of the detector [12].

For polycrystalline CVD diamonds charge collection distances of 275 μm, or 9800 electrons mean signal, have been achieved, where 99% of signals are above approximately 3000e- [13, 14]. Some particularly good samples reach a charge collection distance above 300 μm, as shown in Figure 12.3, which shows CCD versus applied field. From Equation 12.2 it is apparent that CCD will increase with increasing field strength. At high field, the drift velocity will start to saturate [15] as

$$v = \frac{\mu_0 E}{1 + \frac{\mu_0 E}{v_s}} \tag{12.3}$$

where μ_0 denotes the low field mobility for electrons and holes respectively and v_s their respective saturation velocities, which leads to a saturation of measured CCD.

One peculiar feature observed in polycrystalline CVD diamond detectors is that relatively small amounts of radiation increase the measured charge signal. A dose of few tens of Gray is typically sufficient to increase the initial signal (or CCD) by a factor of 1.2 to 1.6, after which the CCD stays constant [1, 16, 17]. This effect is often referred to as "priming" or "pumping" and can be reversed by the exposure

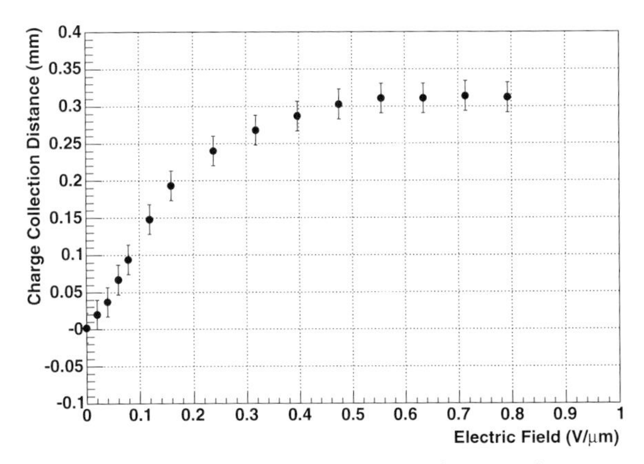

Figure 12.3 Charge collection distance as a function of applied field for a pCVD diamond detector.

to UV light sources and/or heating. It is generally thought to be due to the filling of charge traps. In the initial phase of irradiation, primary free carriers will be trapped and neutralize deep donors and acceptors provided their levels are deep enough to have long lifetimes for charges on defects. During the initial phase of irradiation, the lifetime of free charge carriers increases, hence the CDD increases until an equilibrium is reached. Light will free the charges from the traps and reverse the priming effect.

12.3
Single Crystal CVD Diamond Detectors

In recent years, the first samples of detector grade single crystal CVD diamonds became available. These samples are synthesized with a microwave plasma assisted CVD reactor using <100> oriented single crystal diamond substrates [18]. They promise to resolve many issues commonly associated with polycrystalline diamonds detectors: a uniform signal response due to the absence of a grain structure, little or no charge trapping due to nearly defect free growth, and full charge collection at a lower operation voltage.

Indeed, recent results on samples look very encouraging for their use as particle detectors. Figure 12.4 shows the signal response of a 435 μm thick sample in a CCD measurement using a ^{90}Sr source. The obtained signal is consistent with full charge collection, hence the absence of any significant charge trapping. Furthermore, it shows a much narrower signal distribution compared with pCVD diamonds, and with a FWHM to a most probable signal ratio of ~30%, it has even less variation than silicon detectors in the same measurement. Figure 12.5 shows the most probable charge for samples of different thickness. Their linear increase in signal, again, shows full charge collection. It should be noted that these samples did not exhibit any priming effect during their CCD measurements.

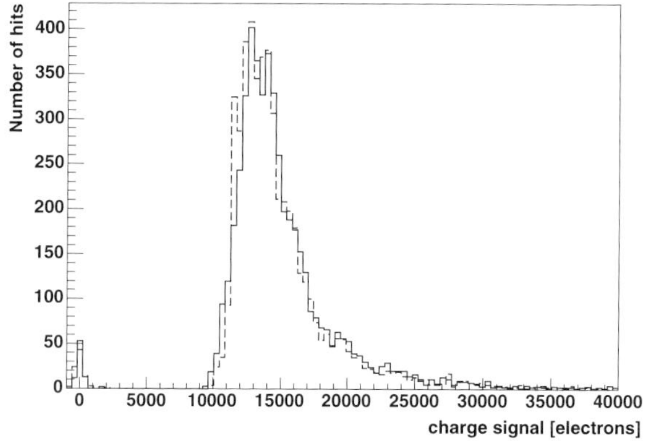

Figure 12.4 Charge signal measured on a single crystal CVD diamond of 435 μm thickness using a ^{90}Sr-source (solid and dashed lines mark positive and negative voltage).

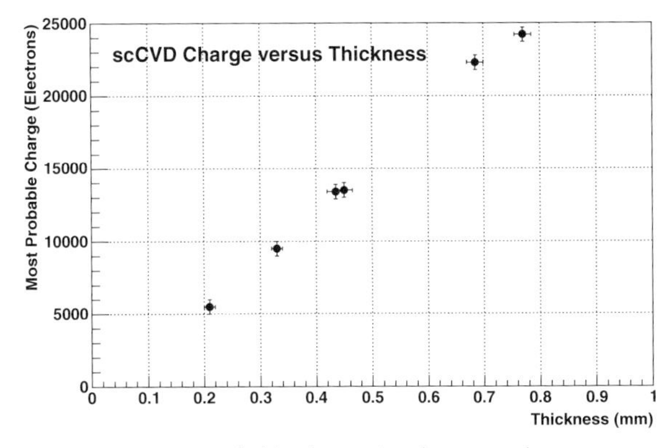

Figure 12.5 Most probable charge signal measured on single crystal CVD diamonds of different thickness using a ^{90}Sr-source.

Given that the signal response shows much less variation than polycrystalline diamonds, single crystal diamond detectors can also be expected to find special spectroscopic applications, an area where thick polycrystalline diamonds are normally unsuitable. First measurements in this direction have been carried out using mixed nuclide α-sources, and demonstrated an energy resolution $\Delta E/E = 0.0035$ [19].

12.4 Charge Carrier Properties in CVD Diamonds

For the development of CVD diamonds and their later operation as charged particle detectors, it is important to have a detailed understanding of the charge carrier transport mechanism and its basic parameters like drift velocity, carrier lifetime, space charge, and field configuration in the detector bulk. The transient current technique (TCT) allows a direct determination of these parameters in a single measurement of the induced current of electrons and holes. The TCT method has been used for the characterization of silicon diodes, irradiated and unirradiated [20–22], and has been extended to the characterization of single crystal CVD diamonds [18, 19, 23, 24].

Also, with the recent availability of single crystal CVD diamond film (scCVD), questions of uniformity and charge lifetime are being addressed. In particular, the high charge carrier mobility allows one to design detectors with very fast signal response, which will be of great interest for future detector systems. The following section will demonstrate the use of TCT measurements for the characterization of diamond bulk properties.

In the TCT method, ionization charge is deposited on one side of the diamond, just beneath the contact. In the applied field, one type of charge carriers, say electrons, will drift through the diamond bulk, while the other type is collected on the contact immediately. In this way, the charge carrier properties of a single carrier type can be measured. Reversing the polarity of the applied voltage on the diamond measures the other carrier type, say holes. Thereby, the properties of electrons and holes can be measured separately, and the field configuration in the bulk can be studied. During the charge drift, the induced current is measured with a high speed current sensitive amplifier. For a detailed description of the method and analysis techniques, see reference [24]. Recent developments of broad band high speed RF amplifiers dedicated to diamond detectors have allowed highly sensitive measurements of this induced current [25, 26].

Depending on the amount and range of injected charge, the actual shape of the measured induced current will look distinctly different. If a high ionization density is created, for example, by lasers, the field will be locally modified by the injected charge leading to a space charge limited (SCL) transient curve as described, for example, in [18, 27]. Using small localized charges, for example, through low-rate α-emitters like ^{241}Am, the measurement is carried out in the space charge free (SCF) regime. Measurements presented here have been carried out in the SCF regime.

Figure 12.6a/b shows the transient curves measured at two different voltages, 100 V and 300 V, on a 470 μm thick sample of single crystal CVD diamond. Plot (a) shows the curves measured for hole drift; plot (b) the respective curves for electron drift. The rising edge marks the start of the drift and the falling edge the arrival of the charge cloud (enlarged by diffusion during the drift) on the opposite side. Using the duration of the pulse, the average drift velocity can be studied for each carrier type as a function of drift field. For a constant field in the diamond, and in absence of charge trapping, the current amplitude will be constant during the drift. However, when the sample contains uniform space charge, like this sample, the electrical field will increase or decrease linearly and the current amplitude will follow a behavior given by

$$i_{e,h}(t) = e^{\frac{t}{\tau_{\mathit{eff}e,h}} - \frac{t}{\tau_{e,h}}} \tag{12.4}$$

$$\tau_{\mathit{eff}e,h} = \frac{\varepsilon\varepsilon_0}{e_0\mu_{e,h}|N_{\mathit{eff}}|} \approx \frac{\varepsilon\varepsilon_0 t_c V}{e_0 d^2 |N_{\mathit{eff}}|} \tag{12.5}$$

where ε_0 and ε denote the permittivity of vacuum and dielectric constant of diamond, $\mu_{e,h}$ the mobility of electrons and holes, V the applied voltage, e_0 the electron charge, N_{eff} the space charge concentration in the bulk and t_c the transit time of charges through the diamond bulk. The first term in the exponent of Equation 12.4 describes the exponential behavior of the current due to effective space charge (increasing current for electrons, decreasing current for holes), the second term accounts for the current's exponential decrease as a result of charge trapping.

The measured drift velocity for electrons and holes is shown in Figure 12.7a/b as a function of electrical field. Solid and open markers denote the measured drift

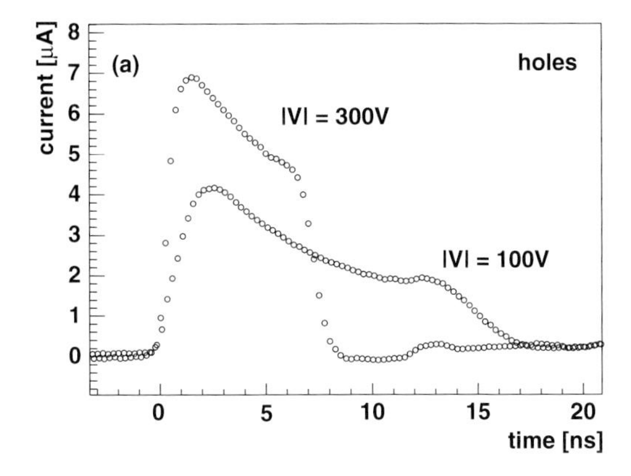

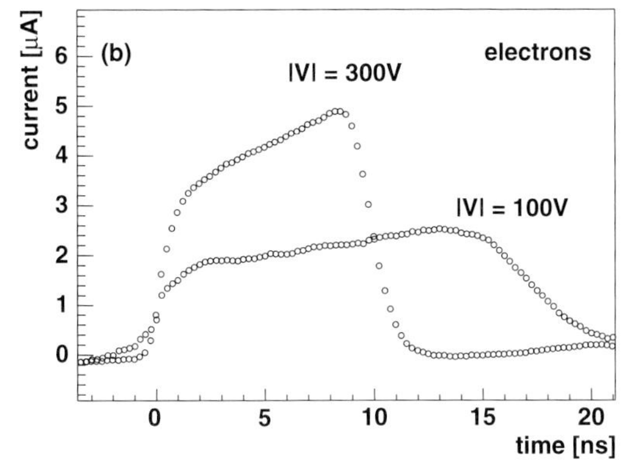

Figure 12.6 Transient current for holes (a) and electrons (b) recorded in TCT measurement of a 470 μm thick single crystal CVD diamond using a ^{241}Am α-source for charge injection.

velocity for electrons and holes respectively. The solid lines represent fits of Equation 12.3 to the data. Both curves clearly deviate from a linear behavior which only exists at low fields. The charge carrier mobility is highest at low field strength, and is gradually reduced. This reduction is generally observed in semiconductors, the dominant scattering mechanism being associated with optical phonons, and leads to a saturation of drift velocity as the field is increased to values where particle detectors are usually operated. For this particular sample, the fits yield low field mobilities $\mu_{0,e} = 1714$ cm^2/Vs and $\mu_{0,h} = 2064$ cm^2/Vs for electrons and holes respectively, and saturation velocities of $v_{s,e} = 9.6 \times 10^6$ cm/s and $v_{s,h} = 14.1 \times 10^6$ cm/s, for electrons and holes respectively. While values appear to show some sample to sample variation, with this sample being on the low side, recent measurements of different samples confirmed the trend of higher mobility for holes than for electrons. Other measurements [19] using the same technique have found mobilities of $\mu_{0,e} = 2071$ cm^2/Vs and $\mu_{0,h} = 2630$ cm^2/Vs for electrons and holes respectively, and, using laser charge injection and operation in the SCL regime,

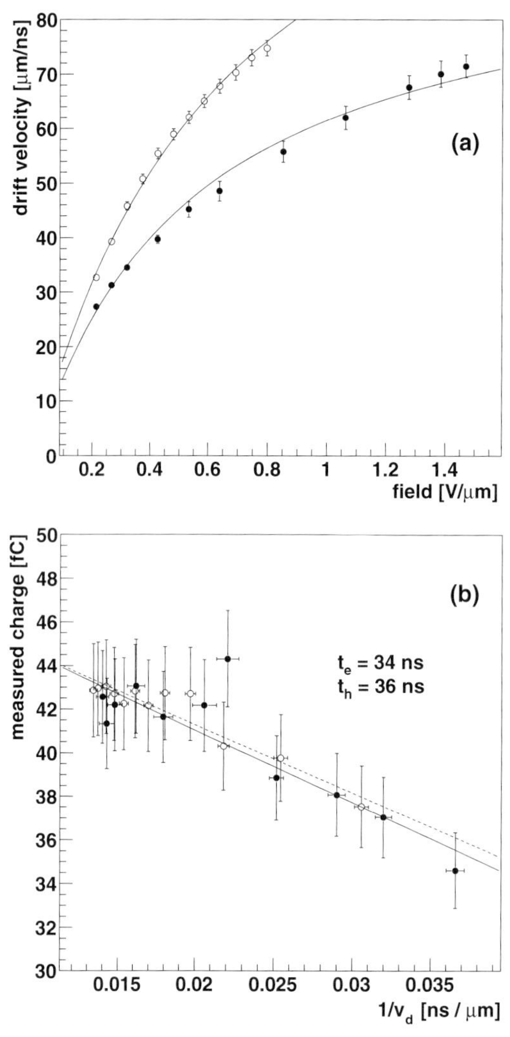

Figure 12.7 Drift velocity for holes and electrons as a function of applied field (a) and total charge, obtained through current integration, as a function of inverse drift velocity (b) as recorded in TCT measurement of a 470 μm thick single crystal CVD diamond using a ^{241}Am α-source for charge injection.

values higher than 3000 cm^2/Vs have been reported [18]. The charge's lifetime is extracted in Figure 12.7b through the linear correlation of total measured charge as a function of the inverse drift velocity. This and other measurements [19] have determined charge lifetimes well in excess of the transit time, with lifetimes reported up to 1 μs. This constitutes a dramatic improvement compared to polycrystalline diamonds, and is a clear advantage for applications where maximum charge signal is desired.

12.5 Future Accelerators and Radiation Hardness

Solid state detectors for high luminosity experiments at present and future accelerators are required to withstand high doses of ionizing and non-ionizing radiation. Inner layers of tracking detectors for experiments at the Large Hadron Collider (LHC) at CERN are typically required to survive around 50 Mrad of ionizing radiation, and to survive the damage due to non-ionizing energy loss of 10^{15} particle/cm^2 accumulated in 10 years of envisaged operation. In addition to this harsh constraint, these tracking detectors have grown into very large systems over the past decade: the silicon detector of the DELPHI experiment at LEP (start of operation 1996 in its final layout) consisted of 1.8 m^2 of silicon detectors with 175,000 channels, up to the silicon detector of the CDF experiment at the Tevatron, the SVX IIa (start of operation 2001), which has 6 m^2 of silicon detectors with 750,000 channels. The present tracking detectors at LHC are due to go into operation in 2008 in the ATLAS (from A Toroidal LHC ApparatuS) and Compact Muon Spectrometer (CMS) experiments with 61 m^2 of silicon and 6 million channels in the case of the ATLAS SCT, and even 220 m^2 of silicon and 10 million channels in case of CMS. It is clear that solid state detectors of this magnitude not only have to have reliable sensor material as their basis, but also that operational and engineering related issues become increasingly more complex, and difficult to overcome. Many of these difficulties are related to practical questions concerning how to power the detectors, and consequently, how to cool them again stably at, in the case of silicon detectors, around −10 °C for their entire lifetime.

The operational requirements will get even more difficult for future upgrades of the LHC to the SuperLHC, where the particle fluence at the innermost layers is expected to increase by another order of magnitude. Silicon detectors like the ones of the ATLAS SCT and CMS tracker are not expected to survive this radiation damage in their present layout. Currently, there is also no alternative to solid state detectors for the inner parts of tracking detectors, given their advantages in spatial resolution, and dense integration. For these reasons, three detector R&D collaborations at CERN are investigating novel and improved solid state detectors: RD39 [28] on the use of cryogenic silicon detectors, RD50 [29] on the use of improved float zone p–n silicon detectors, epitaxial silicon detectors, and novel 3D sensor layouts, and RD42 on the use of CVD diamond detectors. While it seems impractical, and indeed unnecessary, to base outer layers on diamond sensors, CVD diamonds do have attractive properties for the innermost layers, like low leakage currents after irradiation, high thermal conductivity, and the ability to operate at room temperature. To be viable, however, the sensor material needs to provide sufficient signal to the FE electronics even after 10^{16} particle/cm^2.

When diamond is exposed to irradiation, high energetic charged and neutral particles displace atoms, producing vacancies and interstitials, and the displaced atoms themselves may have sufficient energy to further cause damage. The high band gap of 5.5 eV and large displacement energy of 43 eV [30] make diamond attractive in this respect. Reported results indicate, for example, that there is very

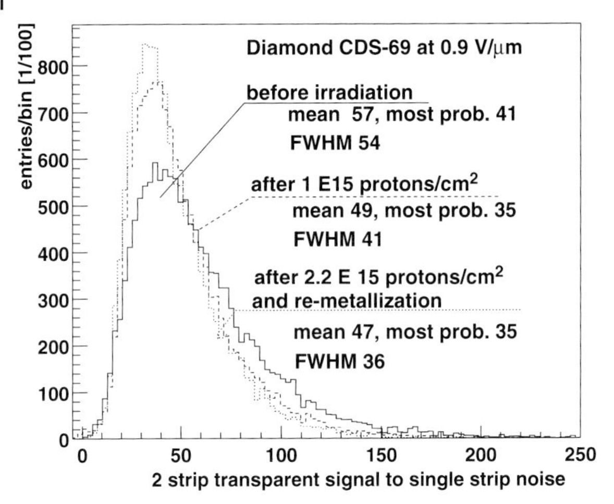

Figure 12.8 Two-strip signal-to-noise distribution for a pCVD diamond strip detector as measured in a beam test after different proton fluences.

little degradation of charge collection distance up to 10^{15} protons/cm^2 [31]. For applications other than tracking detectors, for example as monitoring detectors in nuclear reactors, and for nuclear fuel recycling processes, CVD diamond detectors have been tested with good results up to 250 Mrad of photon irradiation and neutron fluence up to 3×10^{15} neutrons/cm^2 [32].

Further tests have been carried out to extend the irradiation studies on detector grade polycrystalline diamonds up to more than 10^{16} particle/cm^2 and to characterize their CCD as a function of fluence, while, in parallel, similar studies have started on single crystal CVD diamonds. Figure 12.8 shows the signal distribution measured on a polycrystalline CVD diamond detector after 2.2×10^{15} protons/cm^2. The signal is plotted as signal over noise ratio for individual hits with the noise being constant during the measurements (the charge amplifier used in the test was not irradiated). The most probable signal decreases by 15% at 2.2×10^{15} protons/cm^2 when compared to its unirradiated value. The signal response appears more uniform at lower values after irradiation. The diamond has been tested as 50 μm pitch strip detector and has shown a spatial resolution of 11.5 μm before irradiation and 7.4 μm after 2.2×10^{15} protons/cm^2. Figure 12.9 shows the relative signal as a function of fluence up to 1.8×10^{16} protons/cm^2 protons/cm^2. The dashed line indicates a 1/e decrease in relative signal as function of fluence Φ with an exponent of $-0.08\,\Phi$.

12.6
High Resolution Diamond Pixel Detectors

The innermost layers of tracking detectors are currently equipped with high resolution hybrid pixel detectors for tracking at high particle density. In order to

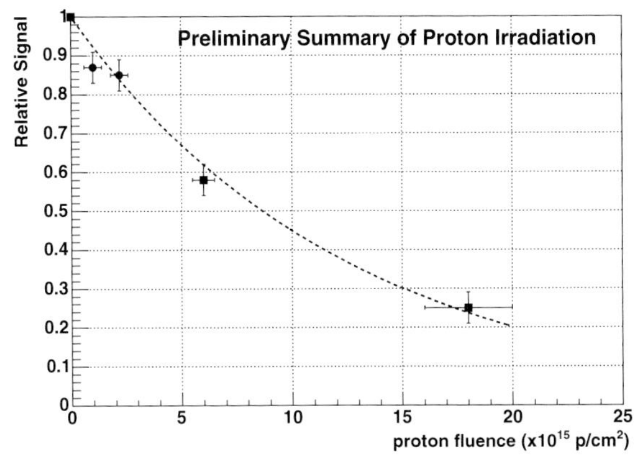

Figure 12.9 Preliminary summary of proton irradiation results for pCVD diamond at fixed field of 1 V/μm.

develop and test the techniques required in the construction of a state-of-the-art pixel detector based on CVD diamonds, a dedicated single chip module, as well as a full ATLAS diamond pixel module, has been constructed. The diamond pixel detectors follow the specification of the ATLAS and CMS pixel detector [33–35]. Several tests have been carried out to develop the bump bonding technique for connecting the diamond sensor to the FE readout pixel chip [36, 37] and prototypes have been examined in beam tests to determine their spatial resolution and efficiency.

Figure 12.2 (middle photograph) shows the electrode pattern on the polycrystalline CVD diamond before bump bonding to the ATLAS pixel readout chip. Each pad of this pixel detector measures $50 \times 400\ \mu m^2$. Beam tests on the single chip module have shown a root mean square spatial resolution of 12 μm along the pixel's short dimension, and a residual distribution with a FWHM of 400 μm along its long dimension as expected for this pixel size. The efficiency for finding hits on tracks reconstructed in the reference telescope was found to be 98% [38]. After these encouraging results, a full size pixel module was constructed, based on a $1.6 \times 6.1\ cm^2$ large pCVD diamond sensor [39]. The diamond sensor is bump bonded to 16 ATLAS pixel readout chips as seen in Figure 12.10 (top) and contains 46,080 readout channels. Beam test studies of this diamond pixel module are currently ongoing. Figure 12.10 (bottom) shows the hitmap during a recent pion beam test with the beam profile clearly visible in the center of the module. The horizontal white line is an artefact of the data presentation. The module is continuously sensitive without a dead area at the chip border.

12.7 Diamond Beam Monitors in High Energy Physics

With their radiation hardness and fast signal response, CVD diamonds lend themselves very well to applications as beam monitors. For example, in these applica-

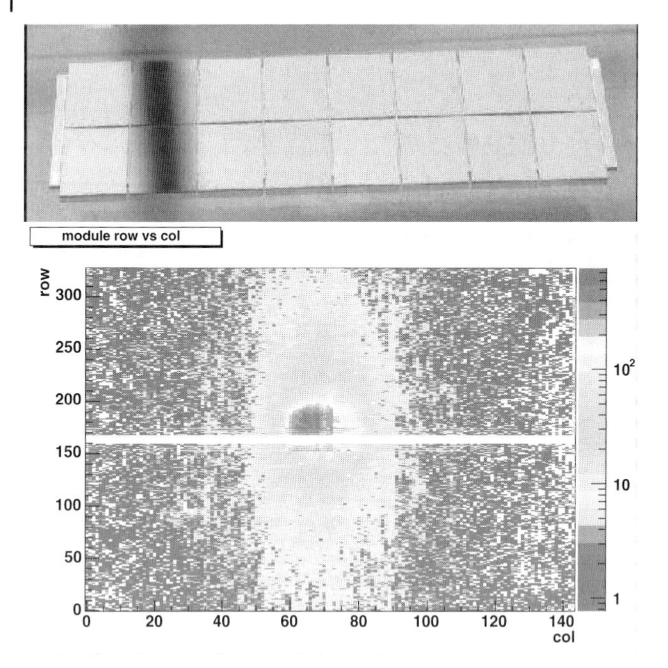

Figure 12.10 pCVD diamond (2×6 cm^2) assembled with 16 ATLAS pixel FE readout chips (top) and beam test hitmap on final diamond pixel modules (bottom) (courtesy SiLab/Bonn).

tions, the diamonds can be used directly in the beam line to monitor the incident beam, or outside the beam line to monitor beam losses. While the examples shown here mainly concern charged particle beams, diamonds have successfully been used for the monitoring of X-ray beams [32] (which, in turn, serves as a nice instrument to investigate diamond uniformity). Their application as visible-blind UV photodetectors is particularly interesting for the monitoring of excimer laser pulses (see e.g. [40, 41]). Similarly, diamond based beam monitors have also been used with heavy ion beams, where they are used as start-veto devices with excellent timing resolution of 29 ps [42], or as α-detectors to monitor plutonium in nuclear recycling processes [43].

As a monitor for beam diagnostics or beam conditions, diamonds can either be used through a measurement of their beam induced "DC current" or as counters for individual particles. Examples of these are given below.

In the DC current mode, the diamond is biased and the current through the diamond is measured continuously. Without beam, only the leakage current (~pA) is measured, while, with beam, a current proportional to the beam intensity (if in line) or beam loss is measured. This method is particularly interesting for high intensity uses, and benefits from the low leakage current, even after heavy irradiation. Beam loss monitors based on this method have been successfully used over the past years in experiments at different high energy physics colliders, like in the BaBar experiment [44] at the Stanford PEP-II storage ring, USA, in the Belle experiment [45] at the KEK B-factory, Japan, and in the CDF experiment [46] at the Fermilab Tevatron, USA. Details of its operation in BaBar can be found in reference [44], where diamonds are used to monitor beam losses and trigger beam aborts, in case losses exceed a threshold where detectors and equipment in the experiment could be damaged. Similar diamond monitors are foreseen at several locations in the CMS experiment close to the beam pipe, where they will serve the same purpose. CMS has tested the function of diamond monitors under worst-case conditions of LHC beam losses, when consecutive proton beam bunches are lost on collimators, at the CERN Super Proton Synchrotron (SPS) accelerator. In the tests, the detectors were exposed to consecutive 40-ns long bunches of proton beams with an intensity of 10^{11} protons/bunch. The response of two diamond monitors to one of those bunches is shown in Figure 12.11. The diamond current matches exactly the gaussian longitudinal beam profile, as monitored separately in SPS beam monitors. CMS also plans to use single crystal CVD diamond detectors bonded to pixel chips, in a telescope type arrangement, as relative luminosity monitors [48].

The method of counting single particles with diamonds to monitor beam conditions is used by the ATLAS experiment. The ATLAS experiment has four identical diamond detector stations on each side of the interaction point as described in [49, 50]. For the ATLAS Beam Conditions Monitor (BCM), the goal is to monitor beam operation conditions by distinguishing proton–proton collisions from background

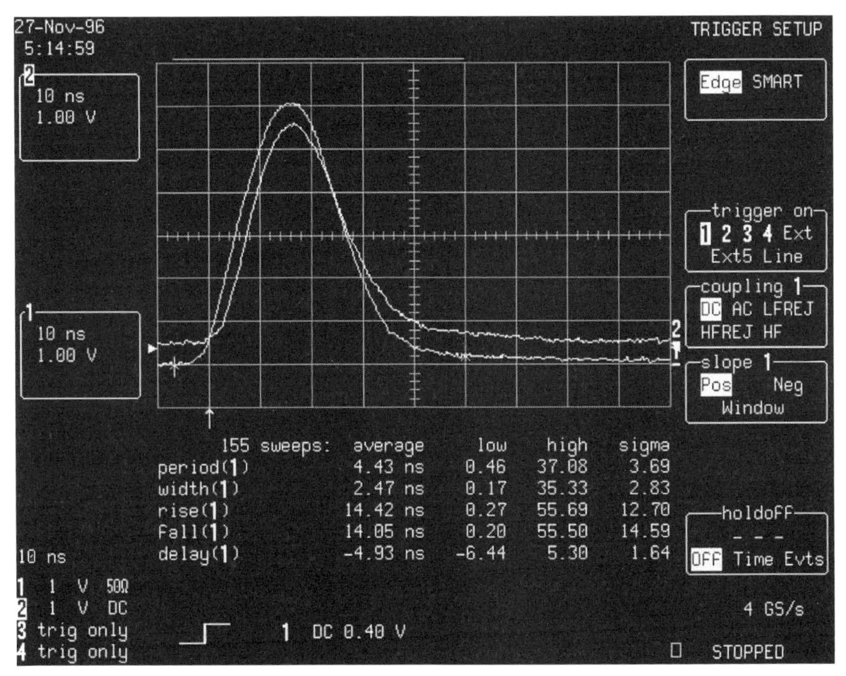

Figure 12.11 Beam spill time structure measured by two pCVD diamond in a high-intensity fast extraction ATLAS/CMS beam test at the SPS.

caused by upstream interactions. The time development of the ratio of background to collisions is monitored and, in case it exceeds a critical threshold, a beam abort can be sent to the accelerator. Collision events are distinguished from background through the diamond time-of-flight measurement between two BCM stations at opposite sides of the interaction point. To achieve this, the BCM diamond system has to be sensitive to single minimum ionizing particles, and provide a signal response with a rise time of 1 ns and a pulse duration of less than 3 ns for single particles. Low noise RF amplifiers have been developed specifically for polycrystalline CVD diamonds to allow this measurement. Test results on first prototypes are shown in Figure 12.12a, which shows the signal of a single minimum ionizing pion traversing the BCM. Signals have an average rise time of 1 ns and a pulse duration (FWHM) of 2.1 ns. Figure 12.12b shows the distribution of signal amplitudes (hatched) in comparison with the noise (open). A signal-to-noise ratio for single minimum ionizing particles in the BCM diamond of 8:1 has been achieved in this measurement, which has recently been improved to 10:1 through noise reduction at the expense of slightly longer rise time.

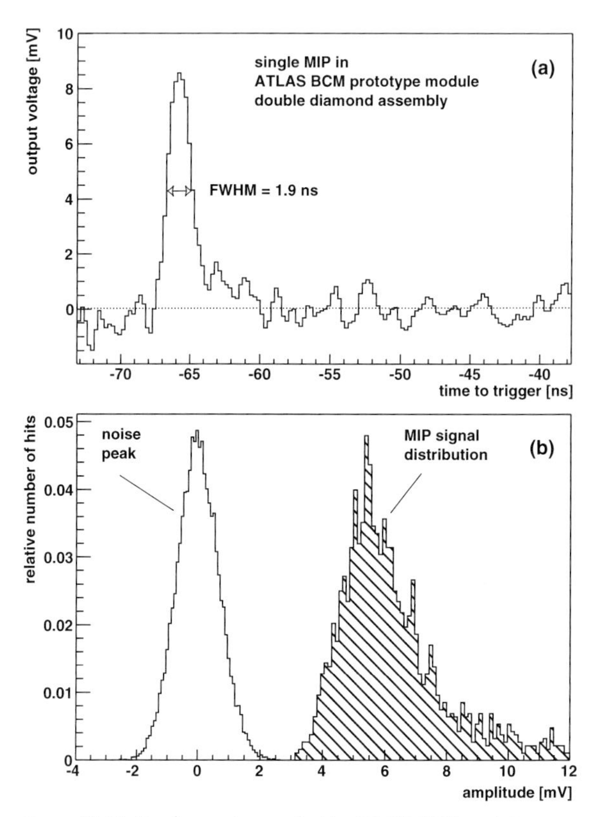

Figure 12.12 Single event recorded in ATLAS BCM prototype module during a pion testbeam at the PS accelerator (a), signal amplitude and noise distribution recorded in the same beam test (b).

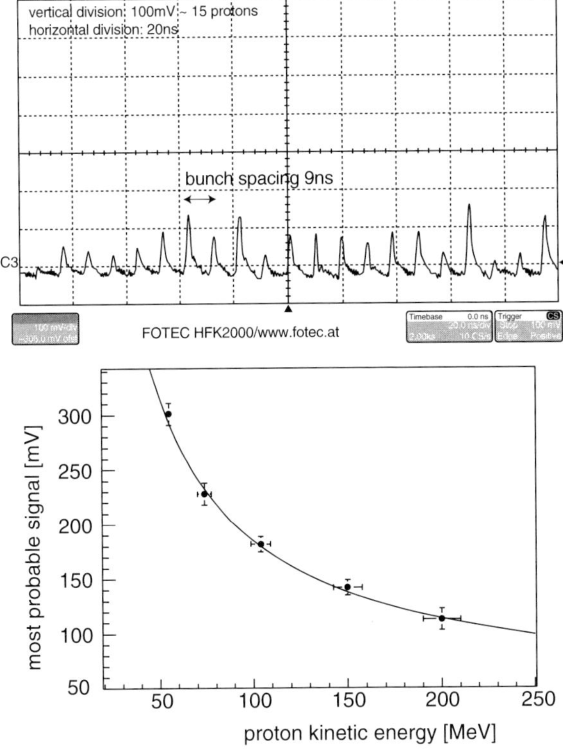

Figure 12.13 Proton beam bunches recorded at NPTC at 200 MeV energy and 2.1×10^9 protons/cm^{2}s (top, courtesy MedAustron)* and diamond most probable signal loss as a function of beam energy recorded at low intensity [26].

12.8
Diamonds for Proton Therapy

The use of diamond beam monitors has recently been investigated in the framework of the MedAustron project [51] where diamonds can be employed as inline beam monitors in proton beam cancer therapy. The MedAustron facility, which is currently under construction, will carry out cancer treatment and non-clinical research with proton beams, and later, carbon beams. Diamond detectors are investigated as high speed counters and beam monitors before the beam is delivered to the patient, in addition to conventional beam monitoring and dosimetry. The aim is to accurately, and in real-time, measure the number of protons delivered with varying intensities and at different beam energy in the range of 70 to 200 MeV. This information, together with a precise measurement of the time structure of the beam, is used for beam diagnostics. In this application, the detector is required to measure individual protons but also larger ns-long bunches. Hence, it requires a large dynamic range, amplifiers in the GHz range, and detec-

*PEG MedAustron Gesellschaft mbH, A – 2700 Wiener Neustadt, Austria.

tors plus electronics need to be radiation hard, to allow device lifetimes from several months up to one year.

Tests have been carried out using 0.5 mm thick polycrystalline CVD diamonds detectors coupled to 2 GHz RF amplifiers [26] at the Northeastern Proton Therapy Center (NPTC), Boston, and the Indiana University Cyclotron Facility to test such a device under comparable conditions. The detectors, $1 \times 1\ cm^2$ in size with an active area of $0.6\ cm^2$ were placed in the beam line with the electronics connected to them. The incoming proton beam intensity was varied in a range of $10^7\ p/cm^2s$ up to several times $10^9\ p/cm^2s$. At the lower intensity settings, individual protons arrive at the detector at 9 ns intervals in case of NPTC. Figure 12.13a shows a train of proton bunches recorded at NPTC at a higher intensity of $2.1 \times 10^9\ p/cm^2s$ and an energy of 200 MeV. Each vertical division corresponds to approximately 15 protons. The plot shows the clear separation of individual bunches. The characteristic energy loss in diamond (dE/dx) has also been measured as a function of beam energy using low intensity running in a previous measurements [26]. The most probable signal per proton as a function of proton energy is shown in Figure 12.13b. The solid line shows the calculated energy loss normalized to the measured one at 107 MeV. Amplifier gain was different in measurements used for plot (a) and (b). The detector operated at a signal-to-noise ratio from 7:1 to 15:1 for single protons depending on the beam energy.

Acknowledgments

The author would like to thank H. Kagan, S. Roe and P. Weilhammer of RD42, E. Griesmayer of Fachhochschule Wiener Neustadt/Fotec, Austria, A. Gorisek and M. Mikuz of the ATLAS BCM collaboration, Alick MacPherson of the CMS BCM collaboration and N. Wermes of University Bonn/SiLab for their help in preparing this manuscript.

References

1 Bauer, C. *et al.* (1996) *Nuclear Instruments and Methods A*, **383**, 64.
2 Manfredotti, C. *et al.* (1998) *Nuclear Instruments and Methods A*, **410**, 96.
3 Kania, D.R. *et al.* (1993) *Diamond and Related Materials*, **2**, 1012.
4 Kozlov, S.F., Konorova, S.A., Hage-Ali, M. and Siffert, P. (1975) *IEEE Transactions on Nuclear Science*, **42**, 160.
5 Canali, C., Gatti, E., Kozlov, S.F., Manfredi, P.F., Manfredotti, C., Nava, F. and Quirini, A. (1979) *Nuclear Instruments and Methods A*, **160**, 73.
6 Zhao, S. (1994) Characterization of the electrical properties of polycrystalline diamond films. PhD Dissertation, State University, Ohio.
7 Pan, L.S., Han, S. and Kania, D.R. (1995) *Diamond: Electronic Properties and Applications*, Kluwer Academic, Dordrecht.
8 Shockley, W. (1938) *Journal of Applied Physics*, **9**, 635–6.
9 Ramo, S. (1939) *Proceedings of the IRE*, **27**, 584–5.
10 The RD42 Collaboration (1994) R&D Proposal, Development of Diamond Tracking Detectors for High Luminosity Experiments at LHC, DRDC/P56, CERN/DRDC 94-21.
11 Manfredotti, C. *et al.* (1998) *Diamond and Related Materials*, **7**, 1338.
12 Adam, W. *et al.* (2003) *Nuclear Instruments and Methods A*, **514**, 79–86.

13 Kagan, H. (2005) *Nuclear Instruments and Methods A*, **546**, 222–7.
14 The RD42 Collaboration (2005) Status Report 2005, Development of Diamond Tracking Detectors for High Luminosity Experiments at LHC, CERN/LHCC 2005-003, LHCC-RD-007.
15 Li, Z. and Kraner, H.W. (1993) *Nuclear Physics B*, **32**, 398.
16 Buttar, C.M., Conway, J., Meyfarth, R., Scarsbrook, G., Sellin, P.J. and Whitehead, A. (1997) *Nuclear Physics A*, **392**, 281.
17 Behnke, T., *et al.* (1998) *Diamond and Related Materials*, **7**, 1553.
18 Isberg, J., Hammersberg, J., Johansson, E., Wikström, T., Twitchen, D.J., Whitehead, A.J., Coe, S.E. and Scarsbrook, G.A. (2002) *Science*, **297**, 1670.
19 Pomorski, M., Berdermann, E., Ciobanu, M., Martemyianov, A., Moritz, P., Rebisz, M. and Marczewska, B. (2005) *physica status solidi (a)*, **202** (11), 2199–205.
20 Eremin, V. *et al.* (1996) *Nuclear Instruments and Methods A*, **372**, 388–98.
21 Li, Z. *et al.* (1997) *Nuclear Instruments and Methods A*, **388**, 297–307.
22 Kramberger, G. (2001) Signal development in irradiated silicon detectors. Doctoral Thesis, University of Ljubljana.
23 Nebel, C.E., Muenz, J. and Stutzmann, M. (1997) *Physical Review B*, **55**, 9786.
24 Pernegger, H. *et al.* (2005) *Journal of Applied Physics*, **97**, 073704.
25 Moritz, P., Berdermann, E., Blasche, K., Stelzer, H. and Voss, B. (2001) *Diamond and Related Materials*, **10**, 1765–9.
26 Frais-Kölbl, H., Griesmayer, E., Kagan, H. and Pernegger, H. (2004) *IEEE Transactions on Nuclear Science*, **51** (6), 3833–7.
27 Lampert, M. and Mark, P. (1970) *Current Injection in Solids*, Academic Press, New York, London.
28 Harkonen, J. *et al.* (2004) *Nuclear Instruments and Methods A*, **535**, 384–8.
29 Bruzzi, M. *et al.* (2005) *Nuclear Instruments and Methods A*, **541**, 189–201.
30 Husson, D. *et al.* (1997) *Nuclear Instruments and Methods A*, **388**, 421.
31 Adam, W. *et al.* (2002) *Nuclear Instruments and Methods A*, **476**, 686–93.
32 Bergonzo, P., Tromson, D., Mer, C., Guizard, B., Foulon, F. and Brambilla, A. (2001) *Physical Review A*, **185** (1), 167181.
33 Rossi, L., Fischer, P., Rohe, T. and Wermes, N. (2006) *Pixel Detectors: From Fundamentals to Applications*, Springer, ISBN 3-540-28332-3.
34 Blanquart, L., Richardson, J., Einsweiler, K., Fischer, P., Mandelli, E., Medeller, G. and Peric, I. (2004) *IEEE Transactions on Nuclear Science*, **51** (4), 1358–64.
35 Wermes, N. (2004) *IEEE Transactions on Nuclear Science*, **51** (3), 1006–15.
36 Adam, W. *et al.* (2001) *Nuclear Instruments and Methods A*, **465**, 88–91.
37 Keil, M. *et al.* (2003) *Nuclear Instruments and Methods A*, **501**, 153–9.
38 Adam, W. *et al.* (2008) *Nuclear Instruments and Methods A*, **565**, 278–83.
39 Velthuis, J.J. *et al.* (2008) Radiation hard diamond pixel detectors. *Nuclear Instruments and Methods A*, in press, doi:10.1016/j.nima.2008.03.061.
40 Whitfield, M.D. *et al.* (2001) *Diamond and Related Materials*, **10**, 650–6.
41 Lansley, S.P. *et al.* (2002) *Diamond and Related Materials*, **11**, 433–6.
42 Berdermann, E., Blasche, K., Moritz, P., Stelzer, H. and Voss, B. (2001) *Diamond and Related Materials*, **10**, 1770–7.
43 Bergonzo, P., Foulon, F., Brambilla, A., Tromson, D., Jany, C. and Han, S. (2000) *Diamond and Related Materials*, **9**, 1003.
44 Bruinsmaa, M. *et al.* (2006) *Nuclear Physics B*, **150**, 164–7.
45 Abashian, A. *et al.* (2002) *The Belle Detector*, NIMA **479**, 117.
46 Abe, F. *et al.* (1988) *The CDF Detector*, NIMA **271**, 487.
47 Fernandez-Hernando, L. *et al.* (2004) *IEEE Instrumentation and Measurement Technology*, **3**, 1855–9.
48 Halkiadakis, E. (2006) A proposed luminosity monitor for CMS based on small angel diamond pixel telescopes. *Nuclear Instruments and Methods A*, **565**, 284–9.
49 Pernegger, H. (2005) *IEEE Transactions on Nuclear Science*, **52** (5), 1590–4.
50 Gorisek, A. (2005) Design and Test of the ATLAS Diamond Beam Conditions Monitor. Proceedings on 9th Conference

on Astroparticle, Particle, Space Physics, Detectors and Medical Physics Applications, Como, Italy, October 2005, World Scientific.

51 Pötter, R. *et al.* (1998) *MedAustron Machbarkeitsstudie*, Eigenverlag der MedAustron Gesellschaft, ISBN 3-9500952-1-7.

13
Superconducting Diamond: An Introduction

Etienne Bustarret
CNRS, Institut Néel, BP 166, 25 rue des Martyrs, 38042 Grenoble Cedex 9, France

13.1
Introduction

13.1.1
The Background to a Discovery

Before introducing the pioneering work of Ekimov *et al.* [1] which added diamond to the long list of superconducting materials and superconductivity to this material's collection of striking properties, it is worth taking a quick look at the relevant background of this discovery, that is, at the contemporary scientific activity in related fields such as superconductivity of column IV elements or superhardness in the B:C:N triangle.

In the former field, beside the stunning success of graphite-like MgB_2, with a superconducting transition temperature T_c of 39 K [2], many graphite intercalation compounds have been studied for some time, for example C_2Na, or C_8K, or more recently C:S composites [3], or CaC_6 where T_c reaches 9 and 11.5 K [4], respectively. In particular, graphite–like BC_3, which was first synthesized a few years ago, has been recently predicted to become superconducting when hole doped [5], under the assumption of an ABC stacking sequence which remains to be fully established [6]. Parallel to this, the superconductivity of silicon based clathrates, where metal atoms are inserted within the cage-like structures, has been observed and quantitatively interpreted [7–10] , showing that the sp^3 hybridization of column IV elements was not incompatible with superconductivity.

Research efforts have been also quite intense in another, apparently quite distant, field: that of superhard materials belonging to the B:C:N chemical composition triangle. While diamond still stands out as the hardest crystal known (80 to 130 GPa depending on the crystallographic orientation of the surface), the hardness of other materials, such as B_4C (30 GPa), BN (62 GPa) or BC_2N (76 GPa [11]), has stimulated creative synthesis strategies. Indeed, in 2004, two groups performed High Pressure High Temperature (HPHT) treatments on a mixture of graphite and of a boron–carbon compound, and reported obtaining carbon borides

Physics and Applications of CVD Diamond. Satoshi Koizumi, Christoph Nebel, and Milos Nesladek

ISBN: 978-3-527-40801-6

together with polycrystalline diamond doped with boron at the 2–3 at.% level. The first study [12], where graphite and CVD BC_3 had been annealed at 2300 K under 20 GPa for 1 min, was focused on the superhardness of the resulting BC_3 phase (88 GPa). The second report, however, which described annealing graphite with B_4C at 2500–2800 K under 8–9 GPa for 5 s, proposed a mechanism for the transformation of graphite into diamond, and made a thorough characterization of the diamond polycrystal [13]. Then, bridging the gap separating the superhardness and superconducting communities, some of the authors of this second paper eventually discovered the superconducting behavior of their diamond sample [1], paving the way for systematic studies of superhard superconducting materials [14].

13.1.2 Back to Basics

The superconducting state of matter, generally observed below a critical temperature T_c and a critical magnetic field H_c, is characterized by an electrical resistance and a magnetic induction both reduced to nil. The external magnetic field is thus fully screened off from the sample. When the magnetic field is increased, in the case of type II superconductors, there is a continuous variation of the magnetic susceptibility χ from −1 in this Meissner state to 0 in the normal state (taken here to be non magnetic). The intermediate region corresponds to a mixed state where quantum flux lines appear in the material as the magnetic field is reached above a first critical value H_{c1}. The density of these vortices increases with the applied field, and their cores, which define local normal regions, merge when the field reaches the upper critical field H_{c2} as the material is restored to the normal state. This puzzling behavior, as well as other related properties, was first explained satisfactorily in the 1950s by Bardeen, Cooper and Schrieffer (BCS) [15], an essential ingredient of the theory being an attractive potential for electrons of opposite spin, leading to the formation of carrier pairs with a binding energy Δ. The resulting bosonic particle, the Cooper pair, has a Ginzburg–Landau coherence length ξ which can be related to the upper critical field H_{c2} extrapolated to zero temperature through the following equation:

$$\xi^2 = \Phi_0 / 2\pi H_{c2}(T = 0) \tag{13.1}$$

where Φ_0 is the flux quantum. The formation of pairs corresponds to a disorder/order phase transition, and implies an abrupt change in the entropy of the system at T_c which can be detected as a jump in the T dependence of the specific heat. Finally, the condensation of superconducting states at the Fermi level opens up an energy gap in the density of states of the (normal) pseudoparticles. The ratio of the width of this gap 2Δ, to the critical thermal energy $k_B T_c$, is called the BCS ratio and rates the strength of the coupling between the two carriers forming a pair. In the case of a weak coupling mechanism, such as that provided by electron/phonon interactions, BCS theory of phonon mediated superconductivity predicts a value of 3.5 for this ratio.

From a practical point of view, it should be emphasized that the observation of zero resistance between two electrical contacts may result solely from the existence of some filamentary percolation paths bypassing the rest of the sample, and that it should not be considered as a proof of its superconductivity. In a similar fashion, negative susceptibility values may be caused by a screening superconducting surface, even when the bulk remains in the normal state. Despite this fact, and the difficulty of evaluating absolute values for χ, especially when dealing with thin films, such observations are generally considered more trustworthy indications of superconductivity. Specific heat measurements are the ultimate judge in such matters, since they probe the bulk of the sample, but they imply the availability of a minimum amount of matter and a good homogeneity which are often difficult to achieve in new materials.

13.1.3 Superconducting Semiconductors

The possibility for a superconducting state to arise in a degenerate semiconductor, or a semimetal, has been theoretically explored in the late 1950s and early 1960s [16–18]. In particular, n-type doping was investigated, and it was stressed that superconductivity should be favored in doped semiconductors having a many-valley band structure, because of the additional attractive interactions provided by intervalley phonons [17]. Experimentally, after many unsuccessful attempts [19], type II superconducting transitions were observed between 50 mK and 500 mK in self doped GeTe [20], SnTe [21] and reduced $SrTiO_3$ [22]. In all three cases, $H_c(T_c)$ phase diagrams were obtained, while specific heat measurements established the bulk character of the transition. The doping dependence of T_c of the two p-type tellurides was also investigated. In particular, the slower uptake of T_c at higher carrier concentrations was attributed to an increased screening of the valley phonon mediated coupling. The same argument was put forward to explain the maximum T_c observed [22] in n-type $SrTiO_3$ for a carrier concentration on the order of 10^{20} cm^{-3}. Finally, tunnel spectroscopy studies [23], performed at various temperatures on GeTe, showed that a $2\Delta/k_B T_c$ ratio of 4.3 could be estimated. At the time, these results were considered to further validate the BCS model for superconductivity [18]. Despite this apparent success, the "superconducting semiconductors" angle had not raised much interest during several decades, until Ekimov's results [1] triggered a renewed excitement, which will probably last for some time, now that superconductivity has also been obtained by heavy boron doping in the parent silicon crystal [24].

13.2
Properties of Heavily Boron Doped Diamond

13.2.1
Chemical Incorporation of Boron at High Levels

Since boron has been identified [25] as the acceptor responsible for the poor insulating properties of natural type IIb diamonds [26], thousands of boron doped single crystals have been synthesized under High Pressure and at High Temperature (HPHT) as well as by Microwave Plasma-assisted Chemical Vapor Deposition (MPCVD) on diamond seeds or substrates. In the 10^{20}–10^{21} cm^{-3} boron concentration range of interest here, polycrystalline p^{++} films grown by MPCVD on Si substrates as well as nanocrystalline layers (which have industrial applications as chemically inert electrodes for advanced electrochemical sensing or processing devices [27–29]) have been found to have electrical properties similar (at room temperature) to those of their single crystal counterparts. Therefore, here, we shall primarily focus on single crystal epilayers grown by MPCVD, which are generally better suited for quantitative studies.

Below about 0.1 at % of boron in diamond (i.e. circa 1.8×10^{20} cm^{-3}), the in situ solid state incorporation of boron scales linearly with the gas phase flow ratio $(B/C)_{gas}$ of boron- to carbon-containing species in the gas phase. More precisely, the efficiency of this linear incorporation depends strongly on the deposition conditions, in particular on the surface quality and orientation and on the purity of the gas mixture. It is generally admitted that, at moderate doping levels, the presence of boron on the growing surface has a positive effect on the MPCVD growth morphology. Moreover, contrary to nitrogen, boron is readily incorporated on substitutional bonding sites of the diamond lattice, and behaves as an electronic acceptor center with an ionization energy of about 370 meV. The presence of oxygen in the gas mixture is known to reduce strongly the incorporation efficiency of boron in the solid state because of the quick formation of volatile boron oxides on the growing surface.

As shown in Figure 13.1, (111)-oriented surfaces pick up 8 times more boron than (100)-oriented facets under otherwise identical plasma conditions [31], and an increase of the methane to hydrogen flow ratio (up to 6% [34]) generally favors the incorporation of boron. Typically, above 0.1 at % in the solid phase, the incorporation becomes superlinear for both crystalline orientations (see Figure 13.1). This indicates a change in the surface chemistry of the adsorbates, probably as a result of stronger boron–boron interactions. In parallel, the growth rate decreases [35], similar to the poisoning effect of dopants observed in the case of silicon epitaxy [36]. Recently, these trends have been tentatively explained by the formation of immobile B—B dimers that become more stable as the p-type doping is increased [37].

The possibility of non-substitutional boron incorporation at high doping levels has been investigated by photoelectron intensity angular distribution (PIAD) circular dichroism [38] and by Nuclear Magnetic Resonance on the ^{11}B isotope [39].

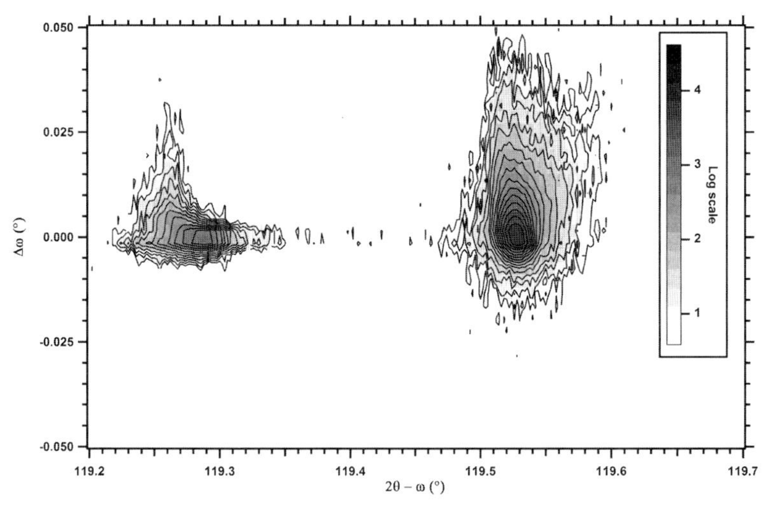

Figure 13.1 Solid phase boron content as measured by SIMS in B doped diamond as a function of the boron-to-carbon atomic concentration ratio in the gas phase during MPCVD growth for various methane-to-hydrogen gas flow ratios: CH4/H2 = 0.15% (111)-oriented epilayers, solid circles [30]; CH_4/H_2 = 1%, (100) and (111) facets of isolated crystallites, empty squares and circles [31]; CH_4/H_2 = 4% (100)-oriented epilayers, solid diamonds [32, 33]; CH_4/H_2 = 6%, (100)-oriented epilayers, solid squares [34]. The broken line corresponds to an incorporation rate of one to one.

Both studies confirmed that the main dopant site for boron was substitutional. However, ^{11}B-NMR studies detected a second boron bonding site of lower symmetry, most abundant in relatively thin (100)-oriented epilayers, and which was proposed to be a local B—H complex [39]. Hydrogen is known to be present in MPCVD diamond layers and to passivate boron acceptors as well as other defects [40, 41]. Interstitial boron as well as boron–vacancy or boron–self-interstitial pairs are also mentioned in the literature [42–47], but the stability of these incorporation sites has been questioned.

13.2.2
Lattice Expansion

As a result of the larger covalent radius of boron (r_B = 0.088 nm) when compared to that of carbon (r_C = 0.077 nm), the introduction of substitutional boron into diamond leads to an expansion of the lattice parameter. This has been found to follow the linear interpolation attributed to Vegard as long as the boron content is lower than 0.2 at.% in MPCVD epilayers [32, 48] or 1.5 at.% in HPHT bulk crystals [49]. In homoepitaxial films, this expansion is severely limited in the film plane by pseudomorphic growth under compressive biaxial stress, but the boron induced positive strain along the growth direction is easily detected in the double crystal X-ray rocking and theta-2theta Bragg reflexion curves [50]. A typical data set obtained on two (100)-oriented homoepitaxial films is represented in Figure

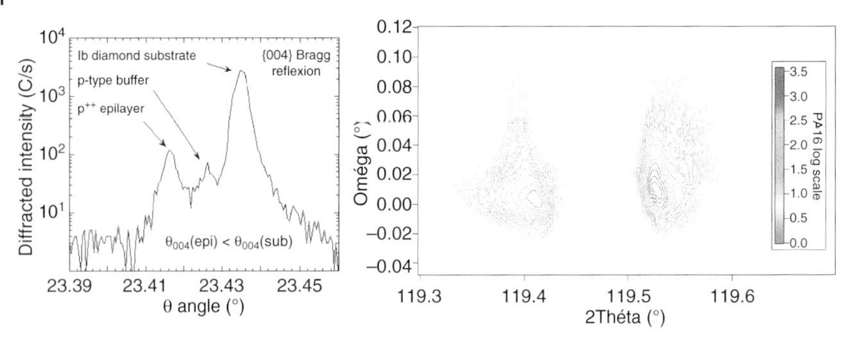

Figure 13.2 (a) θ–2θ diffraction pattern of a 3 μm thick (100)-oriented diamond epilayer doped with about 1021 B/cm^3, including the responses of the Ib substrate, the buffer layer and the p^{++} diamond single crystal. (b) Iso-intensity contours around the {400} diffraction peak of a (100)-oriented 30 μm thick heavily B-doped diamond epilayer. The abscissa axis corresponds to the theta-2 theta scanning direction while the vertical axis corresponds to scanning the omega angle. A double crystal monochromator (channel cut) was used to select the exciting Cu Kα_1 line.

13.2. The downshifted {400} diffraction peak originating from the epilayer has a lineshape (see Figure 13.2a) very similar to that of the type Ib diamond substrate, and if the epilayer is thin enough, an intermediate peak ascribed to the unintentionally doped buffer layer is detected at intermediate angle values. For thicker films, the strain distribution and mosaicity of the epilayer is found to be similar to that of the substrate, as illustrated by the intensity contours in the omega (rocking angle) vs omega minus theta (diffraction angle) map of Figure 13.2b.

Above those threshold concentrations, the expansion is less pronounced [32, 49], contrary to the findings of earlier studies [48, 51] where the boron content had been significantly underestimated [32]. This lower expansion rate has been attributed both to the contribution of free holes [32], to the negative deformation potentials at the valence band maximum [48] and to the occurrence of substitutional boron pairs [49]. Temperature dependent X-ray diffraction measurements have, moreover, established that the thermal expansion coefficient of heavily boron doped diamond was 2–4 times greater than that observed in undoped or moderately B doped crystals [49, 52].

13.2.3 Optical Response

In contrast with the huge amount of optical spectroscopic data available on diamond [53] as a consequence of the economical value of genuine gem stones, relatively few optical studies have been devoted to the darker and often brownish heavily boron doped diamond. For example, infrared transmission spectroscopy, which has been a popular non destructive way of estimating the boron content of polycrystalline or homoepitaxial [54] doped diamond layers grown on infrared-transparent substrates for quite a long time, becomes impractical as the films acquire a metallic opaque character upon heavy doping. Such measurements have thus been restricted to ultrathin homepitaxial films [55]. In a similar manner, the optical emission spectra recorded on p^{++} diamond layers by cathodoluminescence are much weaker and less informative than those measured on the p-type or p^{+} material, particularly in the visible range. As shown in Figure 13.3, even at room temperature, the dominating features are a broad peak around 5.02 eV (5.04 eV at 5 K) with a low energy shoulder around 4.86 eV and another broader and weaker feature extending from 4.1 to 4.6 eV. The latter is probably associated to boron related centers [56], whereas the former feature is usually attributed to a red-shifted and broadened boron-bound excitonic recombination which has been detected in polycrystalline [57, 58] as well as in (001)-oriented [59] and (111)-oriented [30] epilayers. Please note that a peak in the extinction coefficient spectrum (see Figure 13.3) obtained on a p^{++} epilayer by ellipsometry [32] shows up also at 4.86 eV at 300 K, and that the lines around 4.78 eV associated to the 2BD defect [60] attributed to interstitial boron are not seen.

Spectroscopic data obtained by reflectance [32, 55, 61] or ellipsometry [32] have also been reported for (001)- and (111)-oriented heavily doped epilayers, which show as in Figure 13.3 the free carrier reflection edge in the mid infrared range characteristic of a low density plasma. This line shape has been successfully reproduced by the reflectivity of a Drude metal with typical plasmon frequencies around 1 to 2×10^4 cm^{-1} and damping factors γ (or scattering rates) ranging from 3000 to 8000 cm^{-1} [55], and the corresponding optical conductivities extrapolated to zero frequency have been compared to electrical DC conductivities [32]. An exception to this general behavior was reported in the case of a polished polycrystalline sample exhibiting an unusual reflectivity spectrum [62]. In most cases, as remarked in earlier works [32, 55] and illustrated in Figure 13.4, the plasmon edge is strongly modulated by additional contributions in the 450 to 1350 cm^{-1} range correspond-

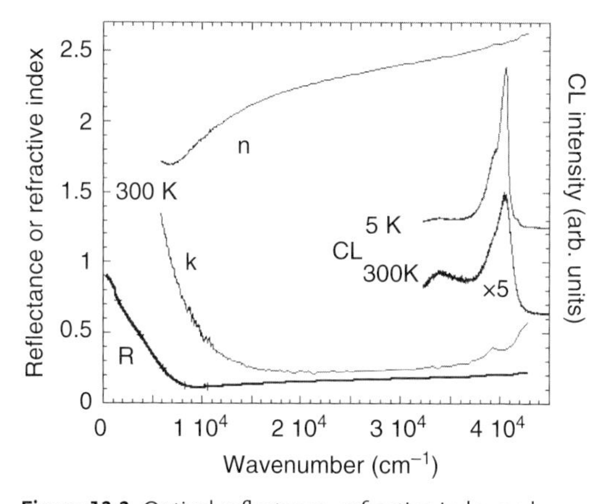

Figure 13.3 Optical reflectance, refractive index and cathodoluminescence (at 5 and 300 K) spectra of a (100)-oriented diamond epilayer doped with about 10^{21} B/cm^3, from the mid infrared to the near UV range.

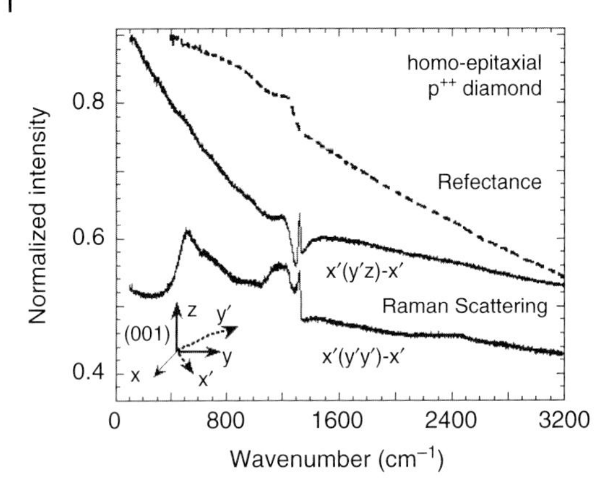

Figure 13.4 IR reflectance and (011)-oriented edge view polarized Raman spectra of a (100)-oriented diamond epilayer doped with about 10^{21} B/cm^3. The Raman spectra have been excited at 633 nm.

ing to one-phonon energies in diamond. These spectral deviations have been recently semi-quantitatively interpreted [61] assuming a frequency dependent scattering rate originating from an electron–phonon coupling spectral function with two peaks around 600 and 1200 cm^{-1}.

Further evidence for both a free hole gas and a strong electron–phonon coupling is provided in Figure 13.4 by the crossed polarized Raman spectrum of a (110)-oriented metallic diamond surface, with both a strong scattering tail of the free holes (probably intraband scattering [32]) and a more complex Fano distorted lineshape indicating an interference between the free carriers continuum and some vibrational discrete levels which might involve also local boron related modes [32]. As can be seen in Figure 13.4, the parallel polarized Raman spectrum is quite different and displays a wide band of excitations starting with a well defined broad peak around 500 cm^{-1} and extending to another broad peak around 1200 cm^{-1} (usually ascribed to both off-center optical modes and Si—B resonant modes) and to a much narrower line around 1320 cm^{-1} attributed to strain-shifted zone-center phonons. According to recent suggestions [63, 64], the most probable assignment for the parallel-polarized Raman peak around 500 cm^{-1} is the A_1 stretching mode of B—B dimers, in agreement with the fact that the frequency of this peak was found not to depend on either the applied magnetic field or the $^{12}C \rightarrow {}^{13}C$ isotopic substitution. The relative strength of these features depends strongly on the excitation wavelength [64, 65], and on the polarization geometry (see Figure 13.4), so that a quantitative lineshape interpretation, and even a comparison of the numerous room temperature Raman spectra of p^{++} diamond published since 1993 [31, 32, 35, 63–71], would be meaningless.

Another way to test the electronic band structure of diamond is to perform X-ray absorption spectroscopy (XAS) and X-ray emission spectroscopy (XES). In the case of polycrystalline p^{++} samples, the spectra [72, 73] demonstrate that the Fermi level lies at the top of the valence band, and that the B2p and C2p states are strongly hybridized. The XAS spectra performed near the C—K edge show two peaks in the partial density of empty states, respectively at 0.3 eV above (probably the boron impurity band) and 1.3 eV below the valence band maximum of the undoped material. More specific to the boron impurity, in gap features show up in the B—K edge XAS spectrum of the most heavily doped sample, indicating the presence of a broad distribution of empty states with features at 1.3 and 3.6 eV above the Fermi level. These bands of localized states have been tentatively attributed to non substitutionally incorporated boron atoms, rather as pairs [37] or clusters than on interstitial sites [72, 73]. In the case of single crystal (111)-oriented layers [74], a high resolution angle-resolved photoemission study has confirmed the metallic character of two layers containing 8.4 and 1.2×10^{21} B/cm^3 and the non metallic character of a sample with 2.9×10^{20} B/cm^3, but failed to observe any additional structure in the occupied density of states at 1.3 eV below E_F. According to the band dispersion, which was measured slightly off but parallel to the Γ—L direction of the Brillouin zone, the experimental band structure parameters were close to those of undoped diamond. The authors claim that their results established that electronic transport in metallic p^{++} diamond takes place in the valence band, and not in the impurity band [74].

13.2.4
Metal-to-Insulator Transition (MIT)

As seen in Figure 13.5, when the boron concentration n_B was increased from 4 to about 5×10^{20} cm^{-3}, the low temperature transport properties of (100)-oriented p^{++}

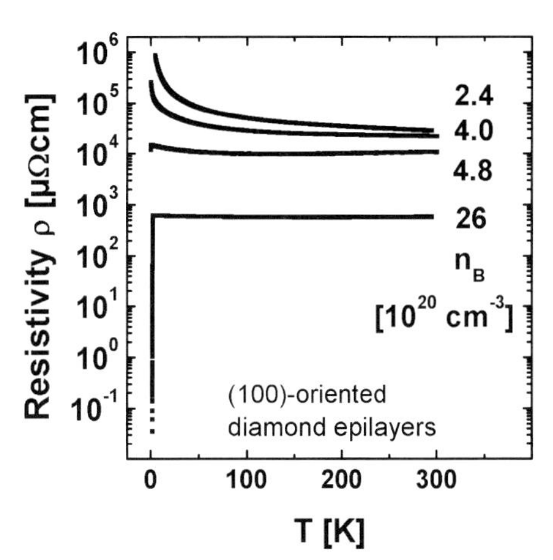

Figure 13.5 Temperature dependence of the AC resistivity (four point) of various (100)-oriented diamond epilayers with different boron contents n_B measured by SIMS. The doping induced MIT is found to occur between 4.0 and 4.8×10^{20} cm^{-3}.

layers changed drastically from an insulating behavior (i.e. the resistivity diverges when $T \rightarrow 0$) to a metallic (and even superconducting) T-dependence which extrapolated to a finite resistivity value at zero temperature. Given the 10% uncertainty of SIMS absolute calibrations, these experimental results are in excellent agreement with the zero temperature model calculations of the vanishing ionization energy or of the chemical potential for both the insulating and the metallic phase [50], which yield a critical boron concentration n_c around 5.2×10^{20} cm^{-3}. A generalized Drude approach [71], taking into account the influence of temperature but neither the hopping transport nor weak localization effects, has also been applied to this system [33] and yielded a similar critical concentration value lying between 4 and 5×10^{20} cm^{-3}. Following Mott, we may assume that this transition results from boron related hydrogenoid states with a Bohr radius $a_H = \varepsilon a_0/m^*$ which overlap when n_B reaches n_c with $a_H n_c^{1/3} = 0.26$ as observed in numerous materials [76]. In this case, taking $\varepsilon = 5.7$ for diamond and $m^* = 0.74$ for the holes, the Bohr radius a_H is estimated at 0.35 nm in fair agreement with values based on the acceptor excited states. The present experimental and theoretical estimates for n_c agree with experimental data recently obtained on free standing polycrystalline p^{++} layers where $n_c < 4.5 \times 10^{20}$ cm^{-3} [77] and 3.4×10^{20} cm^{-3} $< n_c < 5.5 \times 10^{20}$ cm^{-3} [78]. However, they are one order of magnitude lower than those measured in ion implanted diamond [79], where the doping efficiency may have been strongly reduced by the non substitutional incorporation of boron already discussed in Section 13.2.1.

Below the critical boron concentration, the resistivity increases as $\rho = \rho_0 \exp(T_0/T)^m$, where the value of the exponent m depends on the hopping mechanism: $m = 1$ for hopping to the nearest accessible site, $m = 1/4$ for variable range hopping (VRH) assuming [80] a nearly constant density of states at the Fermi level g_F, and $m = 1/2$ when g_F is reduced by a Coulomb gap (ES regime [81]). The T_0 parameter is related to the localization length ξ_{loc} which is of the order of the Bohr radius far from the metal–non metal transition, in a concentration range where the formation of both the impurity band and the valence band tail are expected [82, 83]. Typical values of 10^6 K and crossovers from the VRH to the ES regime and then to a $m = 1$ variation at even lower temperature have been reported for type IIb single crystals with $n_B = 2 \times 10^{19}$ cm^{-3} [84]. Closer to the transition, ξ_{loc} is expected to increase, leading to lower T_0 values, so that the VRH regime extends to lower temperatures, as is the case in Figure 13.5 where T_0 is on the order of a few 10^3 K for $n_B = 2.4 \times 10^{20}$ cm^{-3} and several 10^2 for $n_B = 4 \times 10^{20}$ cm^{-3} with a VRH regime extending down to 10 K [33].

Above the critical concentration, the normal state conductivity can be extrapolated to a finite value σ_0 as $T \rightarrow 0$ K, as seen in Figure 13.5. As a matter of fact, as shown in the inset of Figure 13.6a, for n_B greater than 6×10^{20} cm^{-3} [33], and below a certain temperature at which the inelastic mean free path becomes of the order of the elastic one, the resistivity increases slowly when the temperature is reduced. In this regime where weak localization effects arising from electron–phonon scattering are expected as well as other electronic correlations, the experimental temperature dependence of the conductivity is found in Figure 13.7 to

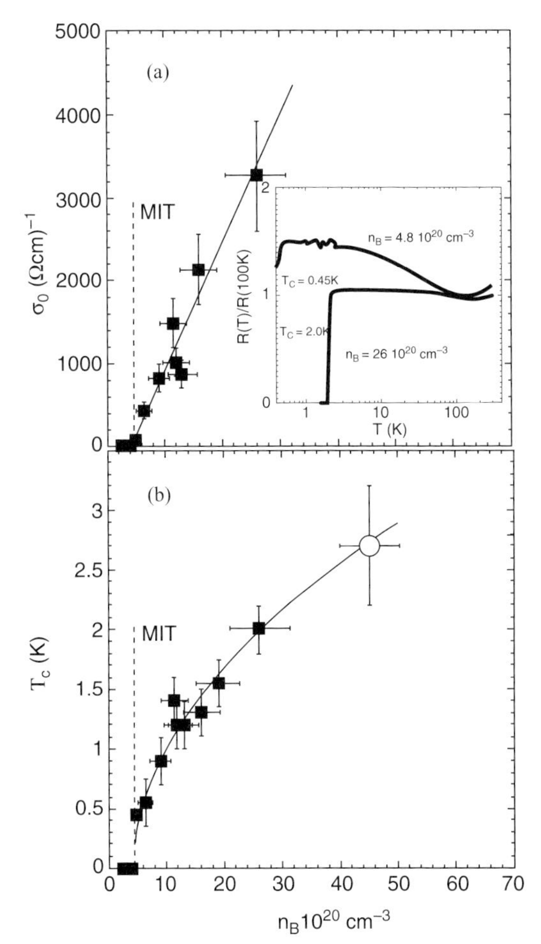

Figure 13.6 (a) Conductivity extrapolated to zero temperature as a function of the boron content, n_B, deduced from SIMS measurements on boron doped diamond (100)-oriented epilayers. The solid line corresponds to the prediction of the scaling theory taking $\nu = 1$ (see text). Insert: T-dependence of the normalized electrical resistance of two samples, illustrating the weak localization regime and the definition of T_c. (b) Critical temperature T_c deduced from resistivity curves (at 90% of the normal state resistance) as a function of the boron content n_B deduced from SIMS measurements. The open circle has been taken from ref. [1]. The solid line corresponds to $T_c \sim (n_B/n_c - 1)^{0.5}$ with $n_c = 4.5 \times 10^{20}$ cm^{-3}.

follow between 3 and 30 K an expression of the type: $\sigma = \sigma_0 + AT^{1/2} + B_{e\text{-}ph}T$. Pronounced weak localization effects have been detected also in heavily boron doped polycrystalline [78] and nanocrystalline [85] diamond, while a $\sigma = \sigma_0 + AT^{1/3}$ variation has been observed in ion-implanted samples [79].

The critical regime of a second order phase transition is generally described by two characteristic exponents ν and η [86]. ν relates the correlation length (here

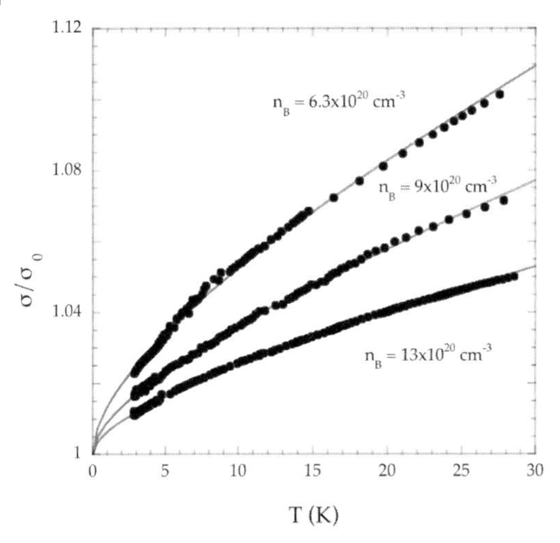

Figure 13.7 Temperature dependence of the conductivity of (100)-oriented diamond epilayers doped with the indicated boron concentrations, on the metallic side of the MIT (adapted from [33]).

ξ_{loc}) to the external parameter driving the transition (here n_B) through $\xi \sim 1/|n_B - n_c|^{\nu}$, whereas η relates the energy and length scales of the system ($E \sim 1/L^{\eta}$). It has been suggested [86] that η ranges from 1 to 3 depending on the relative importance of one electron localization, many body correlations and screening effects, and there are some preliminary indications [33, 79] that for p^{++} diamond $\eta \approx 3$ in agreement with the results obtained on doped silicon [87]. As σ is expected to vary as $1/\xi$, one expects [86] that $\sigma_0 = 0.1(e^2/\hbar)/\xi$ with $a_H/\xi = (n_B/n_c - 1)^{\nu}$. As shown in Figure 13.6a (solid line), σ_0 follows closely the prediction of the scaling theory with $\nu = 1$ and without any other adjustable parameter [33]. In contrast to η, a unique ν value on the order of 1 has been obtained numerically in all systems, independently of the relative importance of the one electron and many body effects. Thus, the $\nu = 1.7$ value reported previously [79] for implanted diamond remains to be explained.

13.3 Superconductivity in Diamond

13.3.1 Experimental Confirmations

The polycrystalline bulk samples synthesized as described in Section 13.1.1 were first shown to be superconducting below about 2.3 K by electrical resistivity and magnetic susceptibility measurements, with an extrapolated upper critical field at zero temperature larger that 3.5 Tesla and a negative dependence of T_c on applied external pressure [1]. Similar results were reported shortly after for polycrystalline [78, 88] and (100)-oriented single crystal [50, 89] diamond films, showing that "zero" resistivity could be observed up to boiling temperature of helium (4.2 K), that superconductivity appeared above about 6×10^{20} B/cm^3, and that T_c increased with the Boron concentration. Other methods, such as Chemical Transport Reaction [90] and Hot Filament-assisted [91] or Electron-assisted [92] chemical vapor deposition, were also shown to yield superconducting layers under the appropriate conditions, confirming the robustness of Ekimov's seminal observation. Heavily boron doped nanocrystalline diamond deposited on seeded glass substrates were also reported to become superconducting below 2 K [93]. Up to now, T_c has been found to reach the 8–10 K range in a few cases [39, 77, 91, 92, 94], the exact values depending not only on the preparation method and conditions but also on the exact experimental characterization and data analysis procedures applied to the transition. In the case of single crystal epilayers, the effect of the substrate orientation has been discussed, (111)-oriented growth being favorable to superconductivity in some cases [39, 94] but not all [37]. As shown on Figure 13.7, for (100)-oriented epilayers, $T_c \sim (n_B - n_c)^{1/2}$, the critical concentration n_c being the same as that of the metal–non metal transition [48, 75, 90], contrary to the case of silicon [24].

13.3.2 Theoretical Calculations

Apart from defect studies, published electronic structure calculations on p^{++} diamond were rather scarce [95] before the superconductivity issue came up. Since then, several studies have appeared involving either a Virtual Crystal Approximation (VCA) [96–99], or a supercell [37, 45–47, 95, 99–101] approach of the doping. In the case of the VCA, Linear Muffin Tin Orbitals (LMTO) [74, 96, 97] as well as *ab initio* pseudopotentials [98] have been used. In the case of the supercell method, both *ab initio* [95, 100] and first principles [37, 45–47, 99, 101] calculations have been performed. Since the substitutional disorder arising from the random incorporation of Boron was not taken into account by the supercell method, its effects have been studied within the Coherent Potential Approximation (CPA) [102, 103]. In the spirit of early studies which pointed to similarities of diamond with MgB_2 [96, 97] and to the strength of the electron–phonon coupling in diamond [104], many of these studies used the calculated electronic and vibrational band structure to compute the electron–phonon spectral function often named after Eliashberg, and to numerically evaluate the coupling strength parameter λ from a sum over the Fermi surface of $g_F V$, where g_F is the density of states at the Fermi level and V the electron–phonon interaction potential. Compared to MgB_2, V is stronger in diamond, but g_F is much smaller because of the three dimensional nature of the sp^3-based carbon network, leading to λ values ranging from 0.15 to 0.55 in the 1 to 5% doping range, depending on the calculation method [96–101, 104], for example, significantly lower than in MgB_2 where $\lambda \approx 1$. Then, assuming that the mechanism for superconductivity was that proposed by the BCS theory [15], and further that the semi-empirical solution of the Eliashberg equations proposed by

McMillan [105] could be applied to diamond, these studies evaluated the critical temperature T_c through:

$$T_c = h\omega_{log}/1.2k_B \exp[-(1.04(1+\lambda))/(\lambda - \mu*(1+0.62\lambda))] \quad (13.2)$$

Where ω_{log} is a logarithmic averaged phonon frequency (about 1020 cm^{-1} in diamond) and $\mu^* = g_F U_C(0)$ the strength of the zero frequency limit of the retarded Coulomb pseudopotential $U_C(0)$. This screening parameter μ^* has generally been taken by the authors to range between 0.1 [96], 0.13 [100] and 0.15 [97, 101], either because these values lead to a good agreement with experiments or because the latter value is typical of usual metals where the Fermi energy is about two orders of magnitude greater than the phonon energy. Within this BCS approach, numerical calculations have also addressed the issue of which phonon were most likely to give rise to superconductivity: works based on the VCA [96–98] insisted on the strong softening of the optical phonons induced by doping and on the similarity of the symmetry of the valence electron and optical phonon states at the center of the Brillouin Zone (BZ), whereas supercell calculations concluded to the significant contribution to the overall electron–phonon coupling coming from localized vibrational modes associated to B atoms [100, 101]. The latter observation has been recently confirmed by a finer sampling of the BZ [99], which also concluded to the significant effect of doping induced disorder (not described by supercell models), as suggested by a previous study [103].

The dominating BCS based explanation of superconductivity in B doped diamond has been challenged by three alternative approaches involving electronic correlations in the impurity band: one dubbed “impurity band resonating valence bond” theory by its author [106] where Mott–Hubbard correlations dominate the physics of the carriers in the impurity band detected the gap by XAS experiments [72, 73]; a second model which insists on the possible role of clustered interstitial Boron and more specifically of composite acceptor states localized on pairs of neighboring Boron interstitials (impurity dumbbells) [43]; a third scenario which insists that the weak localization effects experimentally observed at low temperatures as already mentioned are a precursor to an unconventional pairing mechanism involving spin-flip transitions [85, 107]. Indeed, the fact that T_c values in reasonable agreement with experiments can be obtained from band structure calculations followed by a BCS type estimate of this temperature is not a logical proof that the mechanism is BCS, especially when the doping dependence of T_c is not well reproduced [96]. This theoretical issue may, however, be clarified when a newly developed parameter-free approach where the Coulomb pseudopotential does not need to be adjusted [108] will be applied to the diamond system.

13.3.3
BCS Ratio and Phonon Softening

Apart from the obvious to suggest (but difficult to perform) isotopic substitution effect, there are a few experiments that can shed some light on the nature of the coupling, such as measurements of the lineshape and symmetry of the order parameter of the transition, namely the superconductivity bandgap 2Δ introduced in Section 13.1.2, and of its ratio to the critical temperature (BCS ratio: $2\Delta(T = 0)/T_c$) which measures the strength of the coupling which leads to Cooper pairs formation. Such measurements have been undertaken below T_c and a gap has been observed in the density of states near the Fermi level by point contact tunnelling spectroscopy [50], by optical reflectance in the far infrared [61], by laser excited Photoemission Spectroscopy [109], and by Scanning Tunnel Spectroscopy (STS) with a STM tip [110–113]. When the measurement temperature was comparable to T_c, a V-shaped gap shape has been obtained both on (100)- or (111)-oriented epilayers [50, 109] and on polycrystalline surfaces [112]. However, at lower relative temperatures, a standard BCS lineshape was observed by STS [110, 111] and a ratio of 3.5 typical of weak coupling through phonons was deduced from T dependent spectroscopy [110, 113], as detailed in another work [111].

Another key point is the decrease of the frequency of some specific vibrational modes (phonon softening) and the broadening associated to a strong electron–phonon coupling. This can readily be checked by neutron scattering, or inelastic X-ray scattering (IXS) [114], or even second order Raman scattering under appropriate conditions as described elsewhere [111]. In both cases, evidence was found for a softening of the optical phonons which was smaller than that predicted by VCA calculations, but in good agreement with the values obtained from supercell calculations. Moreover, the IXS data collected along {100} and {111} branches indicated that the softening was strongest close to the Γ point [114]. However, free holes are expected to populate preferentially regions around the {110} direction [104], so that these first IXS results remain to be confirmed and complemented. Additional spectroscopic evidence (not resolved in reciprocal space) came from reflectance spectroscopy in the infrared [61] which showed that the electron–phonon spectral function of B doped diamond had one broad peak around 500 cm^{-1}, and a stronger one near 1200 cm^{-1}, in agreement with supercell calculations [100, 101] which attributed the latter to resonant B related modes.

13.3.4
Boron Concentration Dependence of the Critical Temperature T_c

Although an increase of T_c with either the boron concentration n_B or the free carrier density has been expected on the basis of VCA calculations [95–98] and experimentally observed in some of the early reports [48, 78, 88], systematic studies have remained rather scarce. Within a simplified parabolic band description of the valence band maximum and a BCS model for superconductivity, the curvature of the T_c vs n_B curve may be discussed [33, 78, 88, 104] as function of the λ and μ^* parameters governing Equation 13.2.

Moreover, a comparison between (111)- and (100)-oriented p^{++} diamond epilayers [39, 94] confirmed that above a threshold concentration n_c around 5×10^{20} cm^{-3}, T_c increased sublinearly with n_B in the case of (100)-oriented growth and superlinear for (111)-oriented substrates. The fact that the same boron concentration could lead to quite different T_c values depending on the preparation conditions [94] has raised the question of the doping efficiency of boron atoms depending on their incorporation site. We may thus expect that, in some cases, T_c will reach a maximum value, before decreasing as n_B is increased, if the non substitutional incorporation of boron becomes dominant. For optimized growth conditions, the influence of such extrinsic mechanisms should however remain limited, in particular close to the critical boron concentration n_c [33, 78]. As shown in Figure 13.6b, surprisingly large T_c values (>0.4 K) have been obtained when n_B is only 10% higher than n_c in (100)-oriented epilayers where T_c followed a $(n_B/n_c - 1)^{1/2}$ dependence. Under the assumption that Equation 13.2 still applied despite the fact that close to n_c the Fermi energy becomes smaller than the phonon energy, those relatively high values of T_c could be explained [33] by a slow variation of the electron–phonon coupling parameter λ (typically $\lambda \sim (n_B/n_c - 1)^{0.2}$), together with a pronounced but gradual decrease of the screening parameter μ^* when $n_B \rightarrow n_c$. Such a view where both λ and μ^* are rescaled by the proximity of the metal-to-insulator transition is not so common, although the influence of the proximity of the MIT on the superconducting behavior of disordered metals has been studied extensively over recent years, in order to explain the enhancement of T_c in the vicinity of the MIT [115, 116]. Since such an enhancement has not be seen in diamond so far, the maximum value of T_c is, therefore, most likely to be obtained for very high boron concentration in defect-free substitutional C:B alloys, grown under conditions that remain to be determined.

13.3.5
Is Boron Doped Diamond a "Dirty" Superconductor?

As indicated by the H_c vs T_c phase diagrams published in the first experimental studies [1, 89, 93], one of which is presented in Figure 13.8, p^{++} diamond is a type II superconductor. A striking confirmation of this fact, featuring vortex images obtained by ultra-low temperature STM [110], has been reported elsewhere [111]. The broadening of the transition under magnetic field (see Figure 13.8) is typical of inhomogeneous superconductors. According to Equation 13.1, the coherence length ξ can be estimated either from the low T_c part of such diagrams. In some cases, ξ has been deduced from the dH/dT slope at low field. Typical values for ξ evaluated in one of these ways were 10 nm for polycrystalline HPHT bulk p^{++} diamond [1], 10–30 nm for polycrystalline MPCVD films [78] and 15 and 20 nm for the (100)-oriented epilayers leading to the phase diagrams shown in Figure 13.8. An estimate of the mean free path l_{mfp} for the holes in the normal state has been proposed [78] on the basis of a combination of Hall effect and conductivity measurements at 4.2 K : l_{mfp} was on the order of 0.5 nm in samples where $\xi = 10$ nm. These authors have, moreover, evaluated a London penetration length λ_L of 150 nm for the same film. Seeing that $l_{mfp} \ll \xi \ll \lambda_L$, they concluded that boron-doped superconducting diamond was in the dirty limit. Similar conclusions may be drawn from a free electron description of the room temperature macroscopic

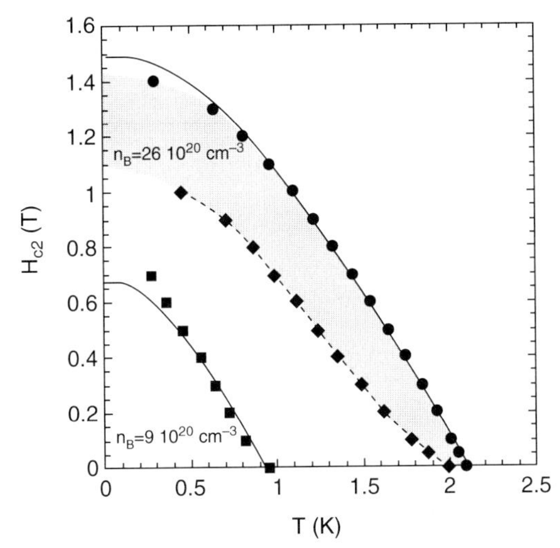

Figure 13.8 *H–T* phase diagram for two (100)-oriented diamond epilayers with different doping levels. The solid circles (diamonds) have been deduced from *T* sweeps of the electrical resistivity for R/R_{Normal} = 90% (10%). The gray area illustrates the width of the transition. The solid squares were defined from the onset of the magnetic susceptibility. The solid lines fit using classical theory, and the dotted line is a guide for the eye.

conductivity values measured on superconductive films (usually in the 200–2000 Ω^{-1} cm^{-1} range), from momentum distribution curves of angle-resolved photoemission around the Fermi level (which yielded l_{mfp} = 0.5 and 0.9 nm in [74]), or from a Drude model analysis of their reflectivity spectrum (microscopic optical conductivity) yielding l_{mfp} = 2.5 and 4 nm respectively for the spectra shown in Figure 13.3. The latter values were obtained on the same epilayers for which a phase diagram is given in Figure 13.8, and again here, $l_{mfp} << \xi$. Moreover, in the thinnest of these boron doped layers, both the room temperature scattering rate γ (from a Drude analysis of the normal state transmittance) and the superconducting gap (2Δ(0), as described elsewhere [111] are known, yielding $\gamma/2\Delta$ = 1700, well into the "dirty" limit of superconductivity.

Clearly, the positive answer to the present subsection title results also from the thorough temperature-dependent infrared reflectivity study down to 5 cm^{-1} region of a superconducting sample with T_c = 6 K [61]. By comparing the spectra above and below T_c, the authors determined not only the value of the superconducting gap 2Δ(0), as mentioned in Section 13.3.3, and the fact that the scattering rate $\gamma(T_c)$ was much greater than 2 Δ(0), but also the field penetration depth λ using the sum rules appropriate in the dirty limit [117]. In the clean limit, λ (λ = 1 μm according to [61]) would coincide with the London penetration length λ_L, while in the dirty limit $\lambda \sim \lambda_L(\gamma/\Delta)^{0.5}$. The γ/Δ ratio of about 400 leads the authors to estimate λ_L~50 nm for their diamond sample, a value somewhat smaller than that proposed above [78].

13.4 Conclusions

The field of superconducting diamond is still young! Although there remains no doubt about the robustness of the effect discovered by Ekimov *et al.* [1], the seemingly conventional pairing mechanism remains to be fully confirmed (and the nature of the pairing phonons to be determined). Many experiments need to be repeated on well characterised and/or specific (for instance isotopically substituted) samples, and slightly higher T_c values (i.e. above 10 K) appear within reach if the material quality and the doping efficiency are further improved.

Because diamond has a simple crystallographic structure and does not involve magnetism, it could become a model system for the study of superconductivity in low dimensional structures of controlled dimensions. This system has the additional benefit of displaying superconductivity close to a metal-to-insulator transition, that is, in a situation where the density of free carriers (and hence the screening) are particularly low, and where disorder can hopefully be restricted to the chemical randomness of an ideal substitutional alloy.

Acknowledgments

The author thanks heartily his colleagues who stimulated his interest in heavily doped diamond epilayers, in the doping induced metal-to-insulator transition, or who introduced him to the fascinating field of superconductivity, and in particular P. Achatz, X. Blase, E. Bourgeois, C. Chapelier, A. Deneuville, E. Gheeraert, J. Kačmarčik, T. Klein, C. Marcenat, L. Ortéga, J. Pernot and B. Sacépé.

References

1 Ekimov, E.A., Sidorov, V.A., Bauer, E.D., Mel'nik, N.N., Curro, N.J., Thompson, J.D. and Stishov, S.M. (2004) *Nature*, **428**, 542.

2 Nagamatsu, J., Nakagawa, N., Muranaka, T., Zenitani, Y. and Akimitsu, J. (2001) *Nature*, **410**, 63.

3 Moehlecke, S., Kopelevich, Y. and Maple, M.B. (2004) *Physical Review B*, **69**, 134519.

4 Emery, N., Hérold, C., d'Astuto, M., Garcia, V., Bellin, C., Marêché, J.F., Lagrange, P. and Loupias, G. (2005) *Physical Review Letters*, **95**, 087003.

5 Ribeiro, F.J. and Cohen, M.L. (2004) *Physical Review B*, **69**, 212507.

6 Sun, H., Ribeiro, F.J., Li, Je-L., Roundy, D., Cohen, M.L. and Louie, S.G. (2004) *Physical Review B*, **69**, 024110.

7 Kawaji, H., Horie, H., Yamanaka, S. and Ishikawa, M. (1995) *Physical Review Letters*, **74**, 1427.

8 Spagnolatti, I., Bernasconi, M. and Benedek, G. (2003) *European Physical Journal B*, **34**, 63.

9 Tanigaki, K., Shimizu, T., Itoh, K.M., Teraoka, J., Moritomo, Y. and Yamanaka, S. (2003) *Nature Materials*, **2**, 653.

10 Connétable, D., Timoshevskii, V., Masenelli, B., Beille, J., Marcus, J., Barbara, B., Saitta, A.M., Rignanese, G.M., Mélinon, P., Yamanaka, S. and Blase, X. (2003) *Physical Review Letters*, **91**, 247001.

11 Solozhenko, V.L., Andrault, D., Fiquet, G., Mezouar, M. and Rubie, D.C. (2001) *Applied Physics Letters*, **78**, 1385.

12 Solozhenko, V.L., Dubrovinskaia, N.A. and Dubrovinsky, L.S. (2004) *Applied Physics Letters*, **85**, 1508.

13 Ekimov, E.A., Sadykov, R.A., Mel'nik, N.N., Presz, A., Tat'yanin, E.V., Slesarev, V.N. and Kuzin, N.N. (2004) *Inorganic Materials*, **40**, 932.

14 Dubitskiy, G.A., Blank, V.D., Buga, S.G., Semenova, E.E., Kul'bachinskii, V.A., Krechetov, A.V. and Kytin, V.G. (2005) *JETP Letters*, **81**, 260.

15 Bardeen, J., Cooper, L.N. and Schrieffer, J.R. (1957) *Physics Review*, **108**, 1175.

16 Pines, D. (1958) *Physics Review*, **109**, 280.

17 Cohen, M.L. (1964) *Physics Review*, **134**, A511.

18 (a) Cohen, M.L. (1964) *Reviews of Modern Physics*, 37, 240. (b) Cohen, M.L (1964) Superconductivity (ed. R.D. Parks), Vol. 1, Marcel Decker, New York, p. 615.

19 Alekseevskii, N.E. and Migenov, L. (1947) *Journal of Physics (USSR)*, **11**, 95.

20 Hein, R.A., Gibson, J.W., Mazelsky, R., Miller, R.C. and Hulm, J.K. (1964) *Physical Review Letters*, **12**, 320.

21 Hein, R.A., Gibson, J.W., Allgaier, R.S., Houston, B.B., Jr, Mazelsky, R. and Miller, R.C. (1965) *Low Temperature Physics, LT9* (eds J.G. Daunt *et al.*), Plenum Press, New York, p. 604.

22 Schooley, J.F., Hosler, W.R. and Cohen, M.L. (1964) *Physical Review Letters*, **12**, 474.

23 Stiles, P.J., Esaki, L. and Schooley, J.F. (1966) *Physics Letters*, **23**, 206.

24 Bustarret, E., Marcenat, C., Achatz, P., Kačmarčik, J., Lévy, F., Huxley, A., Ortéga, L., Bourgeois, E., Blase, X., Débarre, D. and Boulmer, J. (2006) *Nature*, **444**, 465.

25 Collins, A.T. and Williams, A.W.S. (1971) *Journal of Physics C–Solid State Physics*, **4**, 1789.

26 Custers, J.F.H. (1952) *Physica*, **18**, 489.

27 Yagi, I., Notsu, H., Kondo, T., Tryk, D.A. and Fujishima, A. (1999) *Journal of Electroanalytical Chemistry*, **473**, 173.

28 Ndao, A., Zenia, F., Deneuville, A., Bernard, M. and Lévy-Clément, C. (2001) *Diamond and Related Materials*, **10**, 399.

29 Troester, I., Fryda, M., Herrmann, D., Schaefer, L., Haenni, W., Perret, A., Blaschke, M., Kraft, A. and Stadelmann, M. (2002) *Diamond and Related Materials*, **11**, 640.

30 Tavares, C., Omnès, F., Pernot, J. and Bustarret, E. (2006) *Diamond and Related Materials*, **15**, 582.

31 Ushizawa, K., Watanabe, K., Ando, T., Sakaguchi, I., Nishitani-Gamo, M., Sato, Y. and Kanda, H. (1998) *Diamond and Related Materials*, **7**, 1719.

32 Bustarret, E., Gheeraert, E. and Watanabe, K. (2003) *physica status solidi (a)*, **199**, 9.

33 Klein, T., Achatz, P., Kačmarčik, J., Marcenat, C., Gustafsson, F., Marcus, J., Bustarret, E., Pernot, J., Omnès, F., Sernelius, Bo E., Persson, C., Ferreira da Silva, A. and Cytermann, C. (2007) *Physical Review B*, **75**, 165313.

34 Shiomi, H., Nishibayashi, Y. and Fujimori, N. (1991) *Japanese Journal of Applied Physics*, **30**, 1363.

35 Mermoux, M., Jomard, F., Tavares, C., Omnès, F. and Bustarret, E. (2006) *Diamond and Related Materials*, **15**, 572.

36 Mehta, B. and Tao, M. (2005) *Journal of the Electrochemical Society*, **152**, G309

37 Bourgeois, E., Bustarret, E., Achatz, P., Omnès, F. and Blase, X. (2006) *Physical Review B*, **74**, 094509.

38 Kato, Y., Matsui, F., Shimizu, T., Matsushita, T., Guo, F.Z., Tsuno, T. and Daimon, H. (2006) *Science and Technology of Advanced Materials*, **7**, S45.

39 (a) Mukuda, H., Tsuchida, T., Harada, A., Kitaoka, Y., Takenouchi, T., Takano, Y., Nagao, M., Sakaguchi, I., Oguchi, T. and Kawarada, H. (2006) *Science and Technology of Advanced Materials*, **7**, S37. (b) Mukuda, H., Tsuchida, T., Harada, A., Kitaoka, Y., Takenouchi, T., Takano, Y., Nagao, M., Sakaguchi, I., Oguchi, T. and Kawarada, H. (2007) *Physical Review B*, *75*, 033301.

40 Uzan-Saguy, C., Reznik, A., Cytermann, C., Brener, R., Kalish, R., Bustarret, E., Bernard, M., Deneuville, A., Gheeraert, E. and Chevallier, J. (2001) *Diamond and Related Materials*, **10**, 453.

41 Goss, J.P., Briddon, P.R., Jones, R., Teukam, Z., Ballutaud, D., Jomard, F., Chevallier, J., Bernard, M. and Deneuville, A. (2003) *Physical Review B*, **68**, 235209.
42 Walker, J. (1979) *Reports on Progress in Physics*, **42**, 1605.
43 Pogorelov, Yu.G. and Loktev, V.M. (2005) *Physical Review B*, **72**, 075213.
44 Utyuzh, A.N., Timofeev, Yu A. and Rakhmanina, A.V. (2004) *Inorganic Materials*, **40**, 926.
45 Goss, J.P., Briddon, P.R., Sque, S.J. and Jones, R. (2004) *Physical Review B*, **69**, 165215.
46 Goss, J.P. and Briddon, P.R. (2006) *Physical Review B*, **73**, 085204.
47 Oguchi, T. (2006) *Science and Technology of Advanced Materials*, **7**, S67.
48 Brunet, F., Germi, P., Pernet, M., Deneuville, A., Gheeraert, E., Laugier, F., Burdin, M. and Rolland, G. (1998) *Journal of Applied Physics*, **83**, 181.
49 Brazhkin, V.V., Ekimov, E.A., Lyapin, A.G., Popova, S.V., Rakhmanina, A.V., Stishov, S.M., Lebedev, V.M., Katayama, Y. and Kato, K. (2006) *Physical Review B*, **74**, 140502 (R).
50 Kačmarčik, J., Marcenat, C., Cytermann, C., Ferreira da Silva, A., Ortéga, L., Gustafsson, F., Marcus, J., Klein, T., Gheeraert, E. and Bustarret, E. (2005) *physica status solidi (a)*, **202**, 2160.
51 Brunet, F., Deneuville, A., Germi, P., Pernet, M. and Gheeraert, E. (1997) *Journal of Applied Physics*, **81**, 1120.
52 Saotome, T., Ohashi, K., Sato, T., Maeta, H., Haruna, K. and Ono, F. (1998) *Journal of Physics: Condensed Matter*, **10**, 1267.
53 Zaitsev, A.M. (2001) *Optical Properties of Diamond*, Springer-Verlag, Berlin.
54 Gheeraert, E., Deneuville, A. and Mambou, J. (1998) *Diamond and Related Materials*, **7**, 1509.
55 Bustarret, E., Pruvost, F., Bernard, M., Cytermann, C. and Uzan-Saguy, C. (2001) *physica status solidi (a)*, **186**, 303.
56 Sternschulte, H., Horseling, J., Albrecht, T., Thonke, K. and Sauer, R. (1996) *Diamond and Related Materials*, **5**, 585.
57 Sternschulte, H., Albrecht, T., Thonke, K. and Sauer, R. (1996) *The Physics of Semiconductor* (eds M. Scheffler and R. Zimmerman), World Scientific, Singapore, p. 169.
58 Yokota, Y., Tachibana, T., Miyata, K., Hayashi, K., Kobashi, K., Hatta, A., Ito, T., Hiraki, A. and Shintani, Y. (1999) *Diamond and Related Materials*, **8**, 1587.
59 Baron, C., Wade, M., Deneuville, A., Jomard, F. and Chevallier, J. (2006) *Diamond and Related Materials*, **15**, 597.
60 Walker, J. (1979) *Reports on Progress in Physics*, **42**, 1605.
61 Ortolani, M., Lupi, S., Baldassarre, L., Schade, U., Calvani, P., Takano, Y., Nagao, M., Takenouchi, T. and Kawarada, H. (2006) *Physical Review Letters*, **97**, 097002.
62 Wu, D., Ma, Y.C., Wang, Z.L., Luo, Q., Gu, C.Z., Wang, N.L., Li, C.Y., Lu, X.Y. and Jin, Z.S. (2006) *Physical Review B*, **73**, 012501.
63 Bernard, M., Baron, C. and Deneuville, A. (2004) *Diamond and Related Materials*, **13**, 896.
64 Ghodbane, S. and Deneuville, A. (2006) *Diamond and Related Materials*, **15**, 589.
65 Gheeraert, E., Gonon, P., Deneuville, A., Abello, L. and Lucazeau, G. (1993) *Diamond and Related Materials*, **2**, 742.
66 Gonon, P., Gheeraert, E., Deneuville, A., Fontaine, F., Abello, L. and Lucazeau, G. (1995) *Journal of Applied Physics*, **78**, 7059.
67 Ager, J.W., Walukiewicz, W., McCluskey, M., Plano, M.A. and Landstrass, M.I. (1995) *Applied Physics Letters*, **66**, 616.
68 Zhang, R.J., Lee, S.T. and Lam, Y.W. (1996) *Diamond and Related Materials*, **5**, 1288.
69 Pruvost, F., Bustarret, E. and Deneuville, A. (2000) *Diamond and Related Materials*, **9**, 295.
70 Pruvost, F. and Deneuville, A. (2001) *Diamond and Related Materials*, **10**, 531.
71 Bernard, M., Deneuville, A. and Muret, P. (2004) *Diamond and Related Materials*, **13**, 282.
72 Nakamura, J., Kabasawa, E., Yamada, N., Einaga, Y., Saito, D., Isshiki, H., Yugo, S. and Perera, R.C.C. (2004) *Physical Review B*, **70**, 245111.
73 Nakamura, J., Oguchi, T., Yamada, N., Kuroki, K., Okada, K., Takano, Y., Nagao, M., Sakaguchi, I., Kawarada, H., Perera,

R.C.C. and Ederer, D.L. cond-mat/0410144.

74 Yokoya, T., Nakamura, T., Matsushita, T., Muro, T., Takano, Y., Nagao, M., Takenouchi, T., Kawarada, H. and Oguchi, T. (2005) *Nature*, **438**, 647.

75 Ferreira da Silva, A., Sernelius, Bo E., de Souza, J.P., Boudinov, H., Zheng, H.R. and Sarachik, M.P. (1999) *Physical Review B*, **60**, 15824.

76 Edwards, P.P. and Sienko, M.J. (1978) *Physical Review B*, **17**, 2575.

77 Takano, Y., Nagao, M., Takenouchi, T., Umezawa, H., Sakaguchi, I., Tachiki, M. and Kawarada, H. (2005) *Diamond and Related Materials*, **14**, 1936.

78 Winzer, K., Bogdanov, D. and Wild, C. (2005) *Physica C*, **432**, 65.

79 Tshepe, T., Prins, J.F. and Hoch, M.J.R. (1999) *Diamond and Related Materials*, **8**, 1508.

80 Mott, N.F. (1968) *Journal of Non-crystalline Solids*, **1**, 1.

81 (a) Efros, A.L. and Shklovskii, B.I. (1975). *Journal of Physics C – Solid State Physics*, **8**, L49. (b) Efros, A.L. (1976) *Journal of Physics C – Solid State Physics*, **9**, 2021.

82 Fontaine, F. (1999) *Journal of Applied Physics*, **85**, 1409.

83 Serre, J. and Ghazali, A. (1983) *Physical Review B*, **28**, 4704.

84 Sato, T., Ohashi, K., Sugai, H., Sumi, T., Haruna, K., Maeta, H., Matsumoto, N. and Otsuka, H. (2000) *Physical Review B*, **61**, 12970.

85 Mares, J.J., Hubik, P., Nesladek, M., Kindl, D. and Kristofik, J. (2006) *Diamond and Related Materials*, **15**, 1863.

86 McMillan, W.L. (1981) *Physical Review B*, **24**, 2739, and references therein.

87 Castner, T.G. (1997) *Physical Review B*, **55**, 4003, and references therein.

88 Takano, Y., Nagao, M., Sakaguchi, I., Tachiki, M., Hatano, T., Kobayashi, K., Umezawa, H. and Kawarada, H. (2004) *Applied Physics Letters*, **85**, 2851.

89 Bustarret, E., Kačmarčik, J., Marcenat, C., Gheeraert, E., Cytermann, C., Marcus, J. and Klein, T. (2004) *Physical Review Letters*, **93**, 237005.

90 Sidorov, V.A., Ekimov, E.A., Bauer, E.D., Mel'nik, N.N., Curro, N.J., Fritsch, V., Thompson, J.D., Stishov, S.M., Alexenko, A.E. and Spitsyn, B. (2005) *Diamond and Related Materials*, **14**, 335.

91 Wang, Z.L., Luo, Q., Liu, L.W., Li, C.Y., Yang, H.X., Yang, H.F., Li, J.J., Lu, X.Y., Jin, Z.S., Lu, L. and Gu, C.Z. (2006) *Diamond and Related Materials*, **15**, 659.

92 Li, C.Y., Li, B., Lü, X.Y., Li, M.J., Wang, Z.L., Gu, C.Z. and Jin, Z.S. (2006) *Chinese Physical Review Letters*, **23**, 2856.

93 Nesladek, M., Tromson, D., Mer, C., Bergonzo, P., Hubik, P. and Mares, J.J. (2006) *Applied Physics Letters*, **88**, 232111.

94 Umezawa, H., Takenouchi, T., Takano, Y., Kobayashi, K., Nagao, M., Sakaguchi, I., Tachiki, M., Hatano, T., Zhong, G., Tachiki, M. and Kawarada, H. cond-mat/0503303.

95 Barnard, A.S., Russo, S.P. and Snook, I.K. (2003) *Philosophical Magazine*, **83**, 1163.

96 Boeri, L., Kortus, J. and Andersen, O.K. (2004) *Physical Review Letters*, **93**, 237002.

97 Lee, K.-W. and Pickett, W.E. (2004) *Physical Review Letters*, **93**, 237003.

98 Ma, Y., Tse, J.S., Cui, T., Klug, D.D., Zhang, L., Xie, Y., Niu, Y. and Zou, G. (2005) *Physical Review B*, **72**, 014306.

99 Giustino, F., Yates, J.R., Souza, I., Cohen, M.L. and Louie, S.G. (2007) *Physical Review Letters*, **98**, 047005.

100 Blase, X., Adessi, C. and Connétable, D. (2004) *Physical Review Letters*, **93**, 237004.

101 Xiang, H.J., Li, Z., Yang, J., Hou, J.G. and Zhu, Q. (2004) *Physical Review B*, **70**, 212504.

102 Lee, K.-W. and Pickett, W.E. (2006) *Physical Review B*, **73**, 075105

103 Shirakawa, T., Horiuchi, S., Ohta, Y. and Fukuyama, H. (2007) *Journal of the Physical Society of Japan*, **76**, 014711.

104 (a) Cardona, M. (2005) *Solid State Communications*, **133**, 3. (b) Cardona, M. (2006) *Science and Technology of Advanced Materials*, **7**, S60.

105 McMillan, W.L. (1968) *Physics Review*, **167**, 331.

106 Baskaran, G. (2006) *Science and Technology of Advanced Materials*, **7**, S49, cond-mat/0404286; cond-mat/0410296; cond-mat/0611553.

107 Mares, J.J., Nesladek, M., Hubik, P., Kindl, D. and Kristofik, J. (2007) *Diamond and Related Materials*, **16**, 1.

108 Floris, A., Profeta, G., Lathiotakis, N.N., Lüders, M., Marques, M.A.L., Franchini, C., Gross, E.K.U., Continenza, A. and Massidda, S. (2005) *Physical Review Letters*, **94**, 037004.

109 Ishizaka, K., Eguchi, R., Tsuda, S., Yokoya, T., Chainani, A., Kiss, T., Shimojima, T., Togashi, T., Watanabe, S., Chen, C.T., Zhang, C.Q., Takano, Y., Nagao, M., Sakaguchi, I., Takenouchi, T., Kawarada, H. and Shin, S. (2007) *Physical Review Letters*, **98**, 047003.

110 Sacépé, B., Chapelier, C., Marcenat, C., Kačmarčik, J., Klein, T., Bernard, M. and Bustarret, E. (2006) *Physical Review Letters*, **96**, 097006.

111 Sacépé, B., Chapelier, C., Marcenat, C., Kačmarčik, J., Klein, T., Omnès, F. and Bustarret, E. (2006) *physica status solidi (a)*, **203**, 3315.

112 Troyanovskiy, A., Nishizaki, T. and Ekimov, E. (2006) *Science and Technology of Advanced Materials*, **7**, S27.

113 Nishizaki, T., Takano, Y., Nagao, M., Takenouchi, T., Kawarada, H. and Kobayashi, N. (2006) *Science and Technology of Advanced Materials*, **7**, S22.

114 (a) Hoesch, M., Fukuda, T., Takenouchi, T., Sutter, J.P., Tsutsui, S., Baron, A.Q.R., Nagao, M., Takano, Y., Kawarada, H. and Mizuki, J. (2006) *Science and Technology of Advanced Materials*, **7**, S31. (b) Hoesch, M., Fukuda, T., Takenouchi, T., Sutter, J.P., Tsutsui, S., Baron, A.Q.R., Nagao, M., Takano, Y., Kawarada, H. and Mizuki, J. (2007) *Physical Review B*, *75*, 140508 (R).

115 (a) Osofsky, M.S., Soulen, R.J., Jr, Claassen, J.H., Trotter, G., Kim, H. and Horwitz, J.S. (2001) *Physical Review Letters*, **87**, 197004. (b) Osofsky, M.S., Soulen, R.J., Jr, Claassen, J.H., Trotter, G., Kim, H. and Horwitz, J.S. (2002) *Physical Review B*, **66**, 020502 (R).

116 Soulen, R.J., Osofsky, M.S. and Cooley, L.D. (2003) *Physical Review B*, **68**, 094505.

117 Dressel, M. and Grüner, G. (2002) *Electrodynamics of Solids*, Cambridge University Press, Cambridge, UK.

Index

Physics and Applications of CVD Diamond. Satoshi Koizumi, Christoph Nebel, and Milos Nesladek

ISBN: 978-3-527-40801-6

f

n

o